AF879836

Surveys of Modern Mathematics
Volume IV

An Introduction to Rota-Baxter Algebra

Li Guo

International Press
www.intlpress.com

高等教育出版社
HIGHER EDUCATION PRESS

Surveys of Modern Mathematics, Volume IV
An Introduction to Rota-Baxter Algebra

Li Guo
Department of Mathematics, Rutgers University at Newark

2010 Mathematics Subject Classification. Primary: 05C05, 08B20, 18D50, 16W99. Secondary: 05A15, 05E05, 11M32, 16T05, 16T30, 16T25, 11B73, 16Txx, 16S99, 20M99, 68R15, 81T15.

ISBN: 978-1-57146-253-4

Printed in the United States of America.

16 15 14 13 12   1 2 3 4 5 6 7 8 9

SURVEYS OF MODERN MATHEMATICS

Contents

Preface

This is an introduction to Rota-Baxter algebra that aims at presenting basic results of Rota-Baxter algebra with minimal background requirement for the readers.

A Rota-Baxter algebra (first known as a Baxter algebra) is an associative algebra equipped with a linear operator that generalizes the integral operator in analysis. Thus the concept of a Rota-Baxter algebra could have been introduced as an algebraic abstraction of the integral analysis, as in the case of a differential algebra which was studied initiatively by J. F. Ritt in the first third of the twentieth century as an algebraic study of differential equations. Instead Rota-Baxter algebra originated from the 1960 paper [24] of G. Baxter based on his probability study to understand Spitzer's identity in fluctuation theory. Soon afterwards, this concept attracted the attention of well-known mathematicians such as F. V. Atkinson, P. Cartier, and especially G.-C. Rota whose fundamental papers around 1970 brought the subject into the areas of algebra and combinatorics and whose survey articles in the 1990s helped its renascence several years later. While the development of Rota-Baxter (associative) algebra entered a dormant period from the late 1970s to the late 1990s, the corresponding concept for Lie algebra was studied independently by the mathematical physicists, under the name of (the operator form of) the classical Yang-Baxter equation (CYBE), named after the well-known physicists C. N. Yang and R. Baxter.

By coincidence, several discoveries on Rota-Baxter algebra were published just around the turn of the century. After the work of Winkel [173] on Baxter sequences in 1998, Rota-Baxter algebras were found their application in the profound work of Connes and Kreimer [42, 43] in 2000 on the renormalization of perturbative quantum field theory.

Another connection with mathematical physics was also established in year 2000 in the paper [2] of Aguiar that related a Rota-Baxter algebra of weight zero to the associative analog of the classical Yang-Baxter equation [3]. Aguiar further showed that a Rota-Baxter algebra of weight zero naturally carries a structure of a dendriform algebra which had been introduced by Loday [131] in the study of K-theory. Incidentally, a basic example of dendriform algebras is the shuffle product algebra that had been suggested to Loday [134] by Rota and would be shown again coming from Rota-Baxter algebras [53].

Still in year 2000, Keigher and the author, following a suggestion [159] of Rota in an early paper [91], realized free Rota-Baxter algebras as a generalization of the shuffle product [92], called the mixable shuffle product. This did not only facilitated further studies [93, 83, 85] of Rota-Baxter algebras, but also suggested links between these algebras to the many areas related to the shuffle product.

In the same year, a paper of Hoffman [112] introduced another generalization of the shuffle product, called the quasi-shuffle product which had played a fundamental role in the theory of multiple zeta values [111] and quantum field theory [123], and which would turn out to be the same as the mixable shuffle product [53].

Since then, many articles on Rota-Baxter algebras have been completed, in connection with quantum field theory, operads, Hopf algebras, commutative algebra, combinatorics and number theory. An updated list of articles on Rota-Baxter algebra and related areas is maintained at the web site http://andromeda.rutgers.edu/~liguo/rba.html.

The broad connections of Rota-Baxter algebra with many areas in mathematics and mathematical physics are remarkable. Yet the theoretical study of Rota-Baxter algebra is still in its early stage of development in comparison with the rich theory of differential algebra, thus holding the great potential for future researches. These features should make the study of Rota-Baxter algebra attractive to readers with the diverse background. Thus there is a practical need for a monograph on Rota-Baxter algebra to organize the related results spread out in the literature, in addition to the existing survey articles on various aspects of Rota-Baxter algebra. It is the purpose of this monograph to put together the basic properties and constructions of Rota-Baxter algebra as an introduction to two groups of readers who would like to get an overview of Rota-Baxter algebra and who would like to go further to work in this direction. To serve the first group of readers, the main body of the book treats in detail "classical" results that have appeared for several years. The prerequisite is a basic course in abstract algebra either on the undergraduate or the graduate level. Other background will be introduced if need. To serve the second group of readers, a summary has been included at the end of each chapter to briefly discuss more recent developments with the hope to familiarize the readers with some open problems and some active areas of the current research. These summaries are necessarily incomplete, sketchy and assume further background on the part of the readers. A more comprehensive and detailed discussion will have to wait for a future treatise.

The book is divided into three parts, covering three main aspects of Rota-

Baxter algebra, more or less from the author's viewpoint [90]. Each part consists of two chapters. After the basic Section 1.1, the reader can go on with any of the three parts. There is even some flexibility within each part. For example, Chapter 4 is mostly independent of Chapter 3. Overall enough cross references have been provided so that the reader can choose a section and follow the links to refer back to the earlier sections of the book when needed. A Leitfaden is provided for ease of the navigation.

The first part is on the operator aspect of Rota-Baxter algebra that emphasizes the properties of the Rota-Baxter operator, including Spitzer's identity, Atkinson's factorization and its relation to the algebraic Birkhoff decomposition in the Hopf algebra approach of Connes and Kreimer on renormalization of quantum field theory. The concepts, examples and basic properties of Rota-Baxter algebra are introduced in chapter 1. But the main focus of this chapter is to develop various formulations of Spitzer's identity and the related Atkinson's decompositions. Chapter 2 applies Chapter 1 to the special setting where the Rota-Baxter algebra is the convolution algebra of the linear maps from a Hopf algebra to a commutative Rota-Baxter algebra, after the background on connected bialgebras is discussed. This chapter ends with the algebraic Birkhoff decomposition that is applied to renormalization in both physics and mathematics contexts.

The more algebra-oriented readers might find the second part more interesting, since it follows a more traditional treatment of an algebra structure, namely on explicit constructions of the free objects. Because of the rich structure of a Rota-Baxter algebra, we construct free Rota-Baxter algebras in various contexts: commutative versus noncommutative, unitary versus nonunitary, generated by set versus algebras, as well as various representations of the linear bases. Chapter 3 is devoted to the constructions of free commutative Rota-Baxter algebras and their applications to Stirling numbers and partitions. By giving the construction of free commutative Rota-Baxter algebras in terms of the equivalent products of mixable shuffle, quasi-shuffle, stuffle and Delannoy paths, we underline the broad interests from this purely algebraic construction. The original construction of free commutative Rota-Baxter algebras of Rota and Cartier are also discussed, in connection with Spitzer's identity and Waring's formula on symmetric functions. Chapter 4 studies free (noncommutative) Rota-Baxter algebras. We construct the free Rota-Baxter algebra on a set and on an algebra. Further we give the constructions in terms of bracketed words and rooted trees, and in the categories of unitary and nonunitary Rota-Baxter algebras. As an application, we establish the existence and uniqueness of the unitarization of Rota-Baxter algebras.

The third part of the book is on the operad aspect of Rota-Baxter algebra, exploring the connection between Rota-Baxter algebra and operad objects, such as the dendriform algebra. In Chapter 5, we compare free commutative dendriform algebras and free commutative tridendriform algebras with free commutative Rota-Baxter algebras. Then in the non-necessary commutative context, we consider the adjoint functor and universal enveloping algebras related to the functor from Rota-Baxter algebras to (tri-)dendriform algebras. In Chapter 6, we work in the broader context of binary quadratic nonsymmetric operads and consider the relationship between Rota-Baxter operads on such operads and the black square product.

It is my great pleasure to thank all my coauthors, from A(ndrews) to Z(hang), whose collaborations have helped to expose my interests to a wide range of areas in mathematics and physics, and therefore made the writing of this book possible. Particularly, most results in this book came from joint works with them. This book is based on my lectures for graduate courses given at Rutgers University at Newark, Lanzhou University and Zhejiang University. I thank the participants from these courses for their interests. Special appreciation goes to my colleagues and students who have helped to improve the previous drafts, including M. Aggarval, C. Bai, K. Ebrahimi-Fard, X. Gao, H. Kapadia, W. Keigher, P. Lei, F. Li, Z. Lin, J.-L. Loday, Y. Luo, J. Pei, Z. Shi, W. Y. Sit, L. Sun, L. Thrall, B. Vallette, C. Valverde, J. Wang, R. Zhang, Y. Zhang, S. Zheng and C. Zhou, S. Zhou. Thanks also go to my editors Li Zhen Ji and Liping Wang for their encouragement and patience. This book was writing while the author was supported in part by NSF grants DMS-0505643 and DMS-1001855. Finally, I dedicate this book to my family: my parents, my sister, my wife, and my children Audrey and Kevin for their love and support.

Li Guo
November 2011

Leitfaden

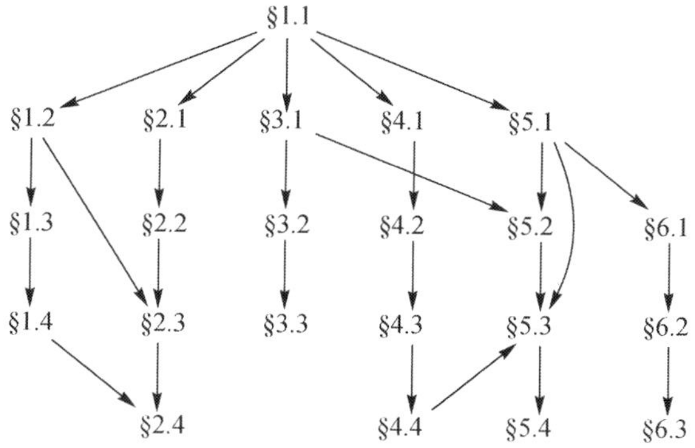

Part I

The Operator Aspect of Rota-Baxter Algebra

Chapter 1

Spitzer's Identity

In this chapter we discuss identities of Rota-Baxter algebras coming from the Rota-Baxter identity, with Spitzer's identity and its generalizations as the centerpiece.

We put our focus on Spitzer's identity for its simplicity, its elegancy, its intricacy, and most importantly, for its surprisingly diverse connections and striking applications. The identity is so simple that its classical form can be explained without any further knowledge beyond the definition of Rota-Baxter algebras. It beautifully encodes an infinite sum in a Rota-Baxter algebra as an exponential function directly generalizing the geometric series formula. Yet its meaning is so involved that its noncommutative generalization was obtained several decades after its commutative counterpart was studied by such leading mathematicians as Cartier and Rota. Despite its complexity and apparently very special form, it appears in many quite unrelated areas of mathematics and physics, such as the Waring's formula in combinatorics, the partition identity of multiple zeta values in number theory, Magnus' expansion formula in matrix differential equations and the Bogoliubov formula in the renormalization of perturbative quantum field theory.

We first give in Section 1.1 the definition of Rota-Baxter algebra and its elementary examples and properties. The first characterization of Rota-Baxter operators is given in Section 1.1.4 as Atkinson's additive decomposition, that expresses a Rota-Baxter algebra as a subdirect sum of subalgebras.

Then we introduce in Section 1.2 the concept of a complete filtered Rota-Baxter algebra as the appropriate framework to study Spitzer's identity and to present its applications in physics (quantum field theory) and number theory (multiple zeta values). Examples from power series and matrices are given. More advanced cases will be discussed in Chapter 2.

In the main Section 1.3, we present different forms of Spitzer's identity, from the classical case to the commutative case and then to the noncommutative case. Then in Section 1.4 we combine Spitzer's identity with Atkinson's multiplicative decomposition theorem to obtain factorization in Rota-Baxter algebras and an algebraic form of the Bogoliubov formula. We end the chapter with a discussion of the related literature and problems.

1.1 Basic definitions, examples and properties

1.1.1 Notations and definitions

We will use $\mathbb{N}_{>0}, \mathbb{N}_{\geqslant 0}, \mathbb{Z}, \mathbb{Q}, \mathbb{R}$ and $\mathbb{C}$ to denote the sets of positive integers, non-negative integers, integers, rational numbers, real numbers and complex numbers, respectively.

We will consider both unitary and nonunitary algebras. To fix the notations and to be self-contained, we briefly recall their definitions. Details can be found in any graduate textbook in algebra, such as [117].

In the following, by a ring we always mean a unitary ring, that is, a set A with binary operations $+$ and $\cdot$ (which will often be suppressed) such that $(A, +)$ is an abelian group, $(A, \cdot)$ is a monoid and $\cdot$ is distributive over $+$. The unit of the monoid is called the identity of A, denoted by $\mathbf{1}_A$ or simply 1. We use nonunitary ring for the corresponding structure where a monoid is replaced by a semigroup, with or without a unit. A ring homomorphism is assumed to preserve the identities. We use $\mathbf{k}$ to denote a commutative ring with identity denoted by $\mathbf{1}$ or simply 1.

Let A be a nonunitary ring. A nonunitary (left) A-module M is an abelian group M together with a *scalar multiplication* $A \times M \to M$ such that

$$a(x+y) = ax + ay, \ (a+b)x = ax + bx, \ (a\,b)x = a(bx), \ \forall a, b \in A, \ x, y \in M.$$

Let A be a ring with identity $\mathbf{1}_A$. A (left) A-module is a nonunitary A-module M with the additional property that $\mathbf{1}_A x = x, \ \forall x \in M$. Particularly, a vector space over a field $\mathbf{k}$ is just a $\mathbf{k}$-module.

Definition 1.1.1. Let $\mathbf{k}$ be a commutative ring. A $\mathbf{k}$-*algebra* (*resp. nonunitary* $\mathbf{k}$-*algebra*) is a ring (resp. nonunitary ring) A together with a unitary $\mathbf{k}$-module structure on the underlying abelian group of A such that

$$k(ab) = (ka)b = a(kb), \ \forall\, k \in \mathbf{k}, \ a, b \in A.$$

For a $\mathbf{k}$-algebra A, there is also the following equivalent definition.

Definition 1.1.2. A $\mathbf{k}$-*algebra* is a ring A together with a ring homomorphism $u : \mathbf{k} \to A$ such that the image of $\mathbf{k}$ is contained in the center of A.

We invite the reader to check that the two definitions of **k**-algebras are indeed equivalent.

The second definition of **k**-algebra can be expressed in terms of commutative diagrams. We will recall it in the next chapter when we introduce the concept of a Hopf algebra.

Now we can define our subject of study in this book.

Definition 1.1.3. Let λ be a given element of **k**. A *Rota-Baxter* **k**-*algebra of weight* λ, or simply an *RBA of weight* λ, is a pair (R, P) consisting of a **k**-algebra R and a linear operator $P : R \to R$ that satisfies the *Rota-Baxter equation*

$$P(x)P(y) = P(xP(y)) + P(P(x)y) + \lambda P(xy), \quad \forall x, y \in R. \tag{1.1}$$

Then P is called a *Rota-Baxter operator (RBO) of weight* λ. If R is only assumed to be a nonunitary **k**-algebra, we call R a nonunitary Rota-Baxter **k**-algebra of weight λ.

We often suppress P in a Rota-Baxter algebra (R, P) when there is no danger of confusion. In the literature, λ is sometimes denoted by $-\theta$ and Eq. (1.1) takes the form

$$P(x)P(y) + \theta P(xy) = P(xP(y)) + P(P(x)y), \quad \forall x, y \in R. \tag{1.2}$$

Basic concepts on **k**-algebras can be similarly defined for Rota-Baxter algebras. Particularly, a *Rota-Baxter subalgebra* (resp. *Rota-Baxter ideal*) of a Rota-Baxter algebra (R, P) is a subalgebra (resp. an ideal) I of R such that $P(I) \subseteq I$. A *Rota-Baxter algebra homomorphism* $f : (R_1, P_1) \to (R_2, P_2)$ between two Rota-Baxter algebras (R_1, P_1) and (R_2, P_2) of the same weight λ is a **k**-algebra homomorphism such that $f \circ P_1 = P_2 \circ f$. We similarly define a homomorphism of nonunitary Rota-Baxter algebras.

Let Alg denote the category of **k**-algebras. Let Alg^0 denote the category of nonunitary **k**-algebras. Similarly let RB_λ (resp. RB^0_λ) denote the category of unitary (resp. nonunitary) Rota-Baxter **k**-algebras of weight λ. The subscript λ will be suppressed if there is no danger of confusion.

Let us compare Rota-Baxter operators with some well-known linear operators. Recall that an *endomorphism* on a **k**-algebra R is a **k**-linear map $f : R \to R$ such that

$$f(xy) = f(x)f(y), \quad \forall x, y \in R.$$

The set $\mathrm{End}(R)$ of endomorphisms on R forms a ring with respect to the addition and composition of functions. On the other hand, a *derivation* on a **k**-algebra R is a **k**-linear map $d : R \to R$ such that

$$d(xy) = d(x)y + xd(y), \quad \forall x, y \in R. \tag{1.3}$$

An algebra with a derivation is called a *differential algebra* [120, 152, 153, 162]. The set $\mathrm{Der}(R)$ of derivations on R forms a Lie algebra with respect to the addition of functions and the Lie bracket

$$[f,g] := f\circ g - g\circ f, \quad \forall f,g\in \mathrm{Der}(R).$$

To the contrary, the set of Rota-Baxter operators on R is not known to possess any algebraic structure.

1.1.2 First examples

We give some simple examples of Rota-Baxter algebras in various contexts. More examples can be found in later chapters.

Example 1.1.4. (Integration) Let R be the $\mathbb{R}$-algebra of continuous functions on $\mathbb{R}$. Define $P: R\to R$ by the integration

$$P(f)(x) = \int_0^x f(t)dt.$$

Then P is an RBO of weight zero. This follows from the integration by parts formula as follows. For $f,g\in R$, let $F(x)=\int_0^x f(t)\,dt$ and $G(x)=\int_0^x g(t)\,dt$. Then $F'(x)=f(x)$ and $G'(x)=g(x)$. So the integration by parts formula

$$\int_0^x F'(t)G(t)\,dt = F(t)G(t)\Big|_0^x - \int_0^x F(t)G'(t)\,dt$$

can be rewritten as

$$\int_0^x f(t)G(t)\,dt = F(t)G(t)\Big|_0^x - \int_0^x F(t)g(t)\,dt = F(x)G(x) - \int_0^x F(t)g(t)\,dt.$$

That is,

$$P(fP(g))(x) = P(f)(x)P(g)(x) - P(P(f)g)(x).$$

Thus

$$P(fP(g)) = P(f)P(g) - P(P(f)g), \quad \forall f,g\in R.$$

Rearranging the terms, we get Eq. (1.1) with $\lambda=0$.

From this example, a Rota-Baxter operator of weight zero is an abstraction of the above integration operator. A Rota-Baxter algebra of weight zero can be regarded as the integral analogue of a differential algebra. More generally, a Rota-Baxter algebra of weight λ can be regarded as the integral analogue of a differential algebra of weight λ [94].

Example 1.1.5. **(Partial sum** [24, 156]) Let R be the set of sequences $(a_n)_{n\geqslant 1}$ with values in $\mathbf{k}$. Then R is a $\mathbf{k}$-algebra with termwise addition, product and scalar product. Define

$$P: R \to R, \quad P(a_1, a_2, \ldots, a_n, \ldots) = (a_1, a_1 + a_2, \ldots, \sum_{1\leqslant i\leqslant n} a_i, \ldots).$$

We will identify $(a_n)_{n\geqslant 1}$ with a function $f: \mathbb{N} \to \mathbb{R}$. Then $P(f)$ is identified with the sequence of partial sums

$$P(f): \mathbb{N} \to \mathbb{R},\ P(f)(n) = \sum_{1\leqslant i\leqslant n} f(i), \quad n \geqslant 1.$$

We show that P is an RBO of weight -1. For $f, g \in R$, have

$$\begin{aligned}(P(f)P(g))(n) &= P(f)(n)P(g)(n)\\ &= \left(\sum_{1\leqslant i\leqslant n} f(i)\right)\left(\sum_{1\leqslant j\leqslant n} g(j)\right)\\ &= \sum_{1\leqslant i,j\leqslant n} f(i)g(j).\end{aligned}$$

The sum is over the set of lattice points

$$\mathrm{I} := \{(i,j)\,|\,1 \leqslant i \leqslant n,\ 1 \leqslant j \leqslant n\} \tag{1.4}$$

which is the union of the subsets

$$\mathrm{II} := \{(i,j)\,|\,1 \leqslant j \leqslant i \leqslant n\} \tag{1.5}$$

and

$$\mathrm{III} := \{(i,j)\,|\,1 \leqslant i \leqslant j \leqslant n\} \tag{1.6}$$

with the overlapping set

$$\mathrm{IV} := \{(i,j)\,|\,1 \leqslant i = j \leqslant n\}. \tag{1.7}$$

We also have

$$P(fP(g))(n) = \sum_{1\leqslant i\leqslant n} f(i)P(g)(i) = \sum_{1\leqslant i\leqslant n} f(i)\left(\sum_{1\leqslant j\leqslant i} g(j)\right) = \sum_{1\leqslant j\leqslant i\leqslant n} f(i)g(j)$$

which is summed over the subset II in Eq. (1.5). Similarly, $P(P(f)g)(n)$ and $P(fg)(n)$ are the sums over the subsets III in Eq. (1.6) and IV in Eq. (1.7), respectively. Since

$$\sum_{(i,j)\in\mathrm{I}} f(i)g(j) = \sum_{(i,j)\in\mathrm{II}} f(i)g(j) + \sum_{(i,j)\in\mathrm{III}} f(i)g(j) - \sum_{(i,j)\in\mathrm{IV}} f(i)g(j),$$

we see that P satisfies Eq. (1.1) with $\lambda = -1$.

Exercise 1.1.6. For the algebra R in Example 1.1.5, define

$$P : R \to R, \quad P(a_1, a_2, \ldots, a_n, \ldots) = (0, a_1, a_1 + a_2, \ldots, \sum_{i<n} a_i, \ldots).$$

Show that P is an RBO of weight 1.

Exercise 1.1.7. (**Scalar product**) Let R be a $\mathbf{k}$-algebra. For any given λ, the operator $P_\lambda : R \to R$ defined by $P_\lambda(a) = -\lambda a$ is an RBO of weight λ. Particularly, the identity map is an RBO of weight -1.

By the last exercise, any $\mathbf{k}$-algebra can be realized as a Rota-Baxter algebra of any given weight. In this sense, the concept of a Rota-Baxter algebra is a generalization of an algebra.

Example 1.1.8. (**q-integral**) Let $R = \mathbf{k}[t]$ and suppose $q \in \mathbf{k}$ is not a root of unity. Formally define

$$P(f)(t) := P_q(f)(t) := \sum_{k=1}^{\infty} f(q^k t). \tag{1.8}$$

More precisely, on the basis $t^n, n \geqslant 1$, of $\mathbf{k}[t]$, we have the formula

$$P(t^n) = \frac{q^n t^n}{1 - q^n}. \tag{1.9}$$

Thus we have

$$\begin{aligned}
&P(t^m P(t^n)) + P(P(t^m)t^n) + P(t^m t^n) \\
&= \frac{q^n q^{m+n}}{(1-q^n)(1-q^{m+n})} t^{m+n} + \frac{q^m q^{m+n}}{(1-q^m)(1-q^{m+n})} t^{m+n} + \frac{q^{m+n}}{1-q^{m+n}} t^{m+n} \\
&= \frac{q^{m+n}}{1-q^{m+n}} \left(\frac{q^n}{1-q^n} + \frac{q^m}{1-q^m} + 1 \right) t^{m+n} \\
&= \frac{q^{m+n}}{1-q^{m+n}} \left(\frac{q^m(1-q^n) + q^n(1-q^m) + (1-q^m)(1-q^n)}{(1-q^m)(1-q^n)} \right) t^{m+n} \\
&= \frac{q^{m+n}}{(1-q^m)(1-q^n)} t^{m+n}.
\end{aligned}$$

This agrees with

$$P(t^m)P(t^n) = \frac{q^m t^m}{1-q^m} \frac{q^n t^n}{1-q^n}.$$

Hence P is an RBO of weight 1.

Exercise 1.1.9. Let R and q be the same as in Example 1.1.8. Show that the operator P on R defined by

$$P(t^n) = \frac{t^n}{1-q^n}, \quad n \geqslant 1$$

is an RBO of weight -1.

The following example plays a key role in the renormalization theory of perturbative quantum field theory [39, 42, 61, 144].

Example 1.1.10. **(Laurent series)** Let R be the algebra of Laurent series:

$$\mathbb{C}[t^{-1},t] := \left\{ \sum_{n=k}^{\infty} a_n t^n \,\middle|\, a_n \in \mathbb{C}, k \leqslant n < \infty, k \in \mathbb{Z} \right\}.$$

Define an operator P on R by

$$P\left(\sum_{n=k}^{\infty} a_n t^n\right) := \sum_{n=k}^{-1} a_n t^n,$$

with the convention that a sum over the empty set is 0. Particularly, $P\left(\sum\limits_{n=0}^{\infty} a_n t^n\right) = 0$. Also define $\tilde{P}\left(\sum\limits_{n=N}^{\infty} a_n t^n\right) = \sum\limits_{n=0}^{\infty} a_n t^n$. Then P and $\tilde{P}$ are Rota-Baxter operators of weight -1. In the case of P, this can be verified as follows (or the reader can apply Theorem 1.1.13). For Laurent series $f(t) := \sum\limits_{m=k}^{\infty} a_m t^m$ and $g(t) := \sum\limits_{n=\ell}^{\infty} b_n t^n$, we have

$$\begin{aligned} P(f)P(g) &= \sum_{m=k}^{-1}\sum_{n=\ell}^{-1} a_m b_n t^m t^n \\ &= \sum_{(m,n)\in \mathrm{I}} a_m b_n t^{m+n}, \end{aligned}$$

where

$$\mathrm{I} := \{(m,n) \in \mathbb{Z}\times\mathbb{Z} \mid m+n<0, k \leqslant m \leqslant -1, \ell \leqslant n \leqslant -1\}. \tag{1.10}$$

We also have

$$\begin{aligned} P(fP(g)) &= P\left(\left(\sum_{m=k}^{-1} a_m t^m + \sum_{m=0}^{\infty} a_m t^m\right)\left(\sum_{n=\ell}^{-1} b_n t^n\right)\right) \\ &= P\left(\left(\sum_{m=k}^{-1} a_m t^m\right)\left(\sum_{n=\ell}^{-1} b_n t^n\right) + \left(\sum_{m=0}^{\infty} a_m t^m\right)\left(\sum_{n=\ell}^{-1} b_n t^n\right)\right) \\ &= \left(\sum_{m=k}^{-1} a_m t^m\right)\left(\sum_{n=\ell}^{-1} b_n t^n\right) + P\left(\left(\sum_{m=0}^{\infty} a_m t^m\right)\left(\sum_{n=\ell}^{-1} b_n t^n\right)\right) \\ &= \sum_{(m,n)\in \mathrm{I}} a_m b_n t^{m+n} + \sum_{(m,n)\in \mathrm{II}} a_m b_n t^{m+n}, \end{aligned}$$

where

$$\text{II} := \{(m,n) \in \mathbb{Z} \times \mathbb{Z} \mid m+n < 0, 0 \leqslant m \leqslant \infty, \ell \leqslant n \leqslant -1\}. \tag{1.11}$$

Similarly,

$$P(P(f)g) = \sum_{(m,n)\in \text{I}} a_m b_n t^{m+n} + \sum_{(m,n)\in \text{III}} a_m b_n t^{m+n},$$

where

$$\text{III} := \{(m,n) \in \mathbb{Z} \times \mathbb{Z} \mid m+n < 0, k \leqslant m \leqslant -1, 0 \leqslant n \leqslant \infty\}. \tag{1.12}$$

Finally, since

$$P\left(\left(\sum_{m=0}^{\infty} a_m t^m\right)\left(\sum_{n=0}^{\infty} b_n t^n\right)\right) = 0,$$

we have

$$P(fg) = \sum_{(m,n)\in \text{I}} a_m b_n t^{m+n} + \sum_{(m,n)\in \text{II}} a_m b_n t^{m+n} + \sum_{(m,n)\in \text{III}} a_m b_n t^{m+n}.$$

Thus

$$P(f)P(g) = P(fP(g)) + P(P(f)g) - P(fg).$$

Example 1.1.11. Let n be a positive integer and R be the $\mathbf{k}$-algebra $M^u_{n\times n}(\mathbf{k})$ of upper triangular $n \times n$-matrices with entries in a ring $\mathbf{k}$. Let $P : R \to R$ be the "collecting operator" that collects all the entries on a row to the corresponding diagonal entry. In other words, for $[a_{ij}]_{ij} \in M^u_{n\times n}(\mathbf{k})$, define

$$P([a_{ij}]_{ij}) = [b_{ij}], \quad b_{ij} := \delta_{ij}\left(\sum_{j=i}^{n} a_{ij}\right), \quad 1 \leqslant i,j \leqslant n.$$

Then P is an RBO of weight -1. In fact, $A = [a_{ij}]_{ij}$ and $B = [b_{ij}]_{ij}$ in R denote $A_i = \sum_{j\geqslant i} a_{ij}$ and $B_i = \sum_{j \geqslant i} b_{ij}$. Then one can check that

$$P(A)P(B) = P(P(A)B) = \operatorname{diag}(A_1B_1, \ldots, A_nB_n)$$

and

$$P(AP(B)) = P(AB) = \left[\delta_{ij}\sum_{k\geqslant i} a_{ik}B_k\right]_{ij}.$$

We will leave the rest of the details to the reader. Another proof is given by the general result in Theorem 1.1.13.

1.1.3 Basic properties

The following elementary properties of a Rota-Baxter algebra follow from its definition.

Proposition 1.1.12. (a) *Let (R, P) be a Rota-Baxter algebra. Then $P(R)$ is a nonunitary subalgebra of R.*

(b) *Let P be an RBO of weight 1. Then λP is an RBO of weight λ. Conversely, if P is an RBO of weight λ and λ is invertible in $\mathbf{k}$, then $\lambda^{-1}P$ is an RBO of weight 1.*

(c) *Let P be an RBO of weight λ. Define $\tilde{P} = -\lambda\mathrm{id} - P$. Then $\tilde{P}$ is also an RBO of weight λ.*

Proof. (a). Since P is $\mathbf{k}$-linear, $P(R)$ is a $\mathbf{k}$-submodule of R. Further, any two elements of $P(R)$ are of the forms $P(x), P(y)$ with $x, y \in R$. Since $P(x)P(y) = P(xP(y) + P(x)y + \lambda xy)$, we see that $P(R)$ is closed under multiplication. Hence $P(R)$ is a nonunitary subalgebra of R.

(b) From

$$P(x)P(y) = P(xP(y)) + P(P(x)y) + P(xy), \quad \forall x, y \in R, \tag{1.13}$$

we obtain

$$(\lambda P)(x)(\lambda P)(y) = (\lambda P)(x(\lambda P)(y)) + (\lambda P)((\lambda P)(x)y) + \lambda(\lambda P)(xy), \quad x, y \in R. \tag{1.14}$$

Hence λP is an RBO of weight λ. Conversely, if λ is invertible, then from Eq. (1.14) we also obtain Eq. (1.13). This is what we want.

(c) Let P be an RBO of weight λ on a $\mathbf{k}$-algebra R. Then for any $x, y \in R$,

$$\begin{aligned}\tilde{P}(x)\tilde{P}(y) &= \lambda^2 xy + \lambda xP(y) + \lambda P(x)y + P(x)P(y)\\ &= \lambda(xP(y) + P(x)y + \lambda xy) + P(xP(y) + P(x)y + \lambda xy)\\ &= -(-\lambda\mathrm{id} - P)(xP(y) + P(x)y + \lambda xy)\\ &= -\tilde{P}(xP(y) + P(x)y + \lambda xy).\end{aligned}$$

On the other hand,

$$\begin{aligned}&\tilde{P}\big(x\tilde{P}(y)\big) + \tilde{P}\big(\tilde{P}(x)y\big) + \tilde{P}\big(\lambda xy\big)\\ &= \tilde{P}\big(x(-\lambda y - P(y)) + (-\lambda x - P(x))y + \lambda xy\big)\\ &= \tilde{P}\big(-xP(y) - P(x)y - \lambda xy\big).\end{aligned}$$

This agrees with $\tilde{P}(x)\tilde{P}(y)$, showing that $\tilde{P}$ is an RBO of weight λ. □

Both Example 1.1.10 and Example 1.1.11 are special cases of a class of Rota-Baxter operators defined by projections.

Theorem 1.1.13. *Let R be a* $\mathbf{k}$*-algebra. A linear operator P on R is an idempotent RBO of weight -1 if and only if there is a* $\mathbf{k}$*-module direct sum decomposition $R = R_1 \oplus R_2$ of R into nonunitary* $\mathbf{k}$*-subalgebras R_1 and R_2 of R such that*

$$P : R \to R_1 \subseteq R \tag{1.15}$$

is the projection of R onto R_1: $P(a_1 + a_2) = a_1$ for $a_1 \in R_1$ and $a_2 \in R_2$.

We note that R is not required to be the algebra direct sum of R_1 and R_2.

Proof. Suppose R has a $\mathbf{k}$-module direct sum decomposition $R = R_1 \oplus R_2$, where R_1 and R_2 are nonunitary $\mathbf{k}$-subalgebras of R. Then the projection P of R onto R_1 is idempotent since, for $a = a_1 + a_2 \in R$ with $a_1 \in R_1$ and $a_2 \in R_2$, we have

$$P^2(a) = P^2(a_1 + a_2) = P(a_1) = a_1 = P(a).$$

Further, for $a = a_1 + a_2$ and $b = b_1 + b_2$ in R with $a_1, b_1 \in R_1$ and $a_2, b_2 \in R_2$, we have $a_1b_1 \in R_1$ and $a_2b_2 \in R_2$. Thus $P(a_1b_1) = a_1b_1$ and $P(a_2b_2) = 0$. Hence

$$\begin{aligned}
P(a)P(b) &= a_1b_1, \\
P(aP(b)) &= P((a_1 + a_2)b_1) = P(a_1b_1 + a_2b_1) = a_1b_1 + P(a_2b_1), \\
P(P(a)b) &= P(a_1(b_1 + b_2)) = P(a_1b_1 + a_1b_2) = a_1b_1 + P(a_1b_2), \\
P(ab) &= P(a_1b_1 + a_1b_2 + a_2b_1 + a_2b_2) = a_1b_1 + P(a_1b_2) + P(a_2b_1).
\end{aligned}$$

In summary, we have

$$P(a)P(b) = P(aP(b)) + P(P(a)b) - P(ab).$$

Thus P is an idempotent RBO of weight -1.

Conversely, suppose $P : R \to R$ is an idempotent RBO of weight -1. Let $R_1 = P(R)$ and $R_2 = (\mathrm{id} - P)(R)$. By Proposition 1.1.12.(a) and (c), both R_1 and R_2 are nonunitary subalgebras of R. Also,

$$a = P(a) + (\mathrm{id} - P)(a), \quad \forall a \in R.$$

Hence $R = R_1 + R_2$. Further, if a is in $R_1 \cap R_2$, then

$$a = P(r_1) = (\mathrm{id} - P)(r_2)$$

for some $r_1, r_2 \in R$. Thus

$$a = P(r_1) = P^2(r_1) = P(\mathrm{id} - P)(r_2) = (P - P^2)(r_2) = 0.$$

Therefore $R = R_1 \oplus R_2$. Finally, since $a = P(a) + (\mathrm{id} - P)(a)$ is the decomposition of $a \in R$ as $a = a_1 + a_2$ with $a_1 \in R_1$ and $a_2 \in R_2$, we see that P is the projection of R onto R_1. □

Exercise 1.1.14. Apply Theorem 1.1.13 to verify that the operators in Examples 1.1.10 and 1.1.11 are RBOs of weight -1.

The following result [7] shows how close it is for an idempotent RBO to have weight -1.

Proposition 1.1.15. *If an idempotent RBO P on R has weight λ, then*

$$(1+\lambda)P(xP(y)) = 0, \tag{1.16}$$
$$(1+\lambda)P(P(x)y) = 0, \tag{1.17}$$
$$(1+\lambda)(P(x)P(y) - \lambda P(xy)) = 0, \quad \forall x, y \in R. \tag{1.18}$$

Thus if $\mathbf{k}$ is a field, then an idempotent RBO must have weight -1.

Proof. Let $x, y \in R$ be given. Since P is an RBO of weight λ, we have

$$P(x)P(y) = P(x \star_P y), \tag{1.19}$$

where $x \star_P y = xP(y) + P(x)y + \lambda xy$. Since P is idempotent, we also have

$$\begin{aligned} P(x)P(y) &= P(x)P^2(y) \\ &= P(xP^2(y)) + P(P(x)P(y)) + \lambda P(xP(y)) \\ &= P(xP^2(y)) + P^2(x \star_P y) + \lambda P(xP(y)) \\ &= P(xP(y)) + P(x \star_P y) + \lambda P(xP(y)). \end{aligned}$$

Comparing this equation with Eq. (1.19), we obtain Eq. (1.16). The proof of Eq. (1.17) is similar. Equation (1.18) follows from Eqs. (1.16) and (1.17) and (1.19). □

Example 1.1.16. (**Maximum**) We now apply Theorem 1.1.13 to introduce "perhaps the most important combinatorial example" [160]. Define a function

$$m : \mathbb{R} \to \mathbb{R}, \quad x \mapsto x^+ := \max\{0, x\}, \quad x \in \mathbb{R}. \tag{1.20}$$

Consider the $\mathbb{R}$-algebra

$$R := \{f : \mathbb{R} \to \mathbb{Q} \,|\, |\mathrm{supp}(f)| < \infty\},$$

where $\mathrm{supp}(f) := \{x \in \mathbb{R} \,|\, f(x) \neq 0\}$ denotes the support of f and where the product is defined by

$$(fg)(x) := \sum_{y,\ z \in \mathbb{R},\ y+z=x} f(y)g(z), \quad x \in \mathbb{R}. \tag{1.21}$$

Define an operator

$$P : R \to R, \quad P(f)(x) := \sum_{y \in \mathbb{R},\ y^+=x} f(y) = \sum_{\max(0,y)=x} f(y), \quad x \in \mathbb{R}, \tag{1.22}$$

with the convention that the sum over the empty set is zero. Explicitly,

$$P(f)(x) = \begin{cases} 0, & x < 0, \\ \sum\limits_{y \leqslant 0} f(y), & x = 0, \\ f(x), & x > 0. \end{cases} \tag{1.23}$$

We then have $\operatorname{supp}(P(f)) \subseteq \mathbb{R}_{\geqslant 0}$. By Eq. (1.23), if $\operatorname{supp}(f) \subseteq \mathbb{R}_{\geqslant 0}$, then $P(f) = f$. Thus

$$P(R) = R_1 := \{f \in R \,|\, \operatorname{supp}(f) \subseteq \mathbb{R}_{\geqslant 0}\}.$$

For $f, g \in R_1$, consider $(fg)(x) = \sum\limits_{y+z=x} f(y)g(z)$. Suppose $x < 0$ and $y+z = x$. Then either $y < 0$ or $z < 0$. Thus $f(y)g(z) = 0$. Hence $\operatorname{supp}(fg) \subseteq \mathbb{R}_{\geqslant 0}$, showing that R_1 is a subalgebra of R. Also define

$$R_2 := \left\{ f \in R \,\middle|\, \operatorname{supp}(f) \subseteq \mathbb{R}_{\leqslant 0}, \sum_{x \in \mathbb{R}} f(x) = 0 \right\}.$$

Then R_2 is a subalgebra of R. Note that, for $f \in R$, we have $f = f_1 + f_2$, where $f_1(x) := P(f)(x)$ and

$$f_2(x) := \begin{cases} 0, & x > 0, \\ -\sum\limits_{y < 0} f(y), & x = 0, \\ f(x), & x < 0. \end{cases}$$

Thus $R = R_1 + R_2$. Clearly $R_1 \cap R_2 = 0$. Therefore $R = R_1 \oplus R_2$ and, by Theorem 1.1.13, P is an RBO of weight -1.

A Rota-Baxter algebra structure can replicate itself as follows.

Theorem 1.1.17. *Let (R, P) be a Rota-Baxter algebra of weight λ. Define another binary operation $\star_P$ on R by*

$$x \star_P y := xP(y) + P(x)y + \lambda xy. \tag{1.24}$$

Then

(a) $P(x)P(y) = P(x \star_P y)$.
(b) *R with the product $\star_P$ is a nonunitary* **k***-algebra.*
(c) *P is an RBO of weight λ on $(R, \star_P)$.*
(d) *P is a homomorphism of nonunitary algebras from $(R, \star_P)$ to R. In fact, it is a homomorphism from the Rota-Baxter algebra $(R, \star_P, P)$ to the Rota-Baxter algebra (R, P).*

The operation $\star_P$ in the proposition is sometimes called the *double product* and the Rota-Baxter algebra structure on $(R, \star_P)$ is called the *double structure* on the original Rota-Baxter algebra R. We can continue this process and obtain a double structure on $(R, \star_P)$. In fact, inductively define

$$x \star_P^{(n+1)} y = x(\star_P^{(n)})_P\, y,\ n \geqslant 1,$$

we obtain a whole hierarchy of Rota-Baxter algebra structures on R:

$$\cdots \xrightarrow{P} (R, \star_P^{(n+1)}) \xrightarrow{P} (R, \star_P^{(n)}) \xrightarrow{P} \cdots \xrightarrow{P} (R, \star_P^{(2)}) \xrightarrow{P} (R, \star_P) \xrightarrow{P} R.$$

Proof. (a) is just the Rota-Baxter equation (1.1).

(b) We just need to verify the associativity of the binary operation $\star_P$. Let x, y, z in R be given. Since R is an associative algebra, by Item (a) we have

$$\begin{aligned} P((x \star_P y) \star_P z) &= P(x \star_P y)P(z) \\ &= P(x)P(y)P(z) \\ &= P(x)P(y \star_P z) \\ &= P(x \star_P (y \star_P z)). \end{aligned}$$

So $\star_P$ is associative if P is injective which might not always be the case. So we instead prove the associativity by the following direct verification.

By Eq. (1.24), we have

$$\begin{aligned} (x \star_P y) \star_P z &= (xP(y) + P(x)y + \lambda xy) \star_P z \\ &= (xP(y)) \star_P z + (P(x)y) \star_P z + (\lambda xy) \star_P z \\ &= xP(y)P(z) + P(xP(y))z + \lambda xP(y)z + P(x)yP(z) \\ &\quad + P(P(x)y)z + \lambda P(x)yz + \lambda xyP(z) + P(\lambda xy)z + \lambda\lambda xyz. \end{aligned}$$

Applying Eq. (1.1) to the first term, we obtain

$$\begin{aligned} (x \star_P y) \star_P z &= xP(yP(z) + P(y)z + \lambda yz) + P(xP(y))z + \lambda xP(y)z \\ &\quad + P(x)yP(z) + P(P(x)y)z + \lambda P(x)yz + \lambda xyP(z) \\ &\quad + \lambda P(xy)z + \lambda^2 xyz \\ &= xP(yP(z)) + xP(P(y)z) + \lambda xP(yz) + P(xP(y))z \\ &\quad + \lambda xP(y)z + P(x)yP(z) + P(P(x)y)z + \lambda P(x)yz \\ &\quad + \lambda xyP(z) + \lambda P(xy)z + \lambda^2 xyz. \end{aligned}$$

Similarly,

$$\begin{aligned}
x \star_P (y \star_P z) &= x \star_P (yP(z) + P(y)z + \lambda yz) \\
&= x \star_P (yP(z)) + x \star_P (P(y)z) + x \star_P (\lambda yz) \\
&= xP(yP(z)) + P(x)yP(z) + \lambda x(yP(z)) + xP(P(y)z) \\
&\quad +P(x)P(y)z + \lambda xP(y)z + xP(\lambda yz) + P(x)\lambda yz + \lambda x(\lambda yz) \\
&= xP(yP(z)) + P(x)yP(z) + \lambda x(yP(z)) + xP(P(y)z) \\
&\quad +(P(xP(y)) + P(P(x)y) + \lambda P(xy))z + \lambda xP(y)z \\
&\quad +\lambda xP(yz) + \lambda P(x)yz + \lambda^2 xyz \\
&= xP(yP(z)) + P(x)yP(z) + \lambda x(yP(z)) + xP(P(y)z) \\
&\quad +P(xP(y))z + P(P(x)y)z + \lambda P(xy)z + \lambda xP(y)z \\
&\quad +\lambda xP(yz) + \lambda P(x)yz + \lambda^2 xyz.
\end{aligned}$$

The associativity of $\star_P$ is verified by comparing the terms in the expression of $(x \star_P y) \star_P z$ and those in the expression of $x \star_P (y \star_P z)$.

(c) By the definition of $\star_P$ and Item (a) we have

$$\begin{aligned}
P(x) \star_P P(y) &= P(x)P^2(y) + P^2(x)P(y) + \lambda P(x)P(y) \\
&= P(x \star_P P(y)) + P(P(x) \star_P y) + \lambda P(x \star_P y),
\end{aligned}$$

as needed.

(d) The first statement follows from Item (a) and the second statement is the tautological statement: $P(P(x)) = P(P(x))$. □

1.1.4 Atkinson's additive decomposition

We now give the additive decomposition of Atkinson [12], published in 1963. His multiplicative decomposition theorem will be presented in Section 1.4.1.

Theorem 1.1.18. (Atkinson [12]) *Let R be a $\mathbf{k}$-algebra. If a linear operator $P : R \to R$ is an RBO, then for any $x, y \in R$, there is a $z \in R$ such that $P(x)P(y) = P(z)$ and $\tilde{P}(x)\tilde{P}(y) = -\tilde{P}(z)$. Suppose that the annihilator of $\lambda \in \mathbf{k}$ in R is zero. Then the converse is also true.*

Proof. Let $P : R \to R$ be an RBO. Then for any $x, y \in R$, we have

$$P(x)P(y) = P(xP(y) + P(x)y + \lambda xy).$$

Let $z = xP(y) + P(x)y + \lambda xy$. Then from the proof of Proposition 1.1.12.(c), we have $\tilde{P}(x)\tilde{P}(y) = -\tilde{P}(z)$.

Now we consider the converse. For any given $x, y \in R$, we verify that the Rota-Baxter relation in Eq. (1.1) holds. Let $z \in R$ be such that $P(x)P(y) = P(z)$ and $\tilde{P}(x)\tilde{P}(y) = -\tilde{P}(z)$. Then by $\tilde{P} = -\lambda \mathrm{id} - P$, we have

$$\begin{aligned} -\lambda z &= P(z) + \tilde{P}(z) \\ &= P(x)P(y) - \tilde{P}(x)\tilde{P}(y) \\ &= P(x)P(y) - \big(\lambda^2 xy + \lambda x P(y) + \lambda P(x)y + P(x)P(y)\big) \\ &= -\lambda\big(xP(y) + P(x)y + \lambda xy\big). \end{aligned}$$

Since λ has trivial annihilator in R, we have

$$z = xP(y) + P(x)y + \lambda xy.$$

So

$$P(x)P(y) = P(z) = P\big(xP(y) + P(x)y + \lambda xy\big),$$

as needed. □

Theorem 1.1.18 can be strengthened as follows. Its formulation as a commutative diagram is also useful when we study the Rota-Baxter property of coalgebras [38]. Let R be a **k**-algebra. Denote $R \pm R$ for the direct product $R \times R$ with component addition and scalar product, and with "signed" product:

$$(x_1, y_1) \pm (x_2, y_2) := (x_1 x_2, -y_1 y_2).$$

Let $P : R \to R$ be a linear map. Define $f := f_P : R \to R \pm R$ by $f(x) = (P(x), \tilde{P}(x))$. Define $\nabla : R \pm R \to R$ by $\nabla(x, y) = x + y$. If P is an RBO, then $\nabla \circ f_P = -\lambda \mathrm{id}$ and $f(R)$ is closed under multiplication in $R \pm R$. So we have the following commutative diagram.

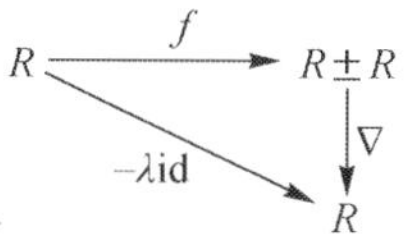

Theorem 1.1.19. *Suppose the annihilator of λ in R is zero. There is a* 1-1 *correspondence between the set of RBOs P of weight λ on R and the set of linear maps $f : R \to R \pm R$ such that $\nabla f = -\lambda\,\mathrm{id}$ and $f(R)$ is closed under the multiplication in $R \pm R$. The correspondence is given by sending P to $f = (P, \tilde{P}) : R \to R \pm R$, and sending $f = (f_1, f_2)$ to $P = f_1$.*

We first prove a result in a more general setting.

Proposition 1.1.20. *For any **k**-algebra R and linear map $f := (f_1, f_2) : R \to R \pm R$, $f(R)$ is closed under the multiplication in $R \pm R$ if and only if*

(a) *$f_1(R)$ and $f_2(R)$ are closed under the multiplication in R, and*
(b) *for $x, y \in R$, if $f_1(x)f_1(y) = f_1(z)$ with $z \in R$, then $f_2(x)f_2(y) = -f_2(z)$.*

Proof. Suppose a subset $f(R)$ of $R \pm R$ is closed under its multiplication. Then for any $x, y \in R$, we have $f(x) \pm f(y) = f(z)$ for some $z \in R$. By the definition of $\pm$, we have

$$\begin{aligned}(f_1(z), f_2(z)) &= f(z) \\ &= f(x) \pm f(y) \\ &= (f_1(x), f_2(x)) \pm (f_1(y), f_2(y)) \\ &= (f_1(x)f_1(y), -f_2(x)f_2(y)).\end{aligned}$$

Thus $f_1(x)f_1(y) = f_1(z)$ and $f_2(x)f_2(y) = -f_2(z)$. This proves (a) and (b).

Conversely, assume that a linear map $f = (f_1, f_2) : R \to R \pm R$ satisfies (a) and (b). For any $x, y \in R$, let $f_1(x)f_1(y) = f_1(z)$ and $f_2(x)f_2(y) = f_2(w)$ by (a). Then $f_2(z) = -f_2(w)$ by (b). Thus

$$\begin{aligned}f(x) \pm f(y) &= (f_1(x), f_2(x)) \pm (f_1(y), f_2(y)) \\ &= \big(f_1(x)f_1(y), -f_2(x)f_2(y)\big) \\ &= (f_1(z), -f_2(w)) \\ &= (f_1(z), f_2(z)) \\ &= f(z).\end{aligned}$$

Thus $f(R)$ is closed under the multiplication in $R \pm R$. □

We now prove Theorem 1.1.19.

Proof. Let $\mathfrak{P}$ be the set of RBOs of weight λ on R and let $\mathfrak{F}$ be the set of linear maps $f : R \to R \pm R$ such that $\nabla \circ f = -\lambda\,\mathrm{id}$ and $f(R)$ is closed under the multiplication of $R \pm R$. For $P \in \mathfrak{P}$, define $\Phi(P) : R \to R \pm R$ by $\Phi(P) = (P, \tilde{P})$. Then by Theorem 1.1.18 and Proposition 1.1.20, $\Phi(P)$ is in $\mathfrak{F}$. Conversely, if $f = (f_1, f_2)$ is in $\mathfrak{F}$, then by $\nabla \circ f = -\lambda\,\mathrm{id}$, we have $f_2 = -\lambda\,\mathrm{id} - f_1$. Since $f(R) \subseteq R \pm R$ is closed under multiplication, by Proposition 1.1.20 and Theorem 1.1.18, $\Psi(f) := f_1$ is an RBO. This gives a map from $\mathfrak{F}$ to $\mathfrak{P}$.

It is easy to verify that $\Phi \circ \Psi$ and $\Psi \circ \Phi$ are the identity maps. Therefore Φ gives a 1-1 correspondence between $\mathfrak{P}$ and $\mathfrak{F}$. □

1.2 Complete Rota-Baxter algebras

We now give the basic algebraic setup to formulate the general form of Spitzer's identity.

Definition 1.2.1. A (decreasing) *filtered* $\mathbf{k}$*-algebra* is a $\mathbf{k}$-algebra R together with nonunitary subalgebras R_n, $n \geqslant 0$, such that

$$R_n \supseteq R_{n+1}, \quad \bigcup_{n\geqslant 0} R_n = R, \quad R_n R_m \subseteq R_{n+m}, \quad m, n \geqslant 0.$$

We let (R, R_n) or simply R denote a filtered $\mathbf{k}$-algebra and call $\{R_n\}_{n\geqslant 0}$ a (decreasing) filtration. Since the filtration is decreasing, we have $R_0 = R$. Since $R_0 R_n \subseteq R_n$ and $R_n R_0 \subseteq R_n$, each R_n is an ideal of R.

In a filtered $\mathbf{k}$-algebra R, we can use the subsets $R_n, n \geqslant 0$, to define a metric on R in the standard way. More precisely, for $x \in R$, define

$$o(x) = \begin{cases} \max\{k \mid x \in R_k\}, & x \notin \cap_{n\geqslant 0} R_n, \\ \infty, & x \in \cap_{n\geqslant 0} R_n. \end{cases}$$

For $x, y \in R$, define

$$d(x,y) = 2^{-o(y-x)}.$$

Equivalently,

$$d(x,y) = \begin{cases} \min\{2^{-n} \,|\, y - x \in R_n\}, & y - x \notin \cap_{n\geqslant 0} R_n, \\ 0, & y - x \in \cap_{n\geqslant 0} R_n. \end{cases}$$

Example 1.2.2. (a) $R := \mathbf{k}[x]$ with $R_n = x^n \mathbf{k}[x]$ is a filtered algebra.
(b) Let p be a prime number. Then $R := \mathbb{Z}$ with $R_n = p^n \mathbb{Z}$ is a filtered algebra.

A filtered algebra (R, R_n) is called *complete* if R is a complete metric space with respect to $d(x, y)$, that is, every Cauchy sequence in R converges. Then we also call the filtration $\{R_n\}_{n\geqslant 0}$ complete. Equivalently, a filtered $\mathbf{k}$-algebra R with filtration $R_n, n \geqslant 0$, is complete if $\cap_{n\geqslant 0} R_n = 0$ and if the resulting injection

$$R \to \bar{R} := \varprojlim R/R_n$$

is surjective. Here, the inverse limit $\varprojlim R/R_n$ is the subalgebra of $\prod_{n\geqslant 0} (R/R_n)$ defined by

$$\varprojlim R/R_n := \left\{ (a_n)_{n\geqslant 0} \in \prod_{n\geqslant 0} (R/R_n) \,\middle|\, \pi_{n+1}(a_{n+1}) = a_n, n \geqslant 0 \right\},$$

where $\pi_{n+1} : R/R_{n+1} \to R/R_n (\cong (R/R_{n+1})/(R_n/R_{n+1}))$ is the quotient map sending $a \bmod R_{n+1}$ to $a \bmod R_n$. For a filtered algebra (R, R_n) with $\cap_{n\geqslant 0} R_n = 0$, $\bar{R}$ is called the *completion* of R with respect to the filtration R_n.

Example 1.2.3. (a) The power series ring $R := \mathbf{k}[[x]]$ with $R_n = x^n\mathbf{k}[[x]]$ is a complete filtered algebra. It is the completion of the filtered algebra $(\mathbf{k}[x], x^n\mathbf{k}[x])$.

(b) Let p be a prime number. The ring of p-adic numbers $R := \mathbb{Z}_p$ with $R_n = p^n\mathbb{Z}_p$ is a complete filtered algebra. It is the completion of the filtered algebra $(\mathbb{Z}, p^n\mathbb{Z})$.

Lemma 1.2.4. *Let (R, R_n) be a complete filtered $\mathbf{k}$-algebra. For $a_n \in R_n, n \geqslant 0$, the series $\sum_{n=0}^{\infty} a_n$ converges in R. Particularly, for $a \in R_1$ and $c_n \in \mathbf{k}, n \geqslant 0$, the series $\sum_{n=0}^{\infty} c_n a^n$ converges in R.*

Proof. By definition, $\sum_{n=0}^{\infty} a_n$ is the limit $\lim_{n\to\infty} s_n$ of the partial sum sequence $s_n := \sum_{k=0}^{n} a_k, n \geqslant 0$, if the limit exists. But the sequence s_n is a Cauchy sequence: for any given $\varepsilon > 0$, let $N = \lceil \log_2 \varepsilon \rceil$. Then for $m, n \geqslant N$, we have $a_m - a_n \in R_N$ and hence $d(a_m, a_n) < 2^{-N} < \varepsilon$. Therefore $\lim_{n\to\infty} s_n$ exists by the completeness of (R, R_n). □

When R is a complete filtered $\mathbb{Q}$-algebra, by Lemma 1.2.4, the functions

$$\exp: R_1 \to 1 + R_1, \ \exp(a) := \sum_{n=0}^{\infty} \frac{a^n}{n!}, \tag{1.25}$$

$$\log: 1 + R_1 \to R_1, \ \log(1 + a) := -\sum_{n=1}^{\infty} \frac{(-a)^n}{n} \tag{1.26}$$

are well-defined.

We also record the following simple fact for later reference.

Lemma 1.2.5. *Let R be a complete filtered algebra. The subset $1 + R_1$ is a group under the multiplication of R.*

Proof. An element of $1 + R_1$ is of the form $1 - a$ with $a \in R_1$. Thus $\sum_{n=0}^{\infty} a^n$ is a well-defined element in $1 + R_1$ and

$$(1 - a)\Big(\sum_{n=0}^{\infty} a^n\Big) = \Big(\sum_{n=0}^{\infty} a^n\Big)(1 - a) = 1.$$

□

Definition 1.2.6. A *complete filtered Rota-Baxter algebra* is defined to be a Rota-Baxter algebra (R, P) with a complete filtration $\{R_n\}_{n\geqslant 0}$ such that

$$P(R_n) \subseteq R_n, \forall\ n \geqslant 0.$$

Exercise 1.2.7. Let (R, P) be a Rota-Baxter algebra. Consider the power series ring $\mathcal{A} := R[[x]]$ on one variable x. Define an operator

$$\mathcal{P} : \mathcal{A} \to \mathcal{A}, \quad \mathcal{P}\Big(\sum_{n=0}^{\infty} a_n x^n\Big) = \sum_{n=0}^{\infty} P(a_n) x^n.$$

Prove that $(\mathcal{A}, \mathcal{P})$, with the filtration $\mathcal{A}_n := \mathcal{A}x^n$, $n \geqslant 0$, is a complete Rota-Baxter algebra with the same weight as that of (R, P).

Another filtered Rota-Baxter algebra comes from matrix algebras. Let (R, P) be a Rota-Baxter algebra and let $\mathcal{M}_n(R)$ be the $n \times n$ matrix algebra with entries in R. We define a linear map $\mathcal{P}$ on $\mathcal{M}_n(R)$ by extending the Rota-Baxter operator P entrywise, i.e. for a matrix $\alpha := [\alpha_{ij}] \in \mathcal{M}_n(R)$, define

$$\mathcal{P}(\alpha) = \big[P(\alpha_{ij})\big]. \tag{1.27}$$

Theorem 1.2.8. *Let (R, P) be a Rota-Baxter algebra. The pair $(\mathcal{M}_n(R), \mathcal{P})$ is a Rota-Baxter algebra with the same weight as (R, P).*

Proof. We only need to verify the Rota-Baxter relation for the operator $\mathcal{P}$. Let $\alpha = [\alpha_{ij}]_{ij}$ and $\beta = [\beta_{ij}]_{ij}$ in $\mathcal{M}_n(R)$ be given. By the entrywise definition of $\mathcal{P}$ in Eq. (1.27) and the Rota-Baxter relation, we have

$$\begin{aligned}
\mathcal{P}(\alpha)\,\mathcal{P}(\beta) &= \big[P(\alpha_{ij})\big]\big[P(\beta_{ij})\big]\\
&= \Big[\sum_k P(\alpha_{ik})P(\beta_{kj})\Big]\\
&= \Big[\sum_k \big(P(P(\alpha_{ik})\beta_{kj}) + P(\alpha_{ik}P(\beta_{kj})) + \lambda P(\alpha_{ik}\beta_{kj})\big)\Big]\\
&= \Big[\sum_k \big(P(P(\alpha_{ik})\beta_{kj})\Big] + \Big[\sum_k P(\alpha_{ik}P(\beta_{kj}))\Big] + \Big[\sum_k \lambda P(\alpha_{ik}\beta_{kj})\Big]\\
&= \Big[P(\sum_k P(\alpha_{ik})\beta_{kj})\Big] + \Big[P(\sum_k \alpha_{ik}P(\beta_{kj}))\Big] + \lambda\Big[P(\sum_k \alpha_{ik}\beta_{kj})\Big]\\
&= \mathcal{P}\Big[\sum_k P(\alpha_{ik})\beta_{kj}\Big] + \mathcal{P}\Big[\sum_k \alpha_{ik}P(\beta_{kj})\Big] + \lambda\mathcal{P}\Big[\sum_k \alpha_{ik}\beta_{kj}\Big]\\
&= \mathcal{P}\big(\mathcal{P}([\alpha_{ij}])[\beta_{ij}]\big) + \mathcal{P}\big([\alpha_{ij}]\mathcal{P}([\beta_{ij}])\big) + \lambda\mathcal{P}\big([\alpha_{ij}]\,[\beta_{ij}]\big)\\
&= \mathcal{P}(\mathcal{P}(\alpha)\beta) + \mathcal{P}(\alpha\mathcal{P}(\beta)) + \lambda\mathcal{P}(\alpha\,\beta),
\end{aligned}$$

as needed. □

To give examples of filtered Rota-Baxter algebras, we consider the subalgebras $\mathcal{M}_n^u(R)$ of $\mathcal{M}_n(R)$, $1 \leqslant n < \infty$, of upper triangular matrices. We also let $\mathcal{M}_\infty^u(R)$ denote the algebra of $\infty \times \infty$ upper triangular matrices.

For each $n \leqslant \infty$, the algebra $\mathcal{M}_n^u(R)$ carries a natural decreasing filtration in terms of the number of zero upper subdiagonals. We denote by $\mathcal{M}_n^u(R)_1$ the upper triangular matrices with the main diagonal being zero, that is, the strict upper triangular matrices. Let $\mathcal{M}_n^u(R)_k$, $k > 1$, denote the ideal of strictly upper triangular matrices with zero in the main diagonal and in the first $k-1$ subdiagonals. We then have the decreasing filtration

$$\mathcal{M}_n^u(R) \supset \mathcal{M}_n^u(R)_1 \supset \cdots \supset \mathcal{M}_n^u(R)_{k-1} \supset \mathcal{M}_n^u(R)_k \supset \cdots, k < n,$$

with

$$\mathcal{M}_n^u(R)_k\, \mathcal{M}_n^u(R)_m \subset \mathcal{M}_n^u(R)_{k+m}.$$

It is easy to see that for any $n \leqslant \infty$, the filtration is complete, that is, $\mathcal{M}_n^u(R)$ is complete with respect to the metric defined by the filtration $\mathcal{M}_n^u(R)_k$, $k \geqslant 0$.

Now let R be a Rota-Baxter algebra with Rota-Baxter operator P. Then By Theorem 1.2.8, we have

Theorem 1.2.9. *For each $1 \leqslant n \leqslant \infty$, the triple $(\mathcal{M}_n^u(R), \mathcal{P}, \{\mathcal{M}_n^u(R)_k\}_{k \geqslant 1})$ forms a complete filtered Rota-Baxter algebra.*

1.3 Formulations of Spitzer's identity, old and new

We now arrive at one of the fundamental theorems of Rota-Baxter algebras, Spitzer's identity. Its importance comes from both its theoretical significance and its surprisingly wide range of applications. Theoretically, it prompted people to find several other related identities, and motivated Rota and Cartier to give their constructions of free commutative Rota-Baxter algebras. It has also found far reaching applications in diverse areas of mathematics and physics such as the renormalization of multiple zeta values and quantum field theory.

1.3.1 Spitzer's identity in the classical case

Spitzer's formula [163] was regarded as a remarkable stepping stone in the theory of sums of independent random variables in the fluctuation theory of probability. It is named after the mathematician Frank Spitzer. Even though Spitzer's identity in its original form describes relations in fluctuation theory in probability, it is better understood in terms of the Rota-Baxter operator. In fact, the very motivation for Glen Baxter [24] to introduce this operator was to give a more conceptual proof of Spitzer's identity.

Kingman [119], Atkinson [12] and Vogel [171] extended and generalized Baxter's work. Later, Rota [160] explored the underlying mathematical structure of Baxter's operator relation, and used free Rota-Baxter algebras to relate Spitzer's identity to Waring's identity. See Section 3.2.4. Recently, a generalization of Spitzer's identity for non-commutative Rota-Baxter algebras has been found and has shown its importance in the Hopf algebraic approach to renormalization in perturbative quantum field theory [62].

The original identity of Spitzer has the following algebraic abstraction [24, 160].

Theorem 1.3.1. (Classical Case of Spitzer's Identity [160]**)** *Let (R,P) be a unitary commutative Rota-Baxter $\mathbb{Q}$-algebra with invertible weight λ. Then for $a \in R$, we have*

$$\exp\left(P\left(\frac{1}{\lambda}\log(1+\lambda a x)\right)\right) = \sum_{n=0}^{\infty} x^n \underbrace{P(P(P(\cdots(P(a)a)a)a))}_{n\text{-iterations}} \tag{1.28}$$

in the ring of power series $R[[x]]$. In other words,

$$\exp\left(-\sum_{k=1}^{\infty} \frac{\lambda^{k-1}(-x)^k P(a^k)}{k}\right) = \sum_{n=0}^{\infty} x^n P_a^n(1). \tag{1.29}$$

Here we have used the notation $P_a(c) := P(ac)$.

The form of Spitzer's identity in Theorem 1.3.1 is an immediate consequence of Theorem 1.3.12 applied to the complete Rota-Baxter algebra $R[[x]]$ in Exercise 1.2.7. A direct proof, using Rota's standard Rota-Baxter algebras, can be found in Section 3.2.4 where we also consider the Rota-Baxter algebra originally considered by Spitzer.

We will generalize the identity in two steps. By Exercise 1.2.7, Spitzer's Theorem 1.3.1 is a statement on the commutative complete Rota-Baxter algebra $R[[x]]$. We first generalize this theorem to any complete commutative Rota-Baxter algebra in Theorem 1.3.12. We then remove the commutativity condition by using the Baker-Campbell-Hausdorff formula, and obtain Theorem 1.3.17.

Before proving Theorem 1.3.1, we first give some examples to get ourselves acquainted with this seemingly unmotivated identity. They can be skipped at the first reading.

Example 1.3.2. (Geometric series) Consider the case where P is the identity map id on a $\mathbf{k}$-algebra R. Recall from Exercise 1.1.7 that id is a Rota-Baxter operator of weight -1. Then the left hand side of Eq. (1.28) becomes the power series

$$\exp\big(-\log(1-ax)\big)=\exp(\log(1-ax)^{-1})=\frac{1}{1-ax}$$

and the right hand side is just

$$\sum_{n=0}^{\infty}x^n a^n=\sum_{n=0}^{\infty}(ax)^n.$$

So we have the familiar geometric series.

In the left hand side of Theorem 1.3.1, we have

$$\frac{1}{\lambda}\log(1+\lambda ax)=\frac{1}{\lambda}(\lambda ax-(\lambda ax)^2/2+\cdots)\xrightarrow{\lambda\to 0}ax.$$

Thus we have the limit form of Classical Spitzer's Identity when $\lambda\to 0$, called the *Classical Spitzer's Identity of weight zero*:

$$\exp\left(P(ax)\right)=\sum_{n=0}^{\infty}x^n\,\underbrace{P\big(aP(aP(\cdots(aP(a)))\big)}_{n\text{-times}}.$$

Note that the right hand side is the solution to the equation

$$y=1+xP(ay)(=1+xP(a)+x^2P(aP(ay))=\cdots).$$

Example 1.3.3. (Integration) Let $A=\mathrm{Cont}(\mathbb{R})$ be the $\mathbb{R}$-algebra of continuous functions on $\mathbb{R}$ and let $P(f)(u)=\int_0^u f(t)dt$. Consider the initial value problem

$$y'(u)=xy(u),y(0)=1.$$

Its solution is equivalent to the solution of

$$y=1+P(xy)$$

as one can check by differentiating the last equation. Thus the Classical Spitzer's Identity of weight zero states that the initial value problem has unique solution

$$\exp\left(x\int_0^u y(t)dt\right).$$

Taking $x=1$, this is a well-known fact in calculus.

Taking the expansion

$$\exp\left(x\int_0^u y(t)dt\right)=1+x\int_0^u y(t)dt+\frac{x^2}{2}\Big(\int_0^u y(t)dt\Big)^2+\cdots$$

and comparing it with the right hand side of the Classical Spitzer's Identity of weight zero

$$1 + x \int_0^u y(t)dt + x^2 \int_0^u \Big(y(t_1) \int_0^{t_1} y(t_2)dt_2\Big)dt_1 + \cdots$$

we obtain

$$\Big(\int_0^u y(t)dt\Big)^2 = 2 \int_0^u y(t_1) \int_0^{t_1} y(t_2)dt_2 dt_1,$$

$$\Big(\int_0^u y(t)dt\Big)^3 = 3! \int_0^u y(t_1) \int_0^{t_1} y(t_2) \int_0^{t_2} y(t_3)dt_3 dt_2 dt_1,$$

and in general,

$$\Big(\int_0^u y(t)dt\Big)^n = n! \underbrace{\int \cdots \int}_{0 \leqslant t_n \leqslant \cdots \leqslant t_1 \leqslant u} y(t_n) \cdots y(t_1)\, dt_n \cdots dt_1.$$

Example 1.3.4. (Waring's formula) Let $T = \{t_n\}_{n \geqslant 1}$ be a set of variables. Important polynomials in $\mathbb{Q}[T]$ are the *power sum symmetric functions*

$$p_k(t_1, \ldots, t_m) := t_1^k + t_2^k + \cdots + t_m^k, \quad k \geqslant 1,$$

and the *elementary symmetric functions*

$$e_n(t_1, \ldots, t_m) := \sum_{1 \leqslant i_1 < \cdots < i_n \leqslant m} t_{i_1} \cdots t_{i_n}, \quad n \geqslant 1,$$

with the convention that $e_0(t_1, \ldots, t_m) = 1$ and $e_n(t_1, \ldots, t_m) = 0$ if $m < n$. See [141, 164] for the role that they play in the study of symmetric functions. The two classes of functions are related by the well-known Waring's formula

$$\exp\Big(- \sum_{k=1}^{\infty} (-1)^k p_k(t_1, \ldots, t_m) x^k / k \Big) = \sum_{n=0}^{\infty} e_n(t_1, \ldots, t_m) x^n, \quad m \geqslant 1.$$

This formula can be derived from Spitzer's identity as follows. Let

$$A = \mathbb{Q}[T]^{\mathbb{N}} = \{(a_n)_{n \geqslant 1} | a_n \in \mathbb{Q}[T], n \geqslant 1\}.$$

Define

$$P((a_n)) = (0, a_1, a_1 + a_2, \ldots, \sum_{k \leqslant n-1} a_k, \ldots)$$

as in Exercise 1.1.6. Then P is a Rota-Baxter operator of weight 1. Taking $a = (t_m)_m$ in Theorem 1.3.1, we have $a^k = (t_m^k)_m$ and hence $P(a^k) =$

$(p_k(t_1, \ldots, t_{m-1}))_m$. Thus the left hand side of Spitzer's identity in Eq. (1.29) is

$$\exp\left(-\sum_{k=1}^{\infty} \frac{(-x)^k (p_k(t_1, \ldots, t_{m-1}))_m}{k}\right).$$

On the other hand, note that elementary symmetric functions can be recursively defined by

$$\begin{aligned} e_1(t_1, \ldots, t_m) &= t_1 + \cdots + t_m, \\ e_{n+1}(t_1, \ldots, t_m) &= \sum_{k=1}^{m-1} e_n(t_1, \ldots, t_k) t_{k+1}, n \geqslant 1, \end{aligned}$$

with the convention that the sum over the empty index set is zero. Then by an inductive argument on n, we obtain

$$P_a^n(1) = (e_n(t_1, \ldots, t_{m-1}))_m, n \geqslant 1, \tag{1.30}$$

in Eq. (1.29) for our choice of $a := (t_m)_m$.

Exercise 1.3.5. Verify Eq. (1.30).

By Eq. (1.30), the right hand side of Eq. (1.29) gives

$$\sum_{n=0}^{\infty} (e_n(t_1, \ldots, t_{m-1}))_m x^n.$$

We then have

$$\exp\left(-\sum_{k=1}^{\infty} \frac{(-x)^k (p_k(t_1, \ldots, t_{m-1}))_m}{k}\right) = \sum_{n=0}^{\infty} (e_n(t_1, \ldots, t_{m-1}))_m x^n. \tag{1.31}$$

For any commutative algebra C, the natural bijection

$$C^{\mathbb{N}}[[x]] \cong (C[[x]])^{\mathbb{N}}, \quad \sum_{n=0}^{\infty} (c_{n,m})_m t^n \mapsto \left(\sum_{n=0}^{\infty} c_{n,m} t^n\right)_m$$

is an algebra homomorphism:

$$\begin{aligned} \left(\sum_{i=0}^{\infty} (c_{i,m})_m t^i\right) \left(\sum_{j=0}^{\infty} (c'_{j,m})_m t^j\right) &= \sum_{k=0}^{\infty} \left(\sum_{i+j=k} (c_{i,m})_m (c'_{j,m})_m\right) t^k \\ &= \sum_{k=0}^{\infty} \left(\sum_{i+j=k} (c_{i,m} c'_{j,m})_m\right) t^k \end{aligned}$$

$$= \left(\sum_{k=0}^{\infty} \left(\sum_{i+j=k} c_{i,m} c'_{j,m} \right) t^k \right)_m$$

$$\mapsto \left(\sum_{k=0}^{\infty} \left(\sum_{i+j=k} c_{i,m} c'_{j,m} \right) t^k \right)_m$$

$$= \left(\left(\sum_{i=0}^{\infty} c_{i,m} t^i \right) \left(\sum_{j=0}^{\infty} c'_{j,m} t^j \right) \right)_m$$

$$= \left(\sum_{i=0}^{\infty} c_{i,m} t^i \right)_m \left(\sum_{j=0}^{\infty} c'_{j,m} t^j \right)_m .$$

Then Eq. (1.31) leads to

$$\left(\exp \left(- \sum_{k=1}^{\infty} \frac{(-x)^k p_k(t_1, \ldots, t_{m-1})}{k} \right) \right)_m = \left(\sum_{n=0}^{\infty} e_n(t_1, \ldots, t_m) x^n \right)_m .$$

Comparing the m-th components gives Waring's formula.

Exercise 1.3.6. Apply the above argument for the RBO P of weight -1 in Example 1.1.5 to prove the "nonstrict" Waring's formula

$$\exp \left(\sum_{k=1}^{\infty} p_k(t_1, \ldots, t_m) x^k / k \right) = \sum_{n=0}^{\infty} \bar{e}_n(t_1, \ldots, t_m) x^n, \quad m \geqslant 1.$$

Here $\bar{e}_n(t_1, \ldots, t_m) = \sum\limits_{1 \leqslant i_1 \leqslant \cdots \leqslant i_n \leqslant m} t_{i_1} \cdots t_{i_n}, n \geqslant 1$ are the nonstrict elementary symmetric functions.

Example 1.3.7. **(Eulerian Identity)** Now consider Spitzer's Identity in the Rota-Baxter $\mathbb{Q}[t]$ with the q-integral in Example 1.1.8 and take $a = t$. Then the right hand side of Eq. (1.28) is

$$1 + \sum_{n=1}^{\infty} \frac{t^n q^{n(n+1)/2}}{(1-q)(1-q^2)\cdots(1-q^n)} x^n .$$

The left hand side of Eq. (1.28) is

$$\exp \left(- \sum_{k=1}^{\infty} \frac{(-x)^k P(t^k)}{k} \right) = \exp \left(- \sum_{k=1}^{\infty} \frac{(-x)^k}{k} \frac{q^k t^k}{1-q^k} \right)$$
$$= \exp \left(- \sum_{n=1}^{\infty} \sum_{k=1}^{\infty} \frac{(-x)^k}{k} q^{kn} t^k \right)$$

$$= \prod_{n=1}^{\infty} \exp\left(-\sum_{k=1}^{\infty} \frac{(-x)^k}{k} (q^n t)^k\right)$$
$$= \prod_{n=1}^{\infty} \exp\left(\log(1 + q^n x t)\right)$$
$$= \prod_{n=1}^{\infty} (1 + q^n x t).$$

Thus we have obtained an Eulerian identity

$$1 + \sum_{n=1}^{\infty} \frac{t^n q^{n(n+1)/2} x^n}{(1-q)(1-q^2)\cdots(1-q^n)} = \prod_{n=1}^{\infty} (1 + q^n x t).$$

Exercise 1.3.8. Applying a similar argument to Exercise 1.1.9 to prove the following Eulerian identity

$$\sum_{n\geqslant 0} \frac{x^n t^n}{(1-q)\cdots(1-q^n)} = \prod_{k\geqslant 0} \frac{1}{1-q^k x t}.$$

Example 1.3.9. (Maximum) We next apply Spitzer's Identity to the "max" operator in Example 1.1.16 for $f : \mathbb{R} \to \mathbb{Q}$ with finite support. By Eqs. (1.22) and (1.21), we have

$$(P(f)f)(t) = \sum_{u+v=t} P(f)(u) f(v)$$
$$= \sum_{u+v=t} \left(\sum_{\max(0,w)=u} f(w) \right) f(v)$$
$$= \sum_{\max(0,w)+v=t} f(w) f(v)$$
$$= \sum_{\max(v,w+v)=t} f(w) f(v).$$

Thus

$$P(P(f)f)(t) = \sum_{\max(0,\max(v,w+v))=t} f(w) f(v)$$
$$= \sum_{\max(0,v,w+v)=t} f(w) f(v).$$

More generally by an inductive proof on n, we obtain

$$P_f^n(1)(u) = \sum_{\max(0,u_1,u_1+u_2,\ldots,u_1+\cdots+u_n)=u} f(u_1)\cdots f(u_n).$$

Thus Spitzer's Identity in Eq. (1.28) yields

$$\exp\left(\sum_{k=1}^{\infty}\frac{x^k}{k}\sum_{\max(0,u_1+\cdots+u_k)=u} f(u_1)\cdots f(u_k)\right)$$
$$=1+\sum_{n=1}^{\infty}x^n \sum_{\max(0,u_1,u_1+u_2,\ldots,u_1+\cdots+u_n)=u} f(u_1)\cdots f(u_n). \qquad (1.32)$$

Example 1.3.10. (Fluctuation theory) We include this example since this is the original case of Spitzer, as presented in [160]. Readers without the probability background can skip this example. But note its similarity to Example 1.3.9.

Define the Banach algebra

$$A := \left\{\varphi(t) := \int_{-\infty}^{\infty} e^{itu}dF(u) \;\Big|\; \int_{-\infty}^{\infty} |dF(u)| < \infty, F(-\infty) \text{ exists}\right\},$$

where $F(-\infty) := \lim_{u\to-\infty} F(u)$. If F is the distribution function of a random variable X, then $\varphi(t)$ is the characteristic functions of X. Define

$$P(\varphi)(t) = \int_0^{\infty} e^{itu}dF(u) + F(0) - F(-\infty).$$

Then $P(\varphi)(t)$ is the characteristic function of the random variable $\max\{0, X\}$. As a consequence, by Example 1.1.16, we conclude that (A, P) is a Rota-Baxter algebra of weight -1.

Let $\{T_k\}$ be a sequence of independent random variables with identical distribution function $F(t)$ and characteristic function

$$\phi(t) = \int_{-\infty}^{\infty} e^{itu}dF(u).$$

Let $S_n = X_1 + \cdots + X_n$ and let $M_n = \max(0, S_1, S_2, \ldots, S_n)$. Let $F_n(u) = \text{Prob}(M_n < u)$ (here Prob stands for probability) be the distribution function of M_n. Applying Spitzer's Identity in Eq. (1.28) to $a = -\phi(t)$, we obtain the *original* identity of Spitzer:

$$\exp\left(\sum_{k=1}^{\infty}\frac{x^k}{k}\left(\int_0^{\infty} e^{itu}dF(u) + F(0)\right)\right) = \sum_{n=0}^{\infty} x^n \int_0^{\infty} e^{itu}dF_n(u).$$

1.3.2 Spitzer's identity in the commutative case

Recall from Theorem 1.1.17 that the double product $\star_P$ from a Rota-Baxter algebra (R, P) is defined by

$$a \star_P b = aP(b) + P(a)b + \lambda ab.$$

Proposition 1.3.11. *Let (R, P) be a Rota-Baxter algebra of weight λ. Then*
(a) $-\lambda a \star_P b = P(a)P(b) - \tilde{P}(a)\tilde{P}(b),\ -\lambda P(a)P(b) = P\big(P(a)P(b) - \tilde{P}(a)\tilde{P}(b)\big).$

(b) *For $n \geqslant 2$,*

$$-\lambda^{n-1}\prod_{i=1}^{n} P(a_i) = \lambda^{n-2} P\left(\prod_{i=1}^{n} P(a_i) - \prod_{i=1}^{n}(-\tilde{P}(a_i))\right), \quad a_i \in R,\ 1 \leqslant i \leqslant n.$$

(c) (Kingman, 1962) [119]

$$-\lambda^{n-1}P(a)^n = \lambda^{n-2}P\big(P(a)^n - (-\tilde{P}(a))^n\big), \quad a \in R. \tag{1.33}$$

(d) *If λ is not a zero divisor of R, then*

$$\lambda P(a)^n = -P\big(P(a)^n - (-\tilde{P}(a))^n\big), \quad a \in R. \tag{1.34}$$

When λ is invertible, Items (b) and (c) can be further simplified by canceling the extra powers of λ on both sides of the equations.

Proof. (a) By Theorem 1.1.18, for given $a, b \in R$, there is a $c \in R$ such that $P(a)P(b) = P(c)$ and $\tilde{P}(a)\tilde{P}(b) = -\tilde{P}(c)$. By the proof of Theorem 1.1.18, one such c can be taken to be $a \star_P b$. With this c we have

$$P(a)P(b) - \tilde{P}(a)\tilde{P}(b) = P(c) + \tilde{P}(c) = -\lambda c.$$

This proves the first equation. Applying P to the first equation, we obtain

$$P(P(a)P(b) - \tilde{P}(a)\tilde{P}(b)) = -\lambda P(a \star_P b) = -\lambda P(a)P(b)$$

by the axiom of the Rota-Baxter operator.

(b) We use induction on $n \geqslant 2$ with $n = 2$ verified in (a). Assume the equation holds for n. Then since $\tilde{P}$ is also an RBO of weight λ with $\tilde{\tilde{P}} = P$, by the induction hypothesis, we have

$$\begin{aligned} -\lambda^{n-1}\prod_{i=1}^{n}\tilde{P}(a_i) &= \lambda^{n-2}\tilde{P}\left(\prod_{i=1}^{n}\tilde{P}(a_i) - \prod_{i=1}^{n}(-P(a_i))\right) \\ &= \lambda^{n-2}(-1)^n\tilde{P}\left(\prod_{i=1}^{n}\tilde{P}(-a_i) - \prod_{i=1}^{n}P(a_i)\right). \end{aligned} \tag{1.35}$$

That is,

$$\lambda^{n-2}\tilde{P}\left(\prod_{i=1}^{n}\tilde{P}(-a_i)-\prod_{i=1}^{n}P(a_i)\right)=(-\lambda)^{n-1}\prod_{i=1}^{n}\tilde{P}(a_i). \tag{1.36}$$

Thus we have

$$\begin{aligned}
-\lambda^n\prod_{i=1}^{n+1}P(a_i) &= \lambda\left(-\lambda^{n-1}\prod_{i=1}^{n}P(a_i)\right)P(a_{n+1})\\
&= \lambda\left(\lambda^{n-2}P\left(\prod_{i=1}^{n}P(a_i)-\prod_{i=1}^{n}(-\tilde{P}(a_i))\right)\right)P(a_{n+1})\\
&\qquad \text{(by induction hypothesis)}\\
&= -\lambda^{n-2}P\left[P\left(\prod_{i=1}^{n}P(a_i)-\prod_{i=1}^{n}(-\tilde{P}(a_i))\right)P(a_{n+1})\right.\\
&\qquad \left.-\tilde{P}\left(\prod_{i=1}^{n}P(a_i)-\prod_{i=1}^{n}(-\tilde{P}(a_i))\right)\tilde{P}(a_{n+1})\right]\\
&\qquad \text{(by Item (a))}\\
&= \lambda^{n-1}P\left(\prod_{i=1}^{n}P(a_i)P(a_{n+1})-(-1)^{n-1}\prod_{i=1}^{n}\tilde{P}(a_i)\tilde{P}(a_{n+1})\right)\\
&\qquad \text{(by induction hypothesis and Eq. (1.36))}\\
&= \lambda^{n-1}P\left(\prod_{i=1}^{n+1}P(a_i)-\prod_{i=1}^{n+1}(-\tilde{P}(a_i))\right).
\end{aligned}$$

This completes the induction.

(c) This follows from (b) by taking $a_i = a$.

(d) This follows directly from Item (c). □

Theorem 1.3.12. (Commutative Case of Spitzer's Identity) *Let* $\mathbf{k}$ *be a commutative unitary* $\mathbb{Q}$*-algebra. Let* (R, R_n, P) *be a commutative complete Rota-Baxter* $\mathbf{k}$*-algebra of weight* λ *such that* λ *is not a zero divisor of* R*. Let* $a \in R_1$*. The equation*

$$b = 1 + P(ba) \tag{1.37}$$

has a unique solution

$$b = \exp\left(P\left(\frac{1}{\lambda}\log(1+\lambda a)\right)\right). \tag{1.38}$$

Further,

$$\exp\left(P\left(\frac{1}{\lambda}\log(1+\lambda a)\right)\right)=\sum_{n=0}^{\infty}\underbrace{P(P(P(\cdots(P(a)a)a)a))}_{n\text{-iterations}}. \tag{1.39}$$

We note that in Eq. (1.38), $\frac{1}{\lambda}\log(1+\lambda a)$ is just a shorthand notation for the series

$$-\sum_{n=1}^{\infty} \frac{\lambda^{n-1}(-a)^n}{n}.$$

Proof. Let $u \in R_1$. Applying Eq. (1.34), we obtain

$$\begin{aligned}
\lambda \exp(P(u)) &= \lambda + \sum_{n=1}^{\infty} \frac{\lambda}{n!} P(u)^n \\
&= \lambda - P\left(\sum_{n=1}^{\infty} \frac{P(u)^n}{n!} - \sum_{n=1}^{\infty} \frac{(-\tilde{P}(u))^n}{n!}\right) \\
&= \lambda - P\left(\sum_{n=0}^{\infty} \frac{P(u)^n}{n!} - \sum_{n=1}^{\infty} \frac{(-\tilde{P}(u))^n}{n!}\right) \\
&= \lambda - P(\exp(P(u)) - \exp(-\tilde{P}(u))) \\
&= \lambda - P\left(\exp(P(u)) - \exp(\lambda u + P(u))\right) \\
&= \lambda - P\left(\exp(P(u)) - \exp(\lambda u)\exp(P(u))\right) \\
&= \lambda - P\big(\exp(P(u))(1 - \exp(\lambda u))\big).
\end{aligned}$$

That is,

$$\lambda \exp(P(u)) = \lambda - P\big(\exp(P(u))(1 - \exp(\lambda u))\big). \tag{1.40}$$

Now for a given $a \in R_1$, denote

$$u := \frac{1}{\lambda}\log(1+\lambda a) \left(= -\sum_{n=1}^{\infty} \lambda^{n-1}\frac{(-a)^n}{n}\right).$$

It is a well-defined element in R_1. Further, solving for λa in the above equation, we obtain

$$-\lambda a = 1 - \exp(\lambda u).$$

Thus from Eq. (1.40) we obtain

$$\lambda \exp(P(u)) = \lambda + P(\exp(P(u))\lambda a).$$

Since λ is not a zero divisor in R, we have

$$\exp(P(u)) = 1 + P(\exp(P(u))a).$$

Thus Eq. (1.38) defines a solution of Eq. (1.37).

Suppose b and b' are both solutions of Eq. (1.37). Then $b-b' = P((b-b')a)$. Suppose $b \neq b'$. Then by the completeness of R as a filtered algebra, we have $\cap_{n\geqslant 0} R_n = \{0\}$. Thus there is an $n \geqslant 1$ such that $b - b' \neq 0$ is in R_n, but not in R_{n+1}. Since a is in R_1, $(b-b')a$, $b-b' = P((b-b')a)$ are in R_{n+1}. This is a contradiction, proving the uniqueness of the solution.

Finally, since a is in R_1, and R_n for $n \geqslant 1$ are Rota-Baxter ideals, we have

$$\underbrace{P(P(P(\cdots P(P(a)a)a\cdots)a)a)}_{n\text{-iterations}} \in R_n$$

and so the series

$$\sum_{n=0}^{\infty} \underbrace{P(P(P(\cdots P(P(a)a)a\cdots)a)a)}_{n\text{-iterations}}$$

converges. Substituting it into Eq. (1.37), we see that it is a solution of the equation. Then Eq. (1.39) follows by the uniqueness of the solution of Eq. (1.37). □

Remark 1.3.13. We note that in the proof we have made essential use of the equation

$$\exp(\lambda u + P(u)) = \exp(\lambda u)\exp(P(u))$$

which requires that λu and $P(u)$ commute with each other. When the two elements do not commute with each other, say, when they are matrices, this relation no longer holds. To overcome this difficulty, we use the Baker-Campbell-Hausdorff formula to be discussed next.

1.3.3 The BCH recursion

We first recall some classical results which can be found in any treatise on Lie groups. See [150], for example, or check the web page:

http://mathworld.wolfram.com/Baker-Campbell-HausdorffSeries.html

Let $M(x,y)$ be the free monoid generated by x and y. Thus

$$M(x,y) = \{1\} \cup \{z_1 \cdots z_k \mid z_i \in \{x,y\}, 1 \leqslant i \leqslant k, k \geqslant 1\}$$

with the concatenation product. Then extending by linearity, this product induces a product on the free $\mathbf{k}$-module

$$\mathbf{k}\langle x,y\rangle := \mathbf{k}M(x,y) = \mathbf{k}\cdot 1 \bigoplus \left(\bigoplus_{w\in M(x,y)} \mathbf{k}w \right),$$

making it into an associative $\mathbf{k}$-algebra. It is in fact the free $\mathbf{k}$-algebra generated by $\{x,y\}$. This follows from the construction of $M(x,y)$ as the free monoid generated by x and y. For $n \geqslant 1$, let I_n be the ideals of $\mathbf{k}\langle x, y\rangle$ generated by the monomials $z_1 \cdots z_n, z_i \in \{x, y\}, 1 \leqslant i \leqslant n$. Define

$$\mathbf{k}\langle\langle x, y\rangle\rangle = \varprojlim_n \mathbf{k}\langle x, y\rangle / I_n.$$

In more concrete terms, $\mathbf{k}\langle\langle x, y\rangle\rangle$ can be identified with the set of formal power series in the noncommutative variables x and y

$$\mathbf{k}\langle\langle x, y\rangle\rangle = \left\{ \sum_{w \in M(x,y)} a_w w \,\middle|\, a_w \in \mathbf{k}, w \in M(x, y) \right\}$$

with product given by the concatenation of the monomials w and by bilinearity. The degree of such a formal sum is defined to be the lowest degree of monomials w with $a_w \neq 0$. So in the sense of Definition 1.2.1, $\mathbf{k}\langle\langle x, y\rangle\rangle$ is a complete filtered algebra with filtration given by the ideals $\bar{I}_n$ of formal power series of degree n for $n \geqslant 1$. Therefore, when $\mathbb{Q} \subseteq \mathbf{k}$, formal exponential and logarithmic functions are well-defined.

Theorem 1.3.14. (Baker-Campbell-Hausdorff formula) *Suppose* $\mathbf{k}$ *contains* $\mathbb{Q}$.

(a) *There is a unique power series* $\mathrm{BCH}(x, y)$ *of degree* 2 *in the noncommutative power series algebra* $\mathbb{Z}\langle\langle x, y\rangle\rangle$ *such that*

$$\exp(x) \exp(y) = \exp\big(x + y + \mathrm{BCH}(x, y)\big).$$

(b) *For any complete filtered* $\mathbf{k}$*-algebra* R *and any* $a, b \in R_1$, *we have*

$$\exp(a) \exp(b) = \exp\big(a + b + \mathrm{BCH}(a, b)\big).$$

The first few terms of $\mathrm{BCH}(x, y)$ are

$$\begin{aligned}\mathrm{BCH}(x, y) = {} & \frac{1}{2}[x, y] + \frac{1}{12}[[x, y], y] - \frac{1}{12}[[x, y], x] \\ & - \frac{1}{48}[y, [x, [x, y]]] - \frac{1}{48}[x, [y, [x, y]]] + \cdots,\end{aligned}$$

where $[x, y] = xy - yx$ denotes the commutator of x and y.

Thus we have

$$\mathrm{BCH}(x, y) = -x - y + \log(\exp(x) \exp(y)).$$

Proposition 1.3.15. *Let* (R, R_n, P) *be a complete filtered algebra equipped with a linear operator* P *that preserves the filtration. Let* $\lambda \in \mathbf{k}$ *be invertible and denote* $\tilde{P} = -\lambda\,\mathrm{id} - P$. *There is a unique map* $\chi := \chi_\lambda : R_1 \to R_1$ *that satisfies the equation*

$$\chi_\lambda(a) = a + \frac{1}{\lambda}\mathrm{BCH}\Big(P(\chi_\lambda(a)), \tilde{P}(\chi_\lambda(a))\Big), \quad a \in R_1. \tag{1.41}$$

χ_λ was introduced in [61] when $\lambda = -1$ and was called the BCH-*recursion*. The general case was considered in [63]. A more conceptual proof of Proposition 1.3.15 in the context of Lie algebras can be found in [144]. To simplify the notations, we usually suppress λ from χ_λ.

Proof. For given $a \in R_1$, define a sequence $\chi_n(a)$ recursively by

$$\begin{aligned}\chi_0(a) &= a,\\ \chi_{n+1}(a) &= a + \frac{1}{\lambda}\mathrm{BCH}(P(\chi_n(a)), \tilde{P}(\chi_n(a))).\end{aligned}$$

We claim that the sequence has a limit that serves as the unique solution to Eq. (1.41). We first define a map

$$\Lambda : R \to R, \quad \Lambda(x) := \frac{1}{\lambda}\mathrm{BCH}(P(x), \tilde{P}(x)), \quad x \in R.$$

Then $\Lambda(x)$ is a degree of two noncommutative power series and

$$\chi_{n+1}(a) = a + \Lambda(\chi_n(a)), \quad \forall a \in R_1.$$

Further, for $s \in R_n, n \geqslant 1$, $\Lambda(a+s)$ is $\Lambda(a)$ plus a sum in which each term has s occurring at least once, and hence is contained in R_{n+1}. Thus we have

$$\Lambda(a \bmod R_n) \equiv \Lambda(a) \bmod R_{n+1}. \tag{1.42}$$

Now we have

$$\chi_1(a) = a + \Lambda(\chi_0(a)) = a + \Lambda(a) \equiv a \equiv \chi_0(a) \bmod R_2.$$

Assume $\chi_n(a) \equiv \chi_{n-1}(a) \bmod R_{n+1}$. Then by Eq. (1.42), we have

$$\begin{aligned}\chi_{n+1}(a) &= a + \Lambda(\chi_n(a))\\ &\equiv a + \Lambda(\chi_{n-1}(a) \bmod R_{n+1})\\ &\equiv a + \Lambda(\chi_{n-1}(a)) \bmod R_{n+2}\\ &\equiv \chi_n(a) \bmod R_{n+2}.\end{aligned}$$

Thus $\lim\limits_{n\to\infty} \chi_n(a)$ exists and is a solution of Eq. (1.41).

Suppose b is another solution of Eq. (1.41). Then, as above, we have

$$\chi_0(a) = a \equiv a + \Lambda(b) \equiv b \bmod R_2.$$

Induction on n gives

$$\begin{aligned}\chi_{n+1}(a) &= a + \Lambda(\chi_n(a))\\ &\equiv a + \Lambda(b \bmod R_{n+2})\\ &\equiv a + \Lambda(b) \bmod R_{n+3}\\ &\equiv b \bmod R_{n+3}.\end{aligned}$$

Thus $b = \lim\limits_{n\to\infty} \chi_n(a)$, proving the uniqueness of the solution of Eq. (1.41). □

As a direct consequence of Proposition 1.3.15, we obtain

Corollary 1.3.16. *Let* (R, R_n, P) *be as in Proposition* 1.3.15. *For* $a \in R_1$, *we have*

$$\exp(-\lambda a) = \exp(P(\chi(a))) \exp(\tilde{P}(\chi(a))). \tag{1.43}$$

Proof. By Proposition 1.3.15 and Theorem 1.3.14, we have

$$\begin{aligned} &\exp(P(\chi(a))) \exp(\tilde{P}(\chi(a))) \\ &= \exp(-\lambda\chi(a) + \mathrm{BCH}(P(\chi(a)), \tilde{P}(\chi(a)))) \\ &= \exp(-\lambda a). \end{aligned}$$

□

1.3.4 Spitzer's identity in the noncommutative case

For $a \in R$, inductively define

$$(Pa)^{[n+1]} := P((Pa)^{[n]}\, a) \text{ and } (Pa)^{\{n+1\}} := P(a\,(Pa)^{\{n\}})$$

with the convention that $(Pa)^{[1]} = P(a) = (Pa)^{\{1\}}$ and $(Pa)^{[0]} = 1 = (Pa)^{\{0\}}$.

Theorem 1.3.17. *Let* (R, R_n, P) *be a complete filtered Rota-Baxter algebra of weight* λ *such that* λ *is not a zero divisor of* R. *Let* $a \in R_1$.

(a) *The equation*

$$b = 1 - P(ba) \tag{1.44}$$

has a unique solution

$$b = \exp\left(-P\left(\chi\left(\frac{\log(1-\lambda a)}{-\lambda}\right)\right)\right). \tag{1.45}$$

(b) *The equation*

$$b = 1 - \tilde{P}(ab) \tag{1.46}$$

has a unique solution

$$b = \exp\left(-\tilde{P}\left(\chi\left(\frac{\log(1-\lambda a)}{-\lambda}\right)\right)\right). \tag{1.47}$$

Proof. We only need to verify for the first equation. The proof of the second equation is similar.

Since a is in R_1 and P preserves the filtration, the series

$$b = 1 - P(a) + P(P(a)a) - \cdots + (-1)^n (Pa)^{[n]} + \cdots,$$

as the limit

$$\lim_{n\to\infty}(1 - P(a) + P(P(a)a) - \cdots + (-1)^n (Pa)^{[n]}),$$

defines a unique element in R. By the recursive definition of $(Pa)^{[n]}$, we obtain

$$\begin{aligned}1 - P(ba) &= 1 - \sum_{n=0}^{\infty} P((-1)^n (Pa)^{[n]} a) \\ &= 1 + \sum_{n=0}^{\infty} (-1)^{n+1} (Pa)^{[n+1]} \\ &= b.\end{aligned}$$

Thus b is a solution of Eq. (1.44). Conversely, if $c \in R$ is a solution of Eq. (1.44), then by iterated substitution, we have

$$\begin{aligned}c &= 1 - P(ca) \\ &= 1 - P\big((1 - P(ca))a\big) \\ &= 1 - P(a) + P(P(ca)a) \\ &= \cdots \\ &= 1 - P(a) + P(P(a)a) - \cdots + (-1)^n (Pa)^{[n]} + \cdots .\end{aligned}$$

Therefore, Eq. (1.44) has a unique solution.

To verify that the b defined in Eq. (1.45) gives the solution, for the given $a \in R_1$, denote

$$u := \frac{\log(1 - \lambda a)}{-\lambda}.$$

Then

$$\lambda a = 1 - \exp(-\lambda u).$$

Applying Eq. (1.33) and Corollary 1.3.16, for our chosen b we have

$$\begin{aligned}\lambda \exp\big(-P(\chi(u))\big) &= \sum_{n=0}^{\infty} \lambda \frac{(-1)^n P(\chi(u))^n}{n!} \\ &= \lambda - \sum_{n=1}^{\infty} (-1)^n P\left(\frac{\big(P(\chi(u))\big)^n}{n!} - \frac{\big(-\tilde{P}(\chi(u))\big)^n}{n!}\right) \\ &= \lambda - P\left(\sum_{n=0}^{\infty} \frac{\big(-P(\chi(u))\big)^n}{n!} - \sum_{n=0}^{\infty} \frac{\big(\tilde{P}(\chi(u))\big)^n}{n!}\right) \\ &= \lambda - P\Big(\exp\big(-P(\chi(u))\big) - \exp\big(\tilde{P}(\chi(u))\big)\Big) \\ &= \lambda - P(\exp(-P(\chi(u))) - \exp(-P(\chi(u)))\exp(-\lambda u))\end{aligned}$$

$$= \lambda - P\left(\exp(-P(\chi(u)))\,(1 - \exp(-\lambda u))\right)$$
$$= \lambda - P\left(\exp(-P(\chi(u)))(\lambda a)\right)$$
$$= \lambda\left(1 - P\left(\exp(-P(\chi(u)))\,a\right)\right).$$

This verifies the first equation. □

Corollary 1.3.18. **(General Case of Spitzer's Identity)** *Let* (R, P, R_n) *be a complete filtered Rota-Baxter algebra. For* $a \in R_1$, *we have*

$$\sum_{n=0}^{\infty} (Pa)^{[n]} = \exp\left(-P\left(\chi\left(\frac{\log(1-\lambda a)}{-\lambda}\right)\right)\right), \tag{1.48}$$

$$\sum_{n=0}^{\infty} (\tilde{P}a)^{\{n\}} = \exp\left(-\tilde{P}\left(\chi\left(\frac{\log(1-\lambda a)}{-\lambda}\right)\right)\right). \tag{1.49}$$

Proof. Both sides of Eq. (1.48) are solutions of Eq. (1.44). This proves Eq. (1.48) by the uniqueness of the solution to Eq. (1.44).

The proof of Eq. (1.49) is the same, considering solutions to Eq. (1.46). □

Exercise 1.3.19. Prove the following variation of Theorem 1.3.17. Let (R, R_n, P) be a complete filtered Rota-Baxter algebra of weight λ such that λ is not a zero divisor of R. Let $a \in R_1$.

(a) The equation
$$b = 1 + P(ab)$$
has a unique solution
$$b = \exp\left(P\left(\chi\left(\frac{\log(1+\lambda a)}{\lambda}\right)\right)\right).$$

(b) The equation
$$b = 1 + \tilde{P}(ba)$$
has a unique solution
$$b = \exp\left(\tilde{P}\left(\chi\left(\frac{\log(1+\lambda a)}{\lambda}\right)\right)\right).$$

1.4 Atkinson's multiplicative decomposition and the algebraic Bogoliubov map

1.4.1 Atkinson's multiplicative decomposition

We begin with proving a lemma.

Lemma 1.4.1. *Let (R, P) be a nonunitary RBA of weight λ. Then we have*

$$P(x)\tilde{P}(y) = P\big(x\tilde{P}(y)\big) + \tilde{P}\big(P(x)y\big) \tag{1.50}$$

$$\tilde{P}(x)P(y) = P\big(\tilde{P}(x)y\big) + \tilde{P}\big(xP(y)\big), \quad x, y \in R. \tag{1.51}$$

Proof. By the Rota-Baxter relation and the definition of $\tilde{P}$, both sides of the first equation yield

$$-\lambda P(x)y - P(x)P(y).$$

This proves the first equation. The proof of the second equation is the same.□

The following is Atkinson's multiplicative decomposition [12].

Theorem 1.4.2. (Atkinson's Multiplicative Decomposition) *Let (R, P) be a Rota-Baxter algebra of weight λ. Let b and b' be the solutions of the equations* (1.44) *and* (1.46) *respectively. Then*

$$b(1 - \lambda a)b' = 1.$$

Proof. By Eq. (1.50) we have

$$\begin{aligned} bb' &= 1 - P(ba) - \tilde{P}(ab') + P(ba)\tilde{P}(ab') \\ &= 1 - P(ba) - \tilde{P}(ab') + P(ba\tilde{P}(ab')) + \tilde{P}(P(ba)ab') \\ &= 1 - P(ba(1 - \tilde{P}(ab'))) - \tilde{P}((1 - P(ba))ab') \\ &= 1 - P(bab') - \tilde{P}(bab') \\ &= 1 + \lambda bab'. \end{aligned}$$

Thus $1 = bb' - \lambda bab' = b(1 - \lambda a)b'$. □

We now prove the algebraic Birkhoff decomposition of complete filtered Rota-Baxter algebras.

Theorem 1.4.3. *Let (R, R_n, P) be a complete filtered Rota-Baxter algebra of weight -1. The following conditions are equivalent.*

(a) *P is idempotent: $P^2 = P$, when restricted to R_1.*

(b) *There is a direct sum decomposition of* $\mathbf{k}$*-modules*

$$R_1 = P(R_1) \oplus \tilde{P}(R_1).$$

(c) *There is a direct product decomposition of groups*

$$(1 + R_1) = (1 + P(R_1)) \times (1 + \tilde{P}(R_1)).$$

Proof. (a) $\Leftrightarrow$ (b): Note that (R_1, P) is a nonunitary Rota-Baxter algebra. If P is idempotent, then by Theorem 1.1.13 we have the direct sum decomposition. Conversely, suppose there is a direct sum decomposition $R_1 = P(R_1) \oplus \tilde{P}(R_1)$. Since

$$P\tilde{P} = P - P^2 = \tilde{P}P,$$

for any $a \in R_1$ we find that $(P - P^2)(a)$ is in $P(R_1) \cap \tilde{P}(R_1)$ which is zero because of the direct sum decomposition. Thus the restriction of P to R_1 is idempotent.

(b) $\Rightarrow$ (c): We just need to show that, for each $a \in R_1$, there is a unique $c \in P(R_1)$ and a unique $\tilde{c} \in \tilde{P}(R_1)$ that gives the factorization

$$1 + a = (1 + c)(1 + \tilde{c}).$$

Let $a \in R_1$ be given, and let b and $\tilde{b}$ be the solution of (1.44) and (1.46) respectively. Then by Theorem 1.4.2, we have

$$b(1 + a)\tilde{b} = 1.$$

By the constructions of $b, \tilde{b}$ and Theorem 1.3.17, we have $b = 1 - b_1$ and $\tilde{b} = 1 - \tilde{b}_1$ for $b_1 \in P(R_1)$ and $\tilde{b}_1 \in \tilde{P}(R_1)$. Thus

$$b^{-1} = 1 + b_1 + b_1^2 + \cdots \in 1 + P(R_1),$$
$$\tilde{b}^{-1} = 1 + \tilde{b}_1 + \tilde{b}_1^2 + \cdots \in 1 + \tilde{P}(R_1).$$

Hence $1 - a = b^{-1}b'^{-1}$. This proves the existence of the factorization.

For the uniqueness of the factorization, suppose we have

$$1 + a = (1 + c)(1 + \tilde{c}) = (1 + d)(1 + \tilde{d})$$

with $c, d \in P(R_1)$ and $\tilde{c}, \tilde{d} \in \tilde{P}(R_1)$. Then

$$(1 + d)^{-1}(1 + c) = (1 + \tilde{d})(1 + \tilde{c})^{-1}$$

which is in $(1 + P(R_1)) \cap (1 + \tilde{P}(R_1))$. If c is in this intersection, then

$$c = 1 + P(d) = 1 + \tilde{P}(d')$$

for some $d, d' \in R_1$. Then $P(d) = \tilde{P}(d')$ is in $P(R_1) \cap \tilde{P}(R_1)$ which is 0 by assumption. Thus $P(d) = 0$ and hence $c = 1$.

(c) $\Rightarrow$ (b): Since $P + \tilde{P} = \mathrm{id}$, we have $R_1 = P(R_1) + \tilde{P}(R_1)$. So we just need to show $P(R_1) \cap \tilde{P}(R_1) = 0$. This follows from $(1 + P(R_1)) \cap (1 + \tilde{P}(R_1)) = \{1\}$. $\square$

1.4.2 The algebraic Bogoliubov map

For $a \in R_1$, let a_- be the unique solution of $b = 1 + P(ba)$ from Theorem 1.3.17 and let

$$\gamma(a) = a_-\, a.$$

Similarly, let $\tilde{a}$ be the unique solution of $b = 1 + \tilde{P}(ab)$ and let

$$\tilde{\gamma}(a) = a\tilde{a}.$$

By Theorem 1.1.17, R_1 with the product $\star_P$ is still a complete algebra. Define

$$\exp_P : R_1 \to 1 + R_1, \ \exp_P(a) := \sum_{n=0}^{\infty} \frac{a^{\star_P n}}{n!}$$

where $a^{\star_P n}$ is the n-th power of a under the product $\star_P$.

Theorem 1.4.4. *Let R be a complete Rota-Baxter algebra of weight -1. The following diagram commutes.*

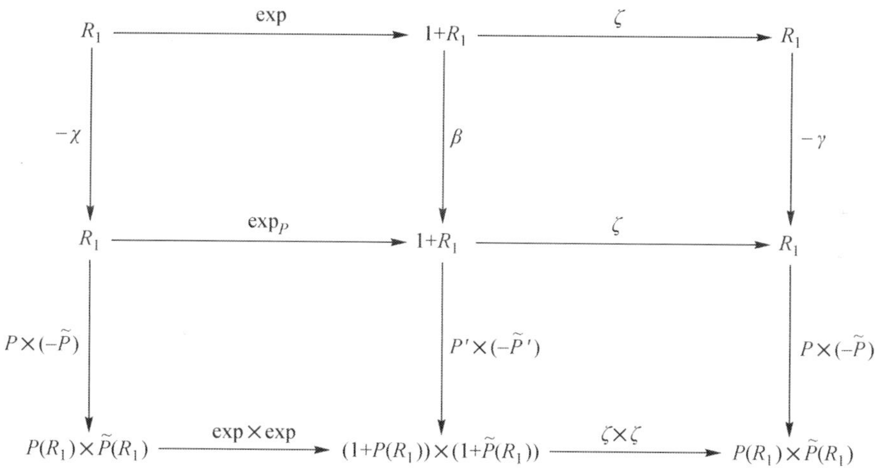

(1.52)

Here $\zeta(x) = x - 1$ and β is defined to be the composition

$$\beta = \zeta^{-1} \circ (-\gamma) \circ \zeta.$$

So $\beta(c) = 1 - \gamma(c-1)$. Similarly define $P' \times (-\tilde{P}')$ to be the composition $(\zeta^{-1} \times \zeta^{-1}) \circ (P \times (-\tilde{P})) \circ \zeta$. So

$$P' : 1 + R_1 \to 1 + R_1, \quad a \mapsto 1 + P(a-1),$$
$$\tilde{P}' : 1 + R_1 \to 1 + R_1, \quad a \mapsto 1 - \tilde{P}(a-1).$$

We call the map β the *algebraic Bogoliubov map* because it gives the Bogoliubov map in the renormalization theory of perturbative quantum field theory. See Section 2.4.2.

Proof. We only need to prove the commutativity of the upper half and the lower half of the diagram. By the way the two maps in the middle column are defined and by the bijectivity of the horizontal maps in the right half of the diagram, it follows that the other squares are also commutative.

Verifying the commutativity of the top half means verifying

$$-\gamma \circ \zeta \circ \exp(u) = \zeta \circ \exp_P \circ(-\chi(u)),$$

that is,

$$-\gamma(\exp(u) - 1) = \exp_P(-\chi(u)) - 1.$$

By Theorem 1.3.17, $\exp(-P(\chi(u)))$ is the unique solution of the equation

$$b = 1 - P(b(1 - \exp(u))) = 1 + P(b(\exp(u) - 1)).$$

Thus

$$\begin{aligned}-\gamma(\exp(u) - 1) &= -\exp\big(-P(\chi(u))\big)\big(\exp(u) - 1\big)\\ &= -\exp\big(-P(\chi(u))\big)\exp(u) + \exp\big(-P(\chi(u))\big).\end{aligned}$$

Hence

$$-\gamma(\exp(u) - 1) = -\exp\big(\tilde{P}(\chi(u))\big) + \exp\big(P(-\chi(u))\big) \tag{1.53}$$

by Corollary 1.3.16. On the other hand, by Theorem 1.1.17 and Theorem 1.1.18, we obtain

$$P(u)^n = P(u^{\star_P n}), \quad \tilde{P}(u)^n = -\tilde{P}((-u)^{\star_P n}).$$

Thus

$$\begin{aligned}P(\exp_P(u)) &= P\left(1 + u + \sum_{n=2}^{\infty} \frac{u^{\star_P n}}{n!}\right)\\ &= P(1) + P(u) + \sum_{n=2}^{\infty} \frac{P(u)^n}{n!}\\ &= \exp(P(u)) + P(1) - 1.\end{aligned}$$

Similarly,

$$\tilde{P}\big(\exp_P(u)\big) = -\exp\big(-\tilde{P}(u)\big) + \tilde{P}(1) + 1. \tag{1.54}$$

Combining these with Eq. (1.53), we obtain

$$\begin{aligned} &-\gamma(\exp(u)-1) \\ &= \tilde{P}\big(\exp_P(-\chi(u))\big) - 1 - \tilde{P}(1) + P\big(\exp_P(-\chi(u))\big) + 1 - P(1) \\ &= \exp_P\big(-\chi(u)\big) - 1, \end{aligned}$$

as desired.

Verifying the commutativity of the lower half of the diagram means verifying the two equations

$$\exp\big(P(u)\big) - 1 = P\big(\exp_P(u) - 1\big), \quad \exp\big(-\tilde{P}(u)\big) - 1 = -\tilde{P}\big(\exp_P(u) - 1\big)$$

which are immediately from Eqs. (1.53) and (1.54). □

1.5 Some related literature and problems

The main references for this chapter are [60, 61, 62, 63].

1.5.1 Spitzer's identity and algebraic Birkhoff decomposition

There are quite a few proofs of the classical case of Spitzer's identity. Other than the original proof of Spitzer [163] in the case of probability distributions, there are the combinatorial proofs of Baxter [24] and Kingman [119], the analytic proofs of Atkinson [12] and Wendel [172], as well as the algebraic proofs of Cartier [33] and Rota-Smith [160] working with free Rota-Baxter algebras and making use of Waring's formula. See also [124].

For the commutative case of Spitzer's Identity, we mainly followed the proof of Kingman since it is the simplest and can be adapted for the noncommutative case in Section 1.3.4.

Further generalizations of Spitzer's identity and Atkinson's factorization can be found in [63], where one can also find applications to the factorization theorem of Barron, Huang and Lepowsky [23] in vertex operator algebra, the plus-minus factorization of Aguiar, Bergeron and Sottile [4] in combinatorial Hopf algebras and Lie algebra polar decomposition of Munthe-Kaas, Quispel and Zanna [148, 175] in the context of the approximation of matrix exponentials in ordinary differential equations. See [50] for Rota-Baxter algebras and combinatorial identities.

1.5.2 Rota-Baxter Lie algebras and the classical Yang-Baxter equation

Let $\mathfrak{g}$ be a Lie algebra over $\mathbb{R}$. Let $r = \sum_i a_i \otimes b_i \in \mathfrak{g} \otimes \mathfrak{g}$. Define (in the universal enveloping algebra $U(\mathfrak{g})$ of $\mathfrak{g}$)

$$r_{12} = \sum_i a_i \otimes b_i \otimes 1, \quad r_{13} = \sum_i a_i \otimes 1 \otimes b_i, \quad r_{23} = \sum_i 1 \otimes a_i \otimes b_i,$$

$$[r_{12}, r_{13}] = \sum_{i,j} [a_i, a_j] \otimes b_i \otimes b_j, \ [r_{13}, r_{23}] = \sum_{i,j} a_i \otimes a_j \otimes [b_i, b_j],$$

$$[r_{23}, r_{12}] = \sum_{i,j} a_j \otimes [a_i, b_j] \otimes b_i.$$

The equation

$$\mathrm{C}(r) := [r_{12}, r_{13}] + [r_{12}, r_{23}] + [r_{13}, r_{23}] = 0 \tag{1.55}$$

is called the *classical Yang-Baxter equation* (CYBE).

Now let $\mathfrak{g}$ be a Lie algebra that is self dual: $\mathfrak{g}^* := \mathrm{Hom}(\mathfrak{g}, \mathbb{R}) \cong \mathfrak{g}$. Then $\mathfrak{g} \otimes \mathfrak{g} \cong \mathfrak{g} \otimes \mathfrak{g}^* \cong \mathrm{End}(\mathfrak{g})$. Let $r \in \mathfrak{g}^{\otimes 2}$ be anti-symmetric (i.e., $\sum_i a_i \otimes b_i = -\sum_i b_i \otimes a_i$). Then r is a solution (r-matrix) of the CYBE (1.55) if and only if the corresponding $P \in \mathrm{End}(\mathfrak{g})$ is a (Lie algebra) *Rota-Baxter operator of weight* 0:

$$[P(x), P(y)] = P[P(x), y] + P[x, P(y)],$$

called the *operator form of the CYBE.*

It is interesting to note [47] that the Baxter in Yang-Baxter equation is the Australian physicist Rodney Baxter, not the American mathematician Glen Baxter in the Rota-Baxter operator. This connection is pursued further by several authors [14, 26, 47, 124, 161]. In [17, 19, 18], the concept of an extended $\mathcal{O}$-operator was introduced that simultaneously generalized the Rota-Baxter operator of arbitrary weight and the $\mathcal{O}$-operator of Kupershmidt [124]. This is applied to the study of Lax pairs, CYBE and Lie bialgebras. Furthermore, behind $\mathcal{O}$-operators there is a natural structure of PostLie algebra coming from recent study of operads (see also Chapter 5).

For an associative algebra A, a less direct connection between the associative analogue of Eq. (1.55) and Rota-Baxter algebras was established by Aguiar [2]. The concept of an extended $\mathcal{O}$-operator on an associative algebra was introduced in [20] that generalizes the concept of a Rota-Baxter algebra [20] and that extends the connection in Lie algebras [14, 17] to associative algebra.

1.5.3 Operators related to the Rota-Baxter operator

Throughout the history of mathematics, mathematical objects have been studied often by operators defined on them. For example, field extensions are studied by their Galois groups and manifolds are studied through their differential fields. In this spirit, Rota [157] long time ago posed the problem of classifying all linear operators that can be defined on associative algebras and satisfy algebraic equations. There has been little progress on this problem [79] in part because there was no precise prescription on what to classify. There are several linear operators related to the Rota-Baxter operator, such as

- the *associative Nijenhuis operator* [48, 121]:
  $P(x)P(y) = P(P(x)y + xP(y) - P(xy))$;
- the (*Leroux*) *TD operator* [128]:
  $P(x)P(y) = P(P(x)y + xP(y) - xP(1)y)$.
- the *average operator* [30, 157]: $P(x)P(y) = P(xP(y))$.

Recent attempts were made in [100, 101] to address Rota's problem by operated algebras [88], computer algebra, rewriting systems [13] and Gröbner-Shirshov [25].

1.5.4 Read the classic

The reader is encouraged to consult the inspiring survey articles of Rota [156, 157, 158] and Rota-Smith [160] for early developments of Rota-Baxter algebras and related problems.

Chapter 2

Connected Hopf Algebras and Rota-Baxter Algebras

We discuss an important class of complete filtered Rota-Baxter algebras that comes from functionals on Hopf algebras with an increasing filtration, laying the mathematical foundation for applications to quantum field renormalization [42, 43, 60, 61, 62]. Particularly, we treat the case of Connes-Kreimer rooted tree Hopf algebra in detail. This is because, even if it is a pure mathematical structure, the rooted tree Hopf algebra carries a hierarchy structure similar to that of Feynman graphs in physics and can be regarded as a toy model for a real quantum field theory. We especially develop the Connes-Kreimer algebraic Birkhoff decomposition and the Bogoliubov map in the context of rooted trees. By doing so we hope to serve well a mathematician who is interested in the physics applications but lacks the physics background, and a physicist who would like to understand the mathematics structure behind the physics applications without having to get into the nitty-gritty of calculational details in Chapter 1.

We first review in Section 2.1 the definitions of modules and algebras by commutative diagrams to motivate the definitions of coalgebras and bialgebras in Section 2.2. Connected bialgebras and convolution (functional) algebras from bialgebras are discussed in Section 2.3, ended with the fundamental result that a connected bialgebra is a Hopf algebra. Similar rooted trees will appear in Chapter 4 for free Rota-Baxter algebras and in Chapter 5 for free dendriform algebras.

We start the study of rooted tree Hopf algebra in Section 2.3.4 as both an interesting example and as a prototype for Feynman graph Hopf algebras. This strong physics flavor is continued in Section 2.4.1 and Section 2.4.2 where we prove the main mathematical theorem in our application to Connes-Kreimer Hopf algebra approach to renormalization.

2.1 Algebras and modules revisited

To motivate the definition of coalgebras and comodules, we give the following interpretations of modules and algebras in terms of commutative diagrams. More details can be found in [117, 118].

2.1.1 Tensor product

Let M and N be $\mathbf{k}$-modules and C an (additive) abelian group. A map $f : M \times N \to C$ is called *bilinear* if

$$f(a_1 + a_2, b) = f(a_1, b) + f(a_2, b),$$

$$f(a, b_1 + b_2) = f(a, b_1) + f(a, b_2),$$

$$f(ka, b) = f(a, kb), \quad \forall a_1, a_2, a \in M, \ b_1, b_2, b \in N, \ k \in \mathbf{k}.$$

The *tensor product* $M \otimes_{\mathbf{k}} N$ of M and N over $\mathbf{k}$ is an abelian group together with a bilinear map $\iota : M \times N \to M \otimes_{\mathbf{k}} N$ satisfying the following universal property: For any $\mathbf{k}$-module C and any bilinear map $f : M \times N \to C$, there is a unique linear map $\tilde{f} : M \otimes_{\mathbf{k}} N \to C$ such that $f = \tilde{f} \circ \iota$

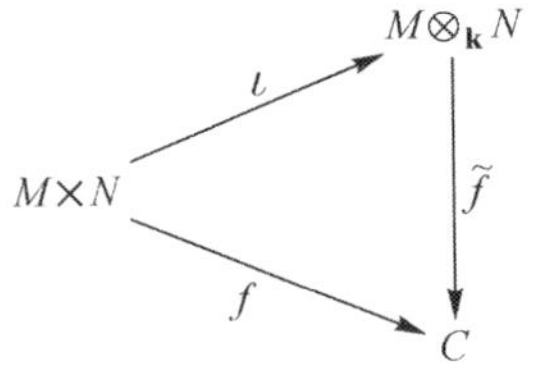

The image of $(a, b) \in M \times N$ in $M \otimes_{\mathbf{k}} N$ is denoted by $a \otimes b$, called a pure tensor. Elements in $M \otimes_{\mathbf{k}} N$ are finite sums of pure tensors. Unless otherwise stated, all the tensor products are taken over $\mathbf{k}$ and will be simply denoted by $M \otimes N$.

Exercise 2.1.1. Let M and N be $\mathbf{k}$-modules. Let F be the free abelian group on the set $M \times N$. Let I be the subgroup of F generated by all elements of F of the forms

(a) $(a_1 + a_2, b) - (a_1, b) - (a_2, b)$;
(b) $(a, b_1 + b_2) - (a, b_1) - (a, b_2)$;
(c) $(ka, b) - (a, kb)$, $\forall a_1, a_2, a \in M, \ b_1, b_2, b \in N, k \in \mathbf{k}$.

Prove that the quotient group F/I with the natural map $M \times N \to F \to F/I$ satisfies the above universal properties for the tensor product. Hence F/I is (isomorphic to) the tensor product $M \otimes N$.

Exercise 2.1.2. Let $\mathbf{k}$ be a field and let M and N be $\mathbf{k}$-modules (i.e., $\mathbf{k}$-vector spaces), with bases X and Y. Let T be the $\mathbf{k}$-vector space on the basis $X \times Y$. Let $\iota : M \times N \to T$ be the linear map induced by $\iota(x, y) = (x, y)$, $x \in X, y \in Y$. Prove that T together with ι is the tensor product of M and N.

Exercise 2.1.3. Let M be a $\mathbf{k}$-module. Then the maps

$$\begin{aligned} \alpha_\ell &: \mathbf{k} \otimes M \to M, k \otimes m \mapsto km, \\ \alpha_r &: M \otimes \mathbf{k} \to M, m \otimes k \mapsto km, \quad k \in \mathbf{k}, m \in M, \end{aligned} \tag{2.1}$$

are isomorphisms. Also show that the inverse maps are

$$\begin{aligned} \beta_\ell &: M \to \mathbf{k} \otimes M, m \mapsto 1 \otimes m, \\ \beta_r &: M \to M \otimes \mathbf{k}, m \mapsto m \otimes 1, \quad k \in \mathbf{k}, m \in M. \end{aligned} \tag{2.2}$$

Let $f_i : M_i \to N_i$, $i = 1, 2$, be linear maps. Then the natural map $M_1 \times N_1 \to M_2 \otimes N_2$ is bilinear and thus induces a linear map $f_1 \otimes f_2 : M_1 \otimes M_2 \to N_1 \otimes N_2$.

2.1.2 Algebras and modules by diagrams

A nonunitary $\mathbf{k}$-algebra A can be equivalently defined as a $\mathbf{k}$-module A together with a $\mathbf{k}$-module homomorphism $m := m_A : A \otimes A \to A$ such that the diagram

$$\text{(Associativity)} \qquad \begin{array}{ccc} A \otimes A \otimes A & \xrightarrow{m \otimes \mathrm{id}_A} & A \otimes A \\ {\scriptstyle \mathrm{id}_A \otimes m} \downarrow & & \downarrow {\scriptstyle m} \\ A \otimes A & \xrightarrow{m} & A \end{array} \tag{2.3}$$

is commutative. Furthermore A is a unitary $\mathbf{k}$-algebra if and only if there is a $\mathbf{k}$-module map $u := u_A : \mathbf{k} \to A$ such that the diagram

$$\text{(Unicity)} \qquad \begin{array}{ccccc} \mathbf{k} \otimes A & \xrightarrow{u \otimes \mathrm{id}_A} & A \otimes A & \xleftarrow{\mathrm{id}_A \otimes u} & A \otimes \mathbf{k} \\ & {\scriptstyle \alpha_\ell} \searrow & \downarrow {\scriptstyle m} & \swarrow {\scriptstyle \alpha_r} & \\ & & A & & \end{array} \tag{2.4}$$

is commutative. Here α_ℓ and α_r are the isomorphisms defined in Eq. (2.1).

In this setup, a linear map $f : A \to B$ between $\mathbf{k}$-algebras is a $\mathbf{k}$-algebra homomorphism meaning the commutativity of the diagrams

$$\begin{array}{ccc} A \otimes A & \xrightarrow{f \otimes f} & B \otimes B \\ {\scriptstyle m_A} \downarrow & & \downarrow {\scriptstyle m_B} \\ A & \xrightarrow{f} & B \end{array} \tag{2.5}$$

and

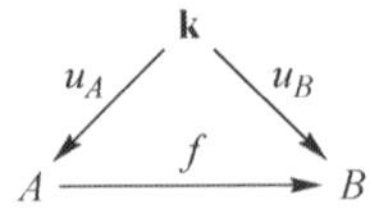

Example 2.1.4. Let A and B be **k**-algebras. The tensor product $A \otimes B$ is a **k**-algebra with the multiplication

$$m_{A\otimes B} : (A \otimes B) \otimes (A \otimes B) \overset{\mathrm{id}_A \otimes \tau \otimes \mathrm{id}_B}{\longrightarrow} A \otimes A \otimes B \otimes B \overset{m_A \otimes m_B}{\longrightarrow} A \otimes B,$$
$$(a_1 \otimes b_1) \otimes (a_2 \otimes b_2) \mapsto (a_1 a_2) \otimes (b_1 b_2),\ a_1, a_2 \in A, b_1, b_2 \in B,$$

with $\tau : B \otimes A \to A \otimes B, b \otimes a \mapsto a \otimes b$ being the switch map, and the unit map

$$u_{A\otimes B} : \mathbf{k} \to A \otimes B, \quad k \mapsto k \otimes k, \quad k \in \mathbf{k}.$$

2.2 Coalgebras and bialgebras

Further details on bialgebras and Hopf algebras can be found in the books [1, 118, 166].

2.2.1 Coalgebra

A **k**-*coalgebra* is obtained by reversing the arrows in the diagrams (2.3) and (2.4) of a **k**-algebra. More precisely, a **k**-coalgebra is a triple (C, Δ, ε) where C is a **k**-module, $\Delta := \Delta_C : C \to C \otimes C$, called the *coproduct*, and $\varepsilon := \varepsilon_C : C \to \mathbf{k}$, called the *counit*, are **k**-linear maps that make the following diagrams commute.

(Coassociativity)

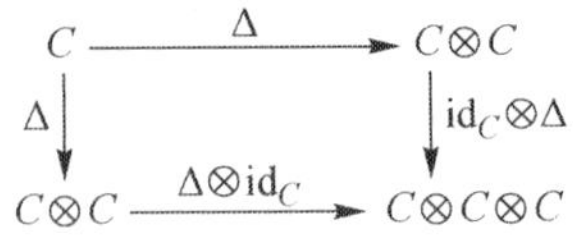

(2.6)

(Counicity)

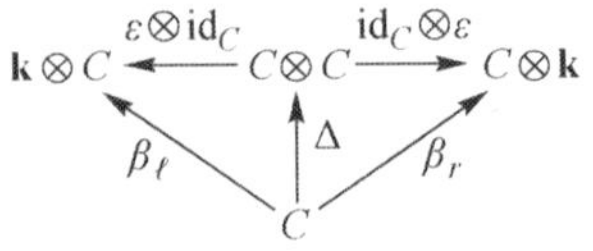

(2.7)

Here β_ℓ and β_r are the isomorphisms defined in Eq. (2.2).

A morphism $f : C \to D$ of coalgebras C and D is defined to be a $\mathbf{k}$-linear map f such that the following diagrams are commutative.

$$\begin{array}{ccc} C & \xrightarrow{\quad f \quad} & D \\ \downarrow{\scriptstyle \Delta_C} & & \downarrow{\scriptstyle \Delta_D} \\ C \otimes C & \xrightarrow{\quad f \otimes f \quad} & D \otimes D \end{array}$$

$$\begin{array}{ccc} C & \xrightarrow{\quad f \quad} & D \\ & {\scriptstyle \varepsilon_C} \searrow \quad \swarrow {\scriptstyle \varepsilon_D} & \\ & \mathbf{k} & \end{array}$$

For $a \in C$, we have $\Delta(a) = \sum_i a_{1\,i} \otimes a_{2\,i}$ which we denote simply by $\sum_{(a)} a_{(1)} \otimes a_{(2)}$ (the Sweedler notation). Then the coassociativity of Δ means

$$\sum_{(a)} \Bigg(\sum_{(a_{(1)})} a_{(1)(1)} \otimes a_{(1)(2)} \Bigg) \otimes a_{(2)} = \sum_{(a)} a_{(1)} \otimes \Bigg(\sum_{(a_{(2)})} a_{(2)(1)} \otimes a_{(2)(2)} \Bigg)$$

for which we use the shorthand notation

$$\Delta^2(a) = \sum_{(a)} a_{(1)} \otimes a_{(2)} \otimes a_{(3)}.$$

Example 2.2.1. The $\mathbf{k}$-algebra $\mathbf{k}$ has a natural structure of a $\mathbf{k}$-coalgebra with

$$\Delta_{\mathbf{k}} : \mathbf{k} \to \mathbf{k} \otimes \mathbf{k}, \;\; c \mapsto c \otimes 1, \;\; c \in \mathbf{k}$$

and

$$\varepsilon_{\mathbf{k}} = \mathrm{id}_{\mathbf{k}} : \mathbf{k} \to \mathbf{k}.$$

We also denote the multiplication in $\mathbf{k}$ by $m_{\mathbf{k}}$.

Example 2.2.2. Let C and D be $\mathbf{k}$-coalgebras. The tensor product $C \otimes D$ is a $\mathbf{k}$-coalgebra with the comultiplication

$$\Delta_{C \otimes D} : C \otimes D \overset{\Delta_C \otimes \Delta_D}{\longrightarrow} C \otimes C \otimes D \otimes D \overset{\mathrm{id}_C \otimes \tau \otimes \mathrm{id}_D}{\longrightarrow} C \otimes D \otimes C \otimes D$$

and the counit

$$\varepsilon_{C \otimes D} : C \otimes D \overset{\varepsilon_C \otimes \varepsilon_D}{\longrightarrow} \mathbf{k} \otimes \mathbf{k} \overset{m_{\mathbf{k}}}{\longrightarrow} \mathbf{k}.$$

2.2.2 Bialgebra

A $\mathbf{k}$-*bialgebra* is a quintuple $(A, m, u, \Delta, \varepsilon)$ where (A, m, u) is a $\mathbf{k}$-algebra and (A, Δ, ε) is a $\mathbf{k}$-coalgebra such that $m : A \otimes A \to A$ and $u : \mathbf{k} \to A$ are morphisms of $\mathbf{k}$-coalgebras, where the coalgebra structures on $A \otimes A$ and $\mathbf{k}$ are defined in Example 2.2.2 and Example 2.2.1 respectively. In other words, we have the commutativity of the following diagrams.

$$\begin{array}{ccc} A\otimes A & \xrightarrow{\quad m \quad} & A \\ {\scriptstyle \Delta_{A\otimes A}}\downarrow & & \downarrow{\scriptstyle \Delta} \\ A\otimes A\otimes A\otimes A & \xrightarrow{\quad m\otimes m \quad} & A\otimes A \end{array} \tag{2.8}$$

$$\begin{array}{ccc} A\otimes A & \xrightarrow{\quad m \quad} & A \\ {\scriptstyle \varepsilon\otimes\varepsilon}\downarrow & & \downarrow{\scriptstyle \varepsilon} \\ \mathbf{k}\otimes\mathbf{k} & \xrightarrow{\quad m_{\mathbf{k}} \quad} & \mathbf{k} \end{array} \tag{2.9}$$

$$\begin{array}{ccc} \mathbf{k} & \xrightarrow{\quad u \quad} & A \\ {\scriptstyle \Delta_{\mathbf{k}}}\downarrow & & \downarrow{\scriptstyle \Delta} \\ \mathbf{k}\otimes\mathbf{k} & \xrightarrow{\quad u\otimes u \quad} & A\otimes A \end{array} \tag{2.10}$$

$$\begin{array}{ccc} \mathbf{k} & \xrightarrow{\quad u \quad} & A \\ & {\scriptstyle \mathrm{id}_{\mathbf{k}}}\searrow \quad \swarrow{\scriptstyle \varepsilon} & \\ & \mathbf{k} & \end{array} \tag{2.11}$$

Exercise 2.2.3. Show that the conditions that m and u are morphisms of coalgebras can be equivalently replaced by the conditions that $\Delta : A \to A \otimes A$ and $\varepsilon : A \to \mathbf{k}$ are homomorphisms of $\mathbf{k}$-algebras. Here $A \otimes A$ is the tensor product algebra in Example 2.1.4.

2.2.3 Examples

2.2.3.1 *Divide power bialgebra*

The *divided power bialgebra* is defined by the quintuple $(A, m, u, \Delta, \varepsilon)$ where

(a) A is the free $\mathbf{k}$-module $\bigoplus\limits_{n=0}^{\infty} \mathbf{k}a_n$ with basis a_n, $n \geqslant 0$;

(b) $m : A \otimes A \to A, a_m \otimes a_n \mapsto \begin{pmatrix} m+n \\ m \end{pmatrix} a_{m+n}$;

(c) $u : \mathbf{k} \to A,\ \mathbf{1} \mapsto a_0$;

(d) $\Delta : A \to A \otimes A, a_n \mapsto \sum\limits_{k=0}^{n} a_k \otimes a_{n-k}$;

(e) $\varepsilon : A \to \mathbf{k},\ a_n \mapsto \delta_{0,n}\mathbf{1}$ where $\delta_{0,n}$ is the Kronecker delta.

Exercise 2.2.4. Verify that this is a bialgebra.

For a generalization of this bialgebra, see [11].

2.2.3.2 *Binomial bialgebra*

The *binomial bialgebra* is defined by the quintuple $(A, m, u, \Delta, \varepsilon)$ where

(a) A is the free $\mathbf{k}$-module $\bigoplus_{n=0}^{\infty} \mathbf{k}a_n$ with basis a_n, $n \geqslant 0$;
(b) $m : A \otimes A \to A, a_m \otimes a_n \mapsto a_{m+n}$;
(c) $u : \mathbf{k} \to A,\ \mathbf{1} \mapsto a_0$;
(d) $\Delta : A \to A \otimes A, a_n \mapsto \sum_{k=0}^{n} \binom{n}{k} a_k \otimes a_{n-k}$;
(e) $\varepsilon : A \to \mathbf{k},\ a_n \mapsto \delta_{0,n}\mathbf{1}$ where $\delta_{0,n}$ is the Kronecker delta.

Exercise 2.2.5. Verify that this is a bialgebra.

2.2.3.3 *Shuffle bialgebra*

More general than the divided power bialgebra is the shuffle product bialgebra. We devote a preliminary discussion here, with more details provided in Chapter 3.

Let V be a free $\mathbf{k}$-module. Define $V^{\otimes n}$ to be the n-th tensor power of V:

$$V^{\otimes n} = \underbrace{V \otimes \cdots \otimes V}_{n-\text{factors}}$$

with the convention that $V^{\otimes 0} = \mathbf{k}$. Let

$$\mathrm{Sh}(V) = \bigoplus_{n \geqslant 0} V^{\otimes n}.$$

Fix pure tensors $\mathfrak{a} := a_1 \otimes \cdots \otimes a_m$ in $V^{\otimes m}$ and $\mathfrak{b} := b_1 \otimes \cdots \otimes b_n$ in $V^{\otimes n}$ where $a_i, b_j \in V, 1 \leqslant i \leqslant m, 1 \leqslant j \leqslant n,$. A *shuffle of* $\mathfrak{a}$ *and* $\mathfrak{b}$ is a tensor list of a_i and b_j that does not change the order of the a_is and b_js. The *shuffle product* $(a_1 \otimes \cdots \otimes a_m) ш (b_1 \otimes \cdots \otimes b_n)$ is the sum of shuffles of $a_1 \otimes \cdots \otimes a_m$ and $b_1 \otimes \cdots \otimes b_n$.

Example 2.2.6. $a_1 ш (b_1 \otimes b_2) = a_1 \otimes b_1 \otimes b_2 + b_1 \otimes a_1 \otimes b_2 + b_1 \otimes b_2 \otimes a_1$.

Expanding by bilinearity, we obtain an associative product on $\mathrm{Sh}(V)$, making $\mathrm{Sh}(V)$ into a unitary $\mathbf{k}$-algebra with $u : \mathbf{k} \to A$ given by $u(k) = k \in \mathbf{k} \subseteq \mathrm{Sh}(V)$.

Define a coproduct $\Delta : \mathrm{Sh}(V) \to \mathrm{Sh}(V) \otimes \mathrm{Sh}(V)$ by

$$\Delta(a_1 \otimes \cdots \otimes a_n) = 1 \bigotimes (a_1 \otimes \cdots \otimes a_n) + a_1 \bigotimes (a_2 \otimes \cdots \otimes a_n) + \cdots$$
$$+ (a_1 \otimes \cdots \otimes a_{n-1}) \bigotimes a_n + (a_1 \otimes \cdots \otimes a_n) \bigotimes 1,\ n \geqslant 1$$

and $\Delta(k) = k\mathbf{1} \otimes \mathbf{1}, k \in \mathbf{k}$. Here we use $\bigotimes$ to denote the tensor product in the coproduct to distinguish it from the tensor product in $\mathrm{Sh}(V)$. Define a counit

$\varepsilon : \mathrm{Sh}(V) \to \mathbf{k}$ by

$$\varepsilon(a_1 \otimes \cdots \otimes a_n) = 0, \quad n \geqslant 1, \quad \text{and} \quad \varepsilon(\mathbf{1}) = \mathbf{1}.$$

Then we obtain a bialgebra, called the *shuffle product bialgebra.*

2.2.3.4 *Group algebra bialgebra*

Let G be a group and let $\mathbf{k}[G]$ be the group algebra over $\mathbf{k}$. So $\mathbf{k}[G]$ is $\oplus_{g\in G}\mathbf{k}g$ as vector space with multiplication induced by that of G. Define a coproduct

$$\Delta : \mathbf{k}[G] \to \mathbf{k}[G] \otimes \mathbf{k}[G], g \mapsto g \otimes g, g \in G,$$

and counit

$$\varepsilon : \mathbf{k}[G] \to \mathbf{k}, g \mapsto \mathbf{1}.$$

Then we have a bialgebra.

2.3 Connected bialgebra and Hopf algebra

In this section we introduce the concept of a Hopf algebra and show that a connected bialgebra is automatically a Hopf algebra.

2.3.1 Connected bialgebra

Definition 2.3.1. A bialgebra H is called a *graded bialgebra* if there are $\mathbf{k}$-submodules $H^{(n)}$, $n \geqslant 0$, of H such that

(a) $H = \bigoplus_{n=0}^{\infty} H^{(n)}$;

(b) $H^{(p)}H^{(q)} \subseteq H^{(p+q)}, p, q \geqslant 0$;

(c) $\Delta(H^{(n)}) \subseteq \bigoplus_{p+q=n} H^{(p)} \otimes H^{(q)}, n \geqslant 0$.

Elements of $H^{(n)}$ are said to have degree n. H is called *connected* if $H^{(0)} = \operatorname{im} u (= \mathbf{k})$.

The term connectedness comes from the fact that the cohomology bialgebra of a connected compact Lie group is a connected bialgebra.

Definition 2.3.2. A bialgebra H is called an (increasing) *filtered bialgebra* if there are $\mathbf{k}$-submodules H^n, $n \geqslant 0$, of H such that

(a) $H^n \subseteq H^{n+1}$;

(b) $\bigcup_{n\geqslant 0} H^n = H$;

(c) $H^pH^q \subseteq H^{p+q}$;
(d) $\Delta(H^n) \subseteq \bigoplus_{p+q=n} H^p \otimes H^q$.

H is called *connected* if $H^0 = \operatorname{im} u (= \mathbf{k})$.

Obviously, a graded bialgebra is a filtered bialgebra with filtration defined by $H^n = \sum_{k \leqslant n} H^{(k)}$.

We note that the map $e := u \circ \varepsilon : H \to H$ has the property that $e(\mathbf{1}_H) = \mathbf{1}_H$ and $e(x) = 0$ for $x \in \ker \varepsilon_H$. So e is idempotent. Therefore

$$H = \operatorname{im} e \bigoplus \ker e = \operatorname{im} u \bigoplus \ker \varepsilon = \mathbf{k}\mathbf{1}_H \bigoplus \ker \varepsilon.$$

In fact, $\ker \varepsilon = \bigoplus_{n>0} H^{(n)}$.

Theorem 2.3.3. *Let H be a connected graded bialgebra. For any $x \in H^{(n)}$ with $n > 0$, the element $\tilde{\Delta}(x) := \Delta(x) - x \otimes 1 - 1 \otimes x$ is in $\bigoplus_{p+q=n, p>0, q>0} H^{(p)} \otimes H^{(q)}$. The map $\tilde{\Delta}$ is coassociative on $\ker \varepsilon$.*

Thus we can use the shorthand notations $\tilde{\Delta}(x) = \sum_{(x)} x' \otimes x''$, $\tilde{\Delta}^2(x) = \sum_{(x)} x' \otimes x'' \otimes x'''$, etc.

Proof. By the connectedness, we have

$$\Delta(x) = y \otimes \mathbf{1}_H + \mathbf{1}_H \otimes z + w$$

with $y, z \in H^{(n)}$ and $w \in \bigoplus_{p+q=n, p>0, q>0} H^p \otimes H^q$. By the counicity (2.7) of Δ, we have

$$x = \beta_\ell^{-1} (\varepsilon \otimes \mathrm{id}_H)\, \Delta(x) = \beta_\ell^{-1} (\varepsilon \otimes \mathrm{id}_H)(y \otimes 1 + 1 \otimes z + w) = \beta_\ell^{-1}(1 \otimes z) = z$$

and similarly

$$x = \beta_r\, (\mathrm{id}_H \otimes \varepsilon)\, \Delta(x) = y.$$

So $\Delta(x) = x \otimes 1 + 1 \otimes x + w$.

To verify the coassociativity of $\tilde{\Delta}$, we note that, for $x \in H^{(n)}$,

$$\begin{aligned}
(\Delta \otimes \mathrm{id}_H)\Delta(x) &= (\Delta \otimes \mathrm{id}_H)(x \otimes \mathbf{1}_H + \mathbf{1}_H \otimes x + \sum_{(x)} x' \otimes x'') \\
&= \Delta(x) \otimes \mathbf{1}_H + \Delta(\mathbf{1}_H) \otimes x + \sum_{(x)} \Delta(x') \otimes x'' \\
&= (x \otimes \mathbf{1}_H + \mathbf{1}_H \otimes x + \tilde{\Delta}(x)) \otimes \mathbf{1}_H + \mathbf{1}_H \otimes \mathbf{1}_H \otimes x \\
&\quad + \sum_{(x)} \left(x' \otimes \mathbf{1}_H + \mathbf{1}_H \otimes x' + \tilde{\Delta}(x') \right) \otimes x''
\end{aligned}$$

$$= (x \otimes \mathbf{1}_H + \mathbf{1}_H \otimes x + \tilde{\Delta}(x)) \otimes \mathbf{1}_H + \mathbf{1}_H \otimes \mathbf{1}_H \otimes x$$
$$+ \sum_{(x)} \left(x' \otimes \mathbf{1}_H \otimes x'' + \mathbf{1}_H \otimes x' \otimes x'' \right) + (\tilde{\Delta} \otimes \mathrm{id}_H)\tilde{\Delta}(x)$$
$$= x \otimes \mathbf{1}_H \otimes \mathbf{1}_H + \mathbf{1}_H \otimes x \otimes \mathbf{1}_H + \mathbf{1}_H \otimes \mathbf{1}_H \otimes x$$
$$+ \sum_{(x)} (x' \otimes x'' \otimes \mathbf{1}_H + x' \otimes \mathbf{1}_H \otimes x'' + \mathbf{1}_H \otimes x' \otimes x'')$$
$$+ (\tilde{\Delta} \otimes \mathrm{id}_H)\tilde{\Delta}(x).$$

Similarly,

$$(\mathrm{id}_H \otimes \Delta)\Delta(x) = x \otimes \mathbf{1}_H \otimes \mathbf{1}_H + \mathbf{1}_H \otimes x \otimes \mathbf{1}_H + \mathbf{1}_H \otimes \mathbf{1}_H \otimes x$$
$$+ \sum_{(x)} (x' \otimes x'' \otimes \mathbf{1}_H + x' \otimes \mathbf{1}_H \otimes x'' + \mathbf{1}_H \otimes x' \otimes x'')$$
$$+ (\mathrm{id}_H \otimes \tilde{\Delta})\tilde{\Delta}(x).$$

So the coassociativity of $\tilde{\Delta}$ follows from that of Δ. □

2.3.2 Convolution product

For a **k**-algebra A and a **k**-coalgebra C, we define the *convolution* of two linear maps f, g in $\mathrm{Hom}(C, A)$ to be the map $f * g \in \mathrm{Hom}(C, A)$ given by the composition

$$C \xrightarrow{\Delta} C \otimes C \xrightarrow{f \otimes g} A \otimes A \xrightarrow{m} A.$$

In other words,

$$(f * g)(a) = \sum_{(a)} f(a_{(1)})\, g(a_{(2)}).$$

Theorem 2.3.4. *Let C be a **k**-coalgebra and A a **k**-algebra. Let $e = u_A \varepsilon_C$.*

(a) *The triple $(\mathrm{Hom}(C, A), *, e)$ is an algebra.*

(b) *Let $H = \cup_{n \geqslant 0} H^n$ be a connected filtered bialgebra. Denote $R = \mathrm{Hom}(H, A)$. Define*

$$R_n = \{ f \in \mathrm{Hom}(H, A) \mid f(H^{n-1}) = 0 \}$$

for $n \geqslant 0$ with the convention that $H^{-1} = \emptyset$. Then (R, R_n) is a complete algebra.

(c) *Under the same hypotheses for Item (b), the set*

$$G = \{ f \in \mathrm{Hom}(H, A) \big| \; f(\mathbf{1}_H) = \mathbf{1}_A \}$$

endowed with the convolution product is a group. The inverse of $f \in G$ is $\sum_{k \geqslant 0} (e - f)^{*k}$.

Proof. (a) By the associativity of m and coassociativity of Δ,

$$\begin{aligned}(f*g)*h &= m((f*g)\otimes h)\Delta\\ &= m(m\otimes \mathrm{id}_A)(f\otimes g\otimes h)(\Delta\otimes \mathrm{id}_C)\Delta\\ &= m(\mathrm{id}_A\otimes m)(f\otimes g\otimes h)(\mathrm{id}_C\otimes \Delta)\Delta\\ &= m(f\otimes (g*h))\Delta\\ &= f*(g*h).\end{aligned}$$

Also,

$$\begin{aligned}f*e &= m(f\otimes u_A\varepsilon_C)\Delta\\ &= m(\mathrm{id}_A\otimes u_A)(f\otimes \mathrm{id}_{\mathbf{k}})(\mathrm{id}_C\otimes \varepsilon_C)\Delta\\ &= \alpha_r(f\otimes \mathrm{id}_{\mathbf{k}})\beta_r\\ &= f,\\ e*f &= m(u_A\varepsilon_C\otimes f)\Delta\\ &= m(u_A\otimes \mathrm{id}_A)(\mathrm{id}_{\mathbf{k}}\otimes f)(\varepsilon_C\otimes \mathrm{id}_C)\Delta\\ &= \alpha_\ell(\mathrm{id}_{\mathbf{k}}\otimes f)\beta_\ell\\ &= f.\end{aligned}$$

This is what we want.

(b) Let $f\in R_p$ and $g\in R_q$. Then for $x\in H^{(k)}$, $k\leqslant p+q-1$, we have

$$(f*g)(x)=\sum_{(x)} f(x_{(1)})g(x_{(2)})$$

with $\deg x_{(1)}+\deg x_{(2)}=k$. So in each term of the sum, either $\deg x_{(1)}\leqslant p-1$ or $\deg x_{(2)}\leqslant q-1$. So either $f(x_{(1)})=0$ or $g(x_{(2)})=0$. Thus $(f*g)(x)=0$. Therefore $R_p*R_q\subseteq R_{p+q}$.

To prove that the filtration is complete, we first note that

$$\bigcap_{n\geqslant 0} R_n=\{f\in R|f(H^n)=0,\ \forall n\}=0,$$

since $H=\bigcup\limits_{n\geqslant 0} H^n$. Next let $\{f_n\}_{n\geqslant 1}$ be a Cauchy sequence in R. Define $f: H\to A$ as follows. Let $x\in H$ be given. Then there is $k\geqslant 0$ such that $x\in H_k$. Since $\{f_n\}_{n\geqslant 1}$ is a Cauchy sequence, there is $N:=N_k$ such that

$$d(f_n,f_N)\leqslant \frac{1}{2^{k+1}},\quad \forall n\geqslant N.$$

This means that f_n-f_N is in R_{k+1}. So $(f_n-f_N)(H^k)=0$ for all $n\geqslant N$. Particularly, $f_n(x)=f_N(x)$ for all $n\geqslant N$ for the given $x\in H^k$. Thus it makes sense to define $f(x) = f_N(x)$ for the given $x \in H$, independent of N for N large enough. This way we obtain a function $f : H \to A$. We note that for the above k and resulting N, we have $d(f, f_N) < 1/2^{k+1}$ since $(f - f_N)(H^k) = 0$. Since $d(f_n, f_N) < 1/2^{k+1}$ for such a k, N and $n \geqslant N$, we get

$$d(f_n, f) \leqslant d(f_n, f_N) + d(f_N, f) < 1/2^k.$$

Therefore $\lim_{n\to\infty} f_n = f$.

(c) We have $f \in \mathrm{Hom}(H, A)$ with $f(\mathbf{1}_H) = \mathbf{1}_A$ if and only if there is $g \in \mathrm{Hom}(H, A)$ with $g(\mathbf{1}_H) = 0$ (say $g = f - e$) such that $f = e + g$. Therefore g is in R_1. So $f(\mathbf{1}_H) = \mathbf{1}_A$ if and only if f is in $e + R_1$. Then by Lemma 1.2.5, f is invertible. □

2.3.3 Antipode and Hopf algebra

Let $(H, m, \eta, \Delta, \varepsilon)$ be a $\mathbf{k}$-bialgebra. A $\mathbf{k}$-linear endomorphism S of H is called an *antipode* for H if it is the inverse of id_H under the convolution product:

$$S * \mathrm{id}_H = \mathrm{id}_H * S = e. \tag{2.12}$$

A *Hopf algebra* is a bialgebra H with an antipode S. In view of Theorem 2.3.4, we have

Theorem 2.3.5. *Any connected graded bialgebra H is a Hopf algebra. The antipode is defined by :*

$$S(x) = \sum_{k\geqslant 0} (e - \mathrm{id}_H)^{*k}(x).$$

It is also defined by $S(\mathbf{1}_H) = \mathbf{1}_H$ and recursively by any of the two formulas for $x \in \ker \varepsilon$:

$$S(x) = -x - \sum_{(x)} S(x')x'',$$

$$S(x) = -x - \sum_{(x)} x'S(x'').$$

Proof. The existence of the antipode and its first formula are consequences of Theorem 2.3.4. (c).

To prove the first recursive formula, we use

$$e = S * \mathrm{id}_H = m(S \otimes \mathrm{id})\Delta.$$

For $x \in \ker \varepsilon$, we have

$$\begin{aligned}
0 &= e(x) \\
&= (S * \mathrm{id}_H)(x) \\
&= m(S \otimes \mathrm{id}_H)\Delta(x) \\
&= m(S \otimes \mathrm{id}_H)(x \otimes 1 + 1 \otimes x + \sum_{(x)} x' \otimes x'') \\
&= \sum S(x')x'' + S(x) + x.
\end{aligned}$$

This gives the first recursive formula of S. The proof of the second recursive formula is similar. □

Remark 2.3.1. (a) S is an algebra anti-homomorphism:

$$S(hk) = S(k)S(h).$$

(b) S is also a coalgebra anti-homomorphism:

$$\Delta \circ S = \tau \circ (S \otimes S) \circ \Delta,$$

where τ is the switch operator.

Exercise 2.3.6. Use Theorem 2.3.5 to find the antipodes of the divided power bialgebra and the binomial bialgebra.

2.3.4 The Hopf algebra of rooted trees

We now introduce one of the primary examples of Hopf algebras for physics applications, the Hopf algebra of rooted trees. It was introduced in [122, 41] as a toy model of the Hopf algebra of Feynman graphs to study the renormalization of perturbative quantum field theory. It is also related to the Hopf algebras of rooted trees of Grossman-Larson [82] and of Loday-Ronco [137].

A rooted tree is a connected and simply-connected set of vertices and oriented edges such that there is precisely one distinguished vertex, called the root, with no incoming edge. A rooted tree is called non-planar if the branches of the same vertex can be permuted. The first few non-planar trees are listed below where the root of a tree is on the top.

Here the term non-planar means different embeddings of a rooted tree into the plane are regarded as being the same.

Remark 2.3.2. For the rest of this chapter, a tree means a non-planar rooted tree unless otherwise stated. The same applies to the concept of a forest introduced below.

Let $\mathcal{T}$ be the set of (isomorphic classes of) rooted trees. Let $\mathcal{H}_{\mathcal{T}} := \mathbb{Q}[\mathcal{T}]$ be the polynomial $\mathbb{Q}$-algebra generated by $\mathcal{T}$. A monomial of trees is called a forest. The *depth* of a leaf in a tree is the length of the path from the leaf to the root. The depth(F) of a forest F is the largest depth of the trees in the forest. A vertex a of a tree is the descendant of another vertex b if a is on the path from b to the root of the tree. We will define a coalgebra structure on $\mathcal{H}_{\mathcal{T}}$. A *subforest* of a tree T consists of a set of vertices of T together with their descendants and edges connecting all these vertices.

The coproduct is then defined as follows. Let $\mathcal{F}_T$ be the set of subforests of the rooted tree T, including the empty subforest which is identified with 1, and the full forest T. Define

$$\Delta(T) = \sum_{F \in \mathcal{F}_T} F \otimes (T/F).$$

Here the quotient T/F is obtained by removing the subforest F and the edges connecting the subforest to the rest of the tree. We use the convention that if F is the empty subforest, then $T/F = T$, and if $F = T$, then $T/F = 1$. For example,

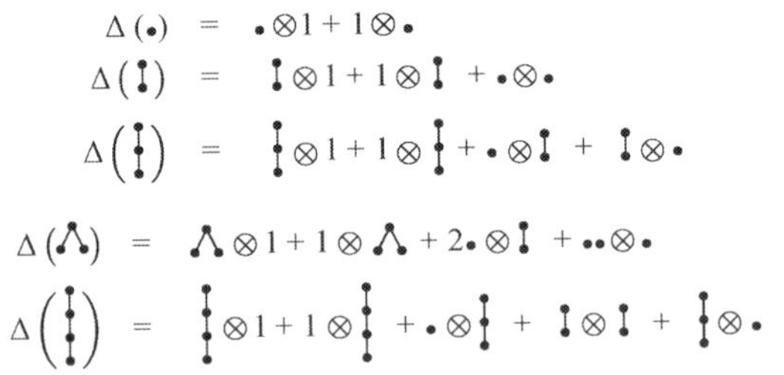
(2.13)

We then extend Δ to $\mathcal{H}_{\mathcal{T}}$ by multiplicativity with the convention $\Delta(1) = 1 \otimes 1$. Then Δ is an algebra homomorphism. Also define

$$\varepsilon : \mathcal{H}_{\mathcal{T}} \to \mathbf{k}$$

by taking $\varepsilon(T) = 0$ for $T \in \mathcal{T}$, $\varepsilon(1) = 1$ and extending by multiplicativity. Then ε is also an algebra homomorphism. It is easy to verify its compatibility with Δ. Further, define the degree of a tree and a forest (i.e., monomial in $\mathbb{C}[\mathcal{T}]$) to be its number of vertices. Then the grading defined by the degree is compatible with both the product and coproduct. It is also connected. Thus to prove that $\mathcal{H}_{\mathcal{T}}$ is a bialgebra and hence a Hopf algebra, we only need to verify the coassociativity of Δ. For this we follow the standard approach. See [75, 122]

Let $B^+ : \mathcal{H}_\mathcal{T} \to \mathcal{H}_\mathcal{T}$ be the linear map given by taking the product $T_1 \cdots T_k$ of k trees $T_1, \ldots, T_k$ to the tree T consisting of a new root, subtrees T_i and an edge from the new root to the root of each T_i. For example

The map B^+ is also called the *grafting operator* in combinatorics.

Proposition 2.3.7. ([41])

$$\Delta B^+ = B^+ \otimes 1 + (\mathrm{id}_H \otimes B^+)\Delta. \tag{2.14}$$

Proof. When $F = 1$, we have

$$\Delta(B^+(F)) = \Delta(\bullet) = \bullet \otimes 1 + 1 \otimes \bullet = B^+(F) \otimes 1 + (\mathrm{id}_H \otimes B^+)\Delta(F).$$

When $F = T_1 \cdots T_k$ is a forest consisting of trees $T_1, \ldots, T_k, k \geqslant 1$, we have

$$\Delta(B^+(F)) = B^+(F) \otimes 1 + 1 \otimes B^+(F) + \sum_{F' \in \mathcal{F}_{B^+(F)}, F' \neq B^+(F), 1} F' \otimes (B^+(F)/F').$$

In sum, one of the F's is F, for the other F's that are not $B^+(F)$ and 1, F' is a proper forest of F and the corresponding $B^+(F)/F'$ is just $B^+(F/F')$. Thus we have

$$\begin{aligned}
\Delta(B^+(F)) &= B^+(F) \otimes 1 + 1 \otimes B^+(F) + F \otimes \bullet + \sum_{F' \in \mathcal{F}_F, F' \neq F, 1} F' \otimes B^+(F/F') \\
&= B^+(F) \otimes 1 + (\mathrm{id}_H \otimes B^+)(1 \otimes F + F \otimes 1 + \sum_{F' \in \mathcal{F}_F, F' \neq F, 1} F' \otimes (F/F')) \\
&= B^+(F) \otimes 1 + (\mathrm{id}_H \otimes B^+)\Delta(F),
\end{aligned}$$

as needed. □

Exercise 2.3.8. For the trees in the list on page 59, compute $\Delta(T)$ by definition and by the recursive formula in Eq. (2.14).

Theorem 2.3.9. *$\mathcal{H}_\mathcal{T}$ is a connected graded Hopf algebra.*

Proof. By the remark before Proposition 2.3.7, it remains to show that the map Δ is coassociative.

Let H^n be the polynomial subalgebra of $\mathcal{H}_\mathcal{T}$ generated by rooted trees with at most n vertices. Then $\mathcal{H}_\mathcal{T} = \cup_{n \geqslant 0} H^n$. We use induction on $n \geqslant 0$ to prove that

$$(\mathrm{id}_H \otimes \Delta)\Delta = (\Delta \otimes \mathrm{id}_H)\Delta$$

on $H^{(n)}$. For $n = 1$, we have

$$(\mathrm{id}_H \otimes \Delta)\Delta(\bullet) = \bullet \otimes 1 \otimes 1 + 1 \otimes \bullet \otimes 1 + 1 \otimes 1 \otimes \bullet = (\Delta \otimes \mathrm{id})\Delta(\bullet). \quad (2.15)$$

Assume that Δ is coassociative on $H^{(n)}$ and let T be a rooted tree with $n+1$ vertices. Then $T = B^+(\overline{T})$ where $\overline{T} = T_1 \cdots T_k$ with T_i the subtrees of T immediately descending from the root of T. Then by Eq. (2.14) and the induction hypothesis, we have

$$\begin{aligned}
&(\mathrm{id}_H \otimes \Delta)\Delta(T)\\
&= (\mathrm{id}_H \otimes \Delta)\Delta(B^+(\overline{T}))\\
&= (\mathrm{id}_H \otimes \Delta)(T \otimes 1 + (\mathrm{id}_H \otimes B^+)\Delta(\overline{T})) \qquad \text{(by Eq. (2.14))}\\
&= T \otimes 1 \otimes 1 + (\mathrm{id}_H \otimes (\Delta B^+))\Delta(\overline{T}) \qquad \text{(by } \Delta(1) = 1 \otimes 1)\\
&= T \otimes 1 \otimes 1 + (\mathrm{id}_H \otimes B^+ \otimes 1 + (\mathrm{id}_H \otimes \mathrm{id}_H \otimes B^+)(\mathrm{id}_H \otimes \Delta))\Delta(\overline{T})\\
&\qquad\qquad \text{(by Eq. (2.14))}\\
&= T \otimes 1 \otimes 1 + ((\mathrm{id}_H \otimes B^+)\Delta(\overline{T})) \otimes 1 + (\mathrm{id}_H \otimes \mathrm{id}_H \otimes B^+)(\mathrm{id}_H \otimes \Delta)\Delta(\overline{T})\\
&= T \otimes 1 \otimes 1 + ((\mathrm{id}_H \otimes B^+)\Delta(\overline{T})) \otimes 1 + (\mathrm{id}_H \otimes \mathrm{id}_H \otimes B^+)(\Delta \otimes \mathrm{id}_H)\Delta(\overline{T})\\
&\qquad\qquad \text{(by induction hypothesis)}\\
&= T \otimes 1 \otimes 1 + \big((\mathrm{id}_H \otimes B^+)\Delta(\overline{T})\big) \otimes 1 + (\Delta \otimes B^+)\Delta(\overline{T})\\
&= (T \otimes 1 + (\mathrm{id}_H \otimes B^+)\Delta(\overline{T})) \otimes 1 + (\Delta \otimes B^+)\Delta(\overline{T})\\
&= \Delta(T) \otimes 1 + (\Delta \otimes B^+)\Delta(\overline{T}) \qquad \text{(by Eq. (2.14))}\\
&= (\Delta \otimes \mathrm{id}_H)\big(T \otimes 1 + (\mathrm{id}_H \otimes B^+)\Delta(\overline{T})\big)\\
&= (\Delta \otimes \mathrm{id}_H)\Delta(T) \qquad \text{(by Eq. (2.14))}.
\end{aligned}$$

This completes the inductive proof of the coassociativity. □

2.4 Algebraic Birkhoff decomposition

In this section, we derive the algebraic Birkhoff decomposition from Atkinson's multiplicative decomposition by regarding a character of the functional algebra $\mathrm{Hom}(H, A)$ as an element in a complete Rota-Baxter algebra.

2.4.1 Characters and derivations

Let H be a Hopf algebra and A an algebra. Then by Theorem 2.3.4, $\mathrm{Hom}(H, A)$ is a **k**-algebra with the convolution product. The map $e := u \circ \varepsilon : H \to A$ is the identity.

We recall that $\mathrm{id}_H \in \mathrm{Hom}(H,H)$ is the identity element with respect to the composition product, but is not the identity element with respect to the convolution product. In fact, the antipode S is the inverse of id_H under the convolution product: $S * \mathrm{id}_H = \mathrm{id}_H * S = e$.

Definition 2.4.1. An element $f \in \mathrm{Hom}(H,A)$ is called a *character* if f is an algebra homomorphism, and is called a *derivation* (or *infinitesimal character*) if

$$f(xy) = \varepsilon(x)f(y) + \varepsilon(y)f(x), \ \forall x, y \in H. \tag{2.16}$$

That is, $f(xy) = e(x)f(y) + e(y)f(x)$. The set of characters (resp. derivations) is denoted by $\mathrm{char}(H,A)$ (resp. $\partial\mathrm{char}(H,A)$).

We note that $f(\mathbf{1}_H) = \mathbf{1}_A$ if f is a character and $f(\mathbf{1}_H) = 0$ if f is a derivation.

We recall the notation $R = \mathrm{Hom}(H,A)$ and

$$R_1 = \{f \in \mathrm{Hom}(H,A) \mid f(\mathbf{1}_H) = 0\}.$$

Proposition 2.4.2. *Let A be a commutative algebra.*

(a) $\mathrm{char}(H,A)$ *forms a group under convolution. The inverse of* $\phi \in \mathrm{char}(H,A)$ *is given by* $\phi^{-1} := \phi \circ S$.

(b) $\partial\mathrm{char}(H,A)$ *forms a Lie algebra under convolution.*

(c) *The bijection* $\exp : R_1 \to e + R_1$ *in Eq.* (1.25) *is restricted to a bijection* $\exp : \partial\mathrm{char}(H,A) \to \mathrm{char}(H,A)$.

Proof. (a) Clearly e is in $\mathrm{char}(H,A)$.

We note that $f \in \mathrm{Hom}(H,A)$ is a character meaning that

$$m_A(f \otimes f) = f\, m_H.$$

Thus for $f, g \in \mathrm{char}(H,A)$, using the fact that Δ is an algebra homomorphism, we have (where $\tau_{2,3}$ is the operator switching the second and third factors)

$$\begin{aligned}
(f * g)m_H &= m_A(f \otimes g)\Delta_H m_H \\
&= m_A(f \otimes g)(m_H \otimes m_H)\tau_{2,3}(\Delta_H \otimes \Delta_H) \\
&= m_A\big((f m_H) \otimes (g m_H)\big)\tau_{2,3}(\Delta_H \otimes \Delta_H) \\
&= m_A\big(m_A(f \otimes f) \otimes m_A(g \otimes g)\big)\tau_{2,3}(\Delta_H \otimes \Delta_H) \\
&= m_A(m_A \otimes m_A)(f \otimes f \otimes g \otimes g)\tau_{2,3}(\Delta_H \otimes \Delta_H) \\
&= m_A(m_A \otimes m_A)(f \otimes g \otimes f \otimes g)(\Delta_H \otimes \Delta_H) \\
&= m_A\big((m_A(f \otimes g)\Delta_H) \otimes (m_A(f \otimes g)\Delta_H)\big) \\
&= m_A\big((f * g) \otimes (f * g)\big)
\end{aligned}$$

proving that $f * g$ is in $\mathrm{char}(H,A)$.

Further, for $f \in \operatorname{char}(H, A)$,

$$\begin{aligned}
(fS) * f &= m_A((fS) \otimes f)\Delta_H \\
&= m_A(f \otimes f)(S \otimes \mathrm{id}_H)\Delta_H \\
&= f m_H(S \otimes \mathrm{id}_H)\Delta \\
&= f(S * \mathrm{id}_H) \\
&= f u_H \varepsilon \\
&= u_A \varepsilon \\
&= e.
\end{aligned}$$

Likewise, $f * (fS) = e$. So $f^{*(-1)} = f\,S$.

(b) As in Item (a), we first note that f is a derivation meaning that

$$f\,m_H = m_A(e \otimes f + f \otimes e).$$

Then for $f, g \in \partial\operatorname{char}(H, A)$, we have

$$\begin{aligned}
(f * g)\,m_H &= m_A(f \otimes g)\Delta_H m_H \\
&= m_A(f \otimes g)(m_H \otimes m_H)\tau_{2,3}(\Delta_H \otimes \Delta_H) \\
&= m_A(f m_H \otimes g m_H)\tau_{2,3}(\Delta_H \otimes \Delta_H) \\
&= m_A\Big(\big(m_A(e \otimes f + f \otimes e)\big) \otimes \\
&\qquad \big(m_A(e \otimes g + g \otimes e)\big)\Big)\tau_{2,3}(\Delta_H \otimes \Delta_H) \\
&= m_A(m_A \otimes m_A)(e \otimes f \otimes e \otimes g + e \otimes f \otimes g \otimes e \\
&\qquad + f \otimes e \otimes e \otimes g + f \otimes e \otimes g \otimes e)\tau_{2,3}(\Delta_H \otimes \Delta_H) \\
&= m_A(m_A \otimes m_A)(e \otimes e \otimes f \otimes g + e \otimes g \otimes f \otimes e \\
&\qquad + f \otimes e \otimes e \otimes g + f \otimes g \otimes e \otimes e)(\Delta_H \otimes \Delta_H) \\
&= m_A\Big((e * e) \otimes (f * g) + (e * g) \otimes (f * e) \\
&\qquad + (f * e) \otimes (e * g) + (f * g) \otimes (e * e)\Big) \\
&= m_A\Big(e \otimes (f * g) + g \otimes f + f \otimes g + (f * g) \otimes e\Big).
\end{aligned}$$

Similarly,

$$(g * f)\,m_H = m_A\Big(e \otimes (g * f) + f \otimes g + g \otimes f + (g * f) \otimes e\Big).$$

Therefore,

$$((f * g) - (g * f))\,m_H = m_A\big(e \otimes (f * g - g * f) + (f * g - g * f) \otimes e\big)$$

as needed.

(c) Let $f \in \partial\mathrm{char}(H, A)$. Using $f(xy) = e(x)f(y) + e(y)f(x)$ and induction, we have

$$f^{*n}(xy) = \sum_{k=0}^{n} \binom{n}{k} f^{*k}(x) f^{*(n-k)}(y), \ n \geqslant 0.$$

Thus

$$\begin{aligned} e^{*f}(xy) &= \sum_{n \geqslant 0} \frac{1}{n!} f^{*n}(xy) \\ &= \sum_{n \geqslant 0} \frac{1}{n!} \sum_{k=0}^{n} \binom{n}{k} f^{*k}(x) f^{*(n-k)}(y) \\ &= \sum_{n \geqslant 0} \sum_{k=0}^{n} \frac{1}{k!} f^{*k}(x) \frac{1}{(n-k)!} f^{*(n-k)}(y) \\ &= e^{*f}(x)\, e^{*f}(y). \end{aligned}$$

So e^{*f} is in $\mathrm{char}(H, A)$. □

2.4.2 Algebraic Birkhoff decomposition

We now derive the algebraic Birkhoff decomposition from [42]. One can think of it as a refinement of Theorem 1.4.3 and Theorem 1.4.4 when the complete Rota-Baxter algebra comes from a connected Hopf algebra.

Theorem 2.4.3. (Algebraic Birkhoff Decomposition) *Let H be a connected graded Hopf algebra and let A be a commutative Rota-Baxter algebra of weight -1 with an idempotent Rota-Baxter operator P. Let R be the complete algebra* $\mathrm{Hom}(H, A)$.

(a) *R is a complete Rota-Baxter algebra with the idempotent operator $\mathcal{P}(f) := Pf$.*

(b) *For $\varphi \in \mathrm{char}(H, A)$, there is unique $\varphi_- \in e + \mathcal{P}(R_1)$ and $\varphi_+ \in e + \tilde{\mathcal{P}}(R_1)$ such that*

$$\varphi = \varphi_-^{*-1} * \varphi_+.$$

(c) *The elements φ_- and φ_+ take the following forms on* $\ker \varepsilon$.

$$\varphi_-(x) = -P(\varphi(x) + \sum_{(x)} \varphi_-(x')\varphi(x'')), \tag{2.17}$$

$$\varphi_+(x) = \tilde{P}(\varphi(x) + \sum_{(x)} \varphi_-(x')\varphi(x'')) \tag{2.18}$$

$$= \tilde{P}(\varphi(x) + \sum_{(x)} \varphi(x')\varphi_+(x'')) \tag{2.19}$$

(d) *The linear maps* φ_- *and* φ_+ *are also algebra homomorphisms.*

The map

$$b(\varphi) : H \to A, x \mapsto \varphi(x) + \sum_{(x)} \varphi_-(x')\varphi(x''), x \in \ker \varepsilon,$$

is called the *Bogoliubov map.* Then we have

$$\varphi_- = -P(b(\varphi)), \quad \varphi_+ = \tilde{P}(b(\varphi)).$$

Example 2.4.4. Let $H := \mathcal{H}_{\mathcal{T}}$ be the Hopf algebra of rooted trees in Section 2.3.4 and let $f \in \operatorname{Hom}(H, A)$ be a character. Then by Eq. (2.17) we have $\varphi_-(\bullet) = -P(\varphi(\bullet))$. Next consider $x := \Lambda$ whose coproduct is given in Eq. (2.13). Then by Eq. (2.17) we have

$$\begin{aligned}
\varphi_-(\Lambda) &= -P\left(\varphi(\Lambda) + 2\varphi_-(\bullet)\varphi(\mathrm{I}) + \varphi_-(\bullet\,\bullet)\varphi(\bullet)\right) \\
&= -P\left(\varphi(\Lambda) - 2P(\varphi(\bullet))\varphi(\mathrm{I}) + (-P(\varphi(\bullet)))^2\varphi(\bullet)\right) \\
&= -P\left(\varphi(\Lambda) - 2P(\varphi(\bullet))\varphi(\mathrm{I}) + P(\varphi(\bullet))^2\varphi(\bullet)\right). \\
\varphi_+(\Lambda) &= \tilde{P}\left(\varphi(\Lambda) - 2P(\varphi(\bullet))\varphi(\mathrm{I}) + P(\varphi(\bullet))^2\varphi(\bullet)\right).
\end{aligned}$$

Proof. (a) It is enough to verify that $\mathcal{P}$ is a Rota-Baxter operator, preserves the filtration and is idempotent. These are all easy to do.

(b) This follows from (a) and Theorem 1.4.3.

(c) The formula of φ_- follows from the proof of Theorem 1.4.3 (i.e., Eq. (1.44)). More precisely, apply Theorem 1.4.3 to the complete Rota-Baxter algebra $R = \operatorname{Hom}(H, A)$ and $a = e + \varphi$. Then a is in R_1. Therefore $\varphi = b^{*-1} * \tilde{b}^{*-1}$ where b, which we now call φ_-, is the unique solution of

$$b = e - \mathcal{P}(b * (e + \varphi)).$$

Thus for $x \in \ker \varepsilon$, using $\Delta(x) = x \otimes \mathbf{1}_H + \mathbf{1}_H \otimes x + \sum_{(x)} x' \otimes x''$, we have

$$\begin{aligned}
\varphi_-(x) &= \big(e - \mathcal{P}(\varphi_- * (e + \varphi))\big)(x) \\
&= e(x) - P\big((\varphi_- * (e + \varphi))(x)\big) \\
&= -P\big(\varphi_-(x)(e + \varphi)(\mathbf{1}_H) + \varphi_-(\mathbf{1}_H)\varphi(x) + \sum_{(x)} \varphi_-(x')\varphi(x'')\big)
\end{aligned}$$

$$= -P(\varphi(x) + \sum_{(x)} \varphi_-(x')\varphi(x'')).$$

Then the first formula for φ_+ follows since

$$\begin{aligned}
\varphi_+(x) &= (\varphi_- * \varphi)(x) \\
&= \varphi(x) + \varphi_-(x) + \sum_{(x)} \varphi_-(x')\varphi(x'') \\
&= \varphi(x) - P(\varphi(x) + \sum_{(x)} \varphi_-(x')\varphi(x'')) + \sum_{(x)} \varphi_-(x')\varphi(x'') \\
&= \tilde{P}(\varphi(x) + \sum_{(x)} \varphi_-(x')\varphi(x'')).
\end{aligned}$$

The second formula for φ_+ is proved in the same way that the formula for φ_- is proved.

(d) Once we prove that φ_- is a character, it is immediate that φ_+ is also character by Proposition 2.4.2 and $\varphi_+ = \varphi_- * \varphi$.

Thus we only need to verify

$$\varphi_-(xy) = \varphi_-(x)\varphi_-(y), x, y \in H.$$

This is immediately checked if either x or y is 1 using

$$\varphi_-(1) = (e + \mathcal{P}(\varphi_- * (e - \varphi)))(1) = e(1) + P(\varphi_-(1)(e(1) - \varphi(1))) = 1.$$

For $x, y \in \ker\varepsilon$, we have

$$\begin{aligned}
\Delta_H(xy) &= \Delta_H(m_H(x \otimes y)) \\
&= (m_H \otimes m_H)\tau_{2,3}(\Delta_H \otimes \Delta_H)(x \otimes y) \\
&= (m_H \otimes m_H)\tau_{2,3}\big((x \otimes 1 + 1 \otimes x + \sum_{(x)} x' \otimes x'') \\
&\quad \otimes (y \otimes 1 + 1 \otimes y + \sum_{(y)} y' \otimes y'')\big) \\
&= (m_H \otimes m_H)(x \otimes y \otimes 1 \otimes 1 + x \otimes 1 \otimes 1 \otimes y + \sum_{(y)} x \otimes y' \otimes 1 \otimes y'' \\
&\quad + 1 \otimes y \otimes x \otimes 1 + 1 \otimes 1 \otimes x \otimes y + \sum_{(y)} 1 \otimes y' \otimes x \otimes y'' \\
&\quad + \sum_{(x)} x' \otimes y \otimes x'' \otimes 1 + \sum_{(x)} x' \otimes 1 \otimes x'' \otimes y \\
&\quad + \sum_{(x),(y)} x' \otimes y' \otimes x'' \otimes y'')
\end{aligned}$$

$$= xy \otimes 1 + x \otimes y + \sum_{(y)} xy' \otimes y'' + y \otimes x + 1 \otimes xy + \sum_{(y)} y' \otimes xy''$$
$$+ \sum_{(x)} x'y \otimes x'' + \sum_{(x)} x' \otimes x''y + \sum_{(x),(y)} x'y' \otimes x''y''.$$

So

$$\begin{aligned}\varphi_-(xy) = &-P(\varphi(xy) + \varphi_-(x)\varphi(y) + \varphi_-(y)\varphi(x))\\ &+ \sum_{(y)} (\varphi_-(y')\varphi(xy'') + \varphi_-(xy')\varphi(y''))\\ &+ \sum_{(x)} (\varphi_-(x'y)\varphi(x'') + \varphi_-(x')\varphi(x''y))\\ &+ \sum_{(x),(y)} \varphi_-(x'y')\varphi(x''y'').\end{aligned}$$

On the other hand, denoting $\varphi_-(x) = -P(X)$, $\varphi_-(y) = -P(Y)$ and using the Rota-Baxter relation of P, we have

$$\begin{aligned}\varphi_-(x)\varphi_-(y) &= P(X)P(Y)\\ &= P(P(X)Y + XP(Y) - XY)\\ &= -P(\varphi_-(x)Y + X\varphi_-(y) + XY)\\ &= -P\Big(\varphi_-(x)\big(\varphi(y) + \sum_{(y)} \varphi_-(y')\varphi(y'')\big)\\ &\quad + \big(\varphi(x) + \sum_{(x)} \varphi_-(x')\varphi(x'')\big)\varphi_-(y)\\ &\quad + \big(\varphi(x) + \sum_{(x)} \varphi_-(x')\varphi(x'')\big)\big(\varphi(y) + \sum_{(y)} \varphi_-(y')\varphi(y'')\big)\Big)\\ &= -P\Big(\varphi_-(x)\varphi(y) + \sum_{(y)} \varphi_-(x)\varphi_-(y')\varphi(y'') + \varphi(x)\varphi_-(y)\\ &\quad + \sum_{(x)} \varphi_-(x')\varphi_-(y)\varphi(x'') + \varphi(x)\varphi(y) + \sum_{(x)} \varphi_-(x')\varphi(x'')\varphi(y)\\ &\quad + \sum_{(y)} \varphi_-(y')\varphi(x)\varphi(y'') + \sum_{(x),(y)} \varphi_-(x')\varphi_-(y')\varphi(x'')\varphi(y'')\Big)\end{aligned}$$

Here we have used the commutativity of A in the last equation. Then by induction on $\deg(x) + \deg(y)$ and the multiplicativity of φ, we see that this equals $\varphi_-(xy)$. □

2.5 Some related literature and problems

2.5.1 Renormalization in mathematical physics

The main references for this chapter are the lecture notes of Figueroa and Gracia-Bondía [75] and Manchon [144]. These references should prepare the reader to understand the original papers of Connes and Kreimer [41, 42, 43], as well as the subsequent monographs [44, 45, 75]. For discussions of quantum field theory renormalization from Rota-Baxter algebra point of view, see the survey articles [55, 68] and reference cited there, such as [51, 59, 60, 61, 62, 63], as well as the more recent developments [49, 50, 67, 71, 72, 73].

2.5.2 Renormalization in mathematics

For applications of the renormalization method to deal with divergence in mathematics, such as in Riemann integrals and multiple zeta values, we refer the reader to the survey articles [89, 96, 97] and the original papers [106, 107, 108, 145, 146, 177].

Part II

The Structure Aspect of Rota-Baxter Algebra

Chapter 3

Free Commutative Rota-Baxter Algebras and Shuffle Products

We will study free commutative Rota-Baxter algebras in this chapter and consider the noncommutative case in the next chapter. We highlight the connection of commutative Rota-Baxter algebras with well-known structures in algebra and combinatorics, such as shuffles, quasi-shuffles, bracketed words, rooted trees and Motzkin paths.

As in the case of any algebraic structure, effective study of commutative Rota-Baxter algebras depends on explicit constructions of free commutative Rota-Baxter algebras. The study of commutative algebras depends critically on the explicit construction of the free objects in terms of polynomials. For commutative Rota-Baxter algebras, a convenient construction is given by the mixable shuffle algebras. We also discuss variations of the mixable shuffle product that have recently appeared in different areas of mathematics.

In fact, the importance of free commutative Rota-Baxter algebras was already noted by Cartier and Rota in the 1960s. As Cartier [33] pointed out forty years ago, "The existence of free (Rota-) Baxter algebras follows from well-known arguments in universal algebra but remains quite immaterial as long as the corresponding word problem is not solved in an explicit way as Rota was the first to do." Both Rota's aforementioned construction [156] and the construction of Cartier himself in the above cited paper dealt with free commutative Rota-Baxter algebras. A third construction was obtained by Guo and Keigher [92] later as a generalization of shuffle product algebras. This construction made a direct connection from commutative Rota-Baxter algebras to the well-known concept of the shuffle product which is basic in many areas of pure and applied mathematics. Applying mixable shuffles to these areas has led to fruitful studies.

Another paper [112] on a generalization of the shuffle product was published in the same year as the work on mixable shuffle products. It was on the quasi-shuffle product by Hoffman. Hoffman's quasi-shuffle product plays a prominent role in the recent studies of harmonic functions, quasi-symmetric functions and multiple zeta values [27, 111, 113] (where it is also called stuffle product or harmonic product) and q-multiple zeta values [29].

We begin this chapter with Section 3.1.1 where we recall the concept of the shuffle product in its various equivalent formulations. Generalizations of these various formulations give various generalizations of the shuffle product that will also turn out to be equivalent. In Section 3.2 we will use the mixable shuffle product to construct free commutative Rota-Baxter algebras and use the connection with quasi-shuffles in Section 3.1 to study their properties. We will in fact consider several variations of free commutative Rota-Baxter algebras, including free on a set and free on another algebra, a concept introduced in [92]. We then discuss the above-mentioned constructions of Rota and Cartier in Section 3.2.3 and Section 3.2.4. In Section 3.3, applications of free commutative Rota-Baxter algebras are given to the study of Stirling numbers, partitions and multinomial coefficients. The chapter ends with a discussion of the related literature and problems in Section 3.4.

Notation. In this chapter, $\mathbf{k}$ is a commutative unitary ring. For a set X, let $\mathbf{k}X$ be the free $\mathbf{k}$-module $\oplus_{x\in X}\mathbf{k}\,x$ generated by X. If X is a semigroup (resp. monoid), $\mathbf{k}X$ is given the natural nonunitary (resp. unitary) $\mathbf{k}$-algebra structures. Let ComAlg be the category of commutative unitary $\mathbf{k}$-algebras whose unit is identified with the unit $\mathbf{1}$ of $\mathbf{k}$ by the structure homomorphism $\mathbf{k} \to A$. Let ComAlg^0 be the category of commutative nonunitary $\mathbf{k}$-algebras. Similarly let CRB_λ (resp. CRB^0_λ) be the category of unitary (resp. nonunitary) commutative Rota-Baxter $\mathbf{k}$-algebras of weight λ. The subscript λ will be suppressed if there is no danger of confusion.

3.1 Shuffle product and generalizations

The shuffle product is an important concept in algebra, combinatorics and topology. Several generalizations of it have been found in recent years that applied to algebra, combinatorics and number theory. We introduce these generalizations separately before showing that they can be identified with one another.

3.1.1 The shuffle product

The shuffle product can be defined in two ways, one is recursive, one is explicit. We will see that Hoffman's quasi-shuffle product is a generalization of the recursive definition and the mixable shuffle product is a generalization of the explicit definition.

3.1.1.1 *The shuffle product*

Let V be a $\mathbf{k}$-module. Consider the $\mathbf{k}$-module

$$T(V) := \bigoplus_{n\geqslant 0} V^{\otimes n} = \mathbf{k} \oplus V \oplus V^{\otimes 2} \oplus \cdots .$$

Here the tensor products are taken over $\mathbf{k}$ and we take $V^{\otimes 0} = \mathbf{k}$.

Let two pure tensors $\mathfrak{a} = a_1 \otimes \cdots \otimes a_m \in V^{\otimes m}$ and $\mathfrak{b} = b_1 \otimes \cdots \otimes b_n \in V^{\otimes n}$ be given. To describe the shuffle product of the two tensors, think of them as two decks of cards. A shuffle of $\mathfrak{a}$ and $\mathfrak{b}$ is just a shuffle of the two decks of cards — a tensor list from the factors of $\mathfrak{a}$ and $\mathfrak{b}$ in which the natural orders of the a_i's and b_j's are preserved. To form the shuffle product of $\mathfrak{a}$ and $\mathfrak{b}$, one puts together all possible shuffles of the two pure tensors.

Example 3.1.1. Let $\mathfrak{a} = a_1$ and $\mathfrak{b} = b_1 \otimes b_2$. Then their shuffles are

$$a_1 \otimes b_1 \otimes b_2, b_1 \otimes a_1 \otimes b_2, b_1 \otimes b_2 \otimes a_1. \tag{3.1}$$

Hence their shuffle product is

$$a_1 \otimes b_1 \otimes b_2 + b_1 \otimes a_1 \otimes b_2 + b_1 \otimes b_2 \otimes a_1.$$

To be precise, the shuffle product on $T(V)$ starts with the shuffles of permutations [150, 166]. Roughly speaking, for $m, n \in \mathbb{N}_+$, an (m, n)-shuffle is a permutation σ of $\{1, \ldots, m, m+1, \ldots, m+n\}$ such that the natural orders of $\{1, \ldots, m\}$ and $\{m+1, \ldots, m+n\}$ are preserved in

$$\{\sigma(1), \ldots, \sigma(m), \sigma(m+1), \ldots, \sigma(m+n)\}.$$

To be precise, define the set of (m, n)-*shuffles* by

$$S(m,n) = \left\{ \sigma \in S_{m+n} \;\middle|\; \begin{array}{l} \sigma^{-1}(1) < \sigma^{-1}(2) < \cdots < \sigma^{-1}(m), \\ \sigma^{-1}(m+1) < \sigma^{-1}(m+2) < \cdots < \sigma^{-1}(m+n) \end{array} \right\}.$$

Here S_{m+n} is the symmetric group on $m+n$ letters.

Example 3.1.2. Consider the case of $m = 1$ and $n = 2$. Then we have

$$\begin{aligned} S(1,2) &= (\sigma \in S_3 \mid \sigma^{-1}(2) < \sigma^{-1}(3)\} \\ &= \left\{ \begin{pmatrix} 1 & 2 & 3 \\ 1 & 2 & 3 \end{pmatrix}, \begin{pmatrix} 1 & 2 & 3 \\ 2 & 1 & 3 \end{pmatrix}, \begin{pmatrix} 1 & 2 & 3 \\ 2 & 3 & 1 \end{pmatrix} \right\}. \end{aligned} \tag{3.2}$$

For a σ in the symmetric group $S_n, n \geqslant 1$, the map

$$f_\sigma : V^n \to V^{\otimes n}, \quad f_\sigma(a_1 \otimes \cdots \otimes a_n) = a_{\sigma(1)} \otimes \cdots \otimes a_{\sigma(n)}, \quad a_1 \otimes \cdots \otimes a_n \in V^n,$$

is multi-linear. Hence it induces a linear map

$$\sigma : V^{\otimes n} \to V^{\otimes n}, \quad \sigma(a_1 \otimes \cdots \otimes a_n) = a_{\sigma(1)} \otimes \cdots \otimes a_{\sigma(n)}.$$

For $\mathfrak{a} = a_1 \otimes \cdots \otimes a_m \in V^{\otimes m}$, $\mathfrak{b} = b_1 \otimes \cdots \otimes b_n \in V^{\otimes n}$ and $\sigma \in S(m,n)$, the *shuffle of* $\mathfrak{a}$ *and* $\mathfrak{b}$ *via* σ is defined by

$$\sigma(\mathfrak{a} \otimes \mathfrak{b}) = u_{\sigma(1)} \otimes u_{\sigma(2)} \otimes \cdots \otimes u_{\sigma(m+n)} \in V^{\otimes(m+n)},$$

where

$$u_k = \begin{cases} a_k, & 1 \leqslant k \leqslant m, \\ b_{k-m}, & m+1 \leqslant k \leqslant m+n. \end{cases}$$

The sum

$$\mathfrak{a} ш \mathfrak{b} := \sum_{\sigma \in S(m,n)} \sigma(a \otimes b) \tag{3.3}$$

is called the *shuffle product* of $\mathfrak{a}$ and $\mathfrak{b}$. Also, by convention, $\mathfrak{a} ш \mathfrak{b}$ is the scalar product if either $m = 0$ or $n = 0$, that is, if either $\mathfrak{a} \in \mathbf{k}$ or $\mathfrak{b} \in \mathbf{k}$. The operation ш extends to a commutative and associative binary operation on $T(V)$, making $T(V)$ into a commutative algebra with identity, called the *shuffle product algebra* generated by V.

Example 3.1.3. Consider the case when $m = 1$ and $n = 2$ as in Example 3.1.2. Then $\mathfrak{a} = a_1$ and $\mathfrak{b} = b_1 \otimes b_2$. From Eq. (3.1) we obtain

$$a_1 ш (b_1 \otimes b_2) = a_1 \otimes b_1 \otimes b_2 + b_1 \otimes a_1 \otimes b_2 + b_1 \otimes b_2 \otimes a_1.$$

The shuffle product on $T(V)$ can also be recursively defined as follows. First define the product with $\mathbf{k} \in V^{\otimes 0} = \mathbf{k}$ to be the scalar product. Particularly, $1_\mathbf{k}$ is the identity. Next for any $m, n \geqslant 1$ and $\mathfrak{a} := a_1 \otimes \cdots \otimes a_m \in V^{\otimes m}$ and $\mathfrak{b} := b_1 \otimes \cdots \otimes b_n \in V^{\otimes n}$, define $\mathfrak{a} ш \mathfrak{b}$ by induction on the sum $m+n \geqslant 2$. When $m+n = 2$, we have $\mathfrak{a}, \mathfrak{b} \in V$ and define

$$\mathfrak{a} ш \mathfrak{b} = \mathfrak{a} \otimes \mathfrak{b} + \mathfrak{b} \otimes \mathfrak{a}.$$

Assume that $\mathfrak{a} ш \mathfrak{b}$ has been defined for $m+n \leqslant k$ with $k \geqslant 2$. Consider $\mathfrak{a}$ and $\mathfrak{b}$ with $m+n = k+1$. Then $m+n \geqslant 3$ and so at least one of m or n is greater than 1. We then define

$$\mathfrak{a} ш \mathfrak{b} = \begin{cases} a_1 \otimes b_1 \otimes \cdots \otimes b_n + b_1 \otimes \big(a_1 ш (b_2 \otimes \cdots \otimes b_n)\big), & m = 1, n \geqslant 2, \\ a_1 \otimes \big((a_2 \otimes \cdots \otimes a_m) ш b_1\big) + b_1 \otimes a_1 \otimes \cdots \otimes a_m, & m \geqslant 2, n = 1, \\ a_1 \otimes \big((a_2 \otimes \cdots \otimes a_m) ш (b_1 \otimes \cdots \otimes b_n)\big) & \\ \quad + b_1 \otimes \big((a_1 \otimes \cdots \otimes a_m) ш (b_2 \otimes \cdots \otimes b_n)\big), & m, n \geqslant 2. \end{cases} \tag{3.4}$$

Here the products by $\sqcup\!\sqcup$ on the right hand side of the equation are well-defined by the induction hypothesis. In short, writing $\mathfrak{a} = a_1 \otimes \mathfrak{a}'$ and $\mathfrak{b} = b_1 \otimes \mathfrak{b}'$, we have

$$\mathfrak{a} \sqcup\!\sqcup \mathfrak{b} = a_1 \otimes (\mathfrak{a}' \sqcup\!\sqcup \mathfrak{b}) + b_1 \otimes (\mathfrak{a} \sqcup\!\sqcup \mathfrak{b}'). \tag{3.5}$$

Example 3.1.4. Consider the case when $m = 1$ and $n = 2$ again. Then $\mathfrak{a} = a_1$ and $\mathfrak{b} = b_1 \otimes b_2$. Then we have

$$\begin{aligned} a_1 \sqcup\!\sqcup (b_1 \otimes b_2) &= a_1 \otimes (1 \sqcup\!\sqcup (b_1 \otimes b_2)) + b_1 \otimes (a_1 \sqcup\!\sqcup b_2) \\ &= a_1 \otimes b_1 \otimes b_2 + b_1 \otimes (a_1 \otimes b_2 + b_2 \otimes a_1) \\ &= a_1 \otimes b_1 \otimes b_2 + b_1 \otimes a_1 \otimes b_2 + b_1 \otimes b_2 \otimes a_1. \end{aligned}$$

This agrees with Example 3.1.3.

Lemma 3.1.5. *For every element $v \in V$, the $\mathbf{k}$-linear map*

$$P_{(v)} : (T(V), \sqcup\!\sqcup) \to (T(V), \sqcup\!\sqcup), \quad P_{(v)}(\mathfrak{a}) := v \otimes \mathfrak{a}, \quad \mathfrak{a} \in T(V),$$

is a Rota-Baxter operator of weight zero.

Proof. Let $\mathfrak{a}$ and $\mathfrak{b}$ in $T(V)$ be given. By the recursive description of $\sqcup\!\sqcup$, we have

$$\begin{aligned} P_{(v)}(\mathfrak{a}) \sqcup\!\sqcup P_{(v)}(\mathfrak{b}) &= (v \otimes \mathfrak{a}) \sqcup\!\sqcup (v \otimes \mathfrak{b}) \\ &= v \otimes (\mathfrak{a} \sqcup\!\sqcup (v \otimes \mathfrak{b})) + v \otimes ((v \otimes \mathfrak{a}) \sqcup\!\sqcup \mathfrak{b}) \\ &= P_{(v)}(\mathfrak{a} \sqcup\!\sqcup P_{(v)}(\mathfrak{b})) + P_{(v)}(P_{(v)}(\mathfrak{a}) \sqcup\!\sqcup \mathfrak{b}), \end{aligned}$$

as needed. □

3.1.1.2 *The Hopf algebra structure*

Define deconcatenation coproduct

$$\begin{aligned} &\Delta : T(V) \to T(V) \bigotimes T(V), \\ &\quad a_1 \otimes \cdots \otimes a_k \mapsto 1 \bigotimes (a_1 \otimes \cdots \otimes a_k) + a_1 \bigotimes (a_2 \otimes \cdots \otimes a_k) \\ &\quad + \cdots + (a_1 \otimes \cdots \otimes a_k) \bigotimes 1 \end{aligned}$$

and counit

$$\epsilon : T(V) \to \mathbf{k}, \quad \begin{cases} \mathfrak{a} \mapsto 0, & \mathfrak{a} \in V^{\otimes n}, n \geqslant 1, \\ \mathfrak{a} \mapsto \mathfrak{a}, & \mathfrak{a} \in \mathbf{k}. \end{cases}$$

Then $T(V)$ becomes a connected graded bialgebra, therefore a Hopf algebra by Theorem 2.3.5. It is in fact the Hopf algebra from a free commutative dendriform algebra that we will introduce in Chapter 5.

3.1.2 The quasi-shuffle product

3.1.2.1 *The quasi-shuffle product of Hoffman*

We recall the construction of quasi-shuffle algebras [112]. Let X be a locally finite set, that is, X is the disjoint union of finite sets X_n, $n \geqslant 1$. The elements of X_n are defined to have degree n. Elements in X are called letters and monomials in the letters are called words. For a commutative ring $\mathbf{k}$, consider the $\mathbf{k}$-module underlying the noncommutative polynomial algebra $\mathfrak{A} = \mathbf{k}\langle X\rangle$, that is, the free $\mathbf{k}$-algebra generated by X. The identity 1 of $\mathfrak{A}$ is called the empty word. Define $\bar{X} = X \cup \{0\}$. Suppose that there is a product

$$[\cdot,\cdot] : \bar{X} \times \bar{X} \to \bar{X} \tag{3.6}$$

with the properties

S0. $[a, 0] = 0$ for all $a \in \bar{X}$;

S1. $[a, b] = [b, a]$ for all $a, b \in \bar{X}$;

S2. $[[a, b], c] = [a, [b, c]]$ for all $a, b, c \in \bar{X}$;

S3. either $[a, b] = 0$ for all $a, b \in \bar{X}$, or $\deg([a, b]) = \deg(a) + \deg(b)$ for all $a, b \in \bar{X}$.

We define a *Hoffman set* to be a locally finite set X with a pairing in Eq. (3.6) that satisfies conditions S0 — S3.

Definition 3.1.6. *Let X be a Hoffman set. The quasi-shuffle product $*$ on $\mathfrak{A}$ is defined recursively by*

(a) $1 * w = w * 1 = w$ *for any word* w;

(b) $(aw_1) * (bw_2) = a(w_1 * (bw_2)) + b((aw_1) * w_2) + [a, b](w_1 * w_2)$, *for any words* w_1, w_2 *and letters* a, b.

Then $*$ is extended to a binary operation on $\mathfrak{A}$ by bilinearity.

Example 3.1.7. For letters a_1 and b_1, b_2, we have

$$\begin{aligned} a_1 * (b_1b_2) &= a_1(1 * (b_1b_2)) + b_1(a_1 * b_2) + [a_1, b_1](1 * b_2) \\ &= a_1(b_1b_2) + b_1(a_1 * b_2) + [a_1, b_1]b_2 \\ &= a_1b_1b_2 + b_1a_1b_2 + b_1b_2a_1 + b_1[a_1, b_2] + [a_1, b_1]b_2. \end{aligned}$$

When $[\cdot,\cdot]$ is identically zero, $*$ is the usual shuffle product ш defined in Eq. (3.4).

Theorem 3.1.8 ((Hoffman)[112]). (a) *$(\mathfrak{A}, *)$ is a commutative graded $\mathbf{k}$-algebra.*

(b) *When $[\cdot,\cdot] \equiv 0$, $(\mathfrak{A}, *)$ is the shuffle product algebra $(T(V), \text{ш})$, where V is the free $\mathbf{k}$-module generated by X.*

Proof. We only need to prove that the product $*$ is associative for which we only need to show

$$(u * v) * w = u * (v * w) \tag{3.7}$$

for any words $u = u_1 \cdots u_m$, $v = v_1 \cdots v_n$ and $w = w_1 \cdots w_\ell$. Here we use the convention that $u = \mathbf{1}$ if $m = 0$. For this we apply induction on the sum $i := m + n + \ell$ of the lengths of the words. If one of u, v, w is $\mathbf{1}$, then Eq. (3.7) is clearly true. This verifies Eq. (3.7) when $0 \leqslant i \leqslant 2$. Assume that it has been verified for $i \leqslant k$ with $k \geqslant 2$ and consider u, v, w with $i = k + 1$. So $i \geqslant 3$. With the remark just above, we can further assume $m, n, \ell \geqslant 1$. Thus we can apply Definition 3.1.6 (b). Let $u = au_1, v = bv_1, w = cw_1$ with $a, b, c \in X$ and u_1, v_1, w_1 being words. Then we have

$$\begin{aligned}(u * v) * w &= \big(a(u_1 * v) + b(u * v_1) + [a, b](u_1 * v_1)\big) * w \\ &= a\big((u_1 * v) * w\big) + c\big((a(u_1 * v)) * w_1\big) + [a, c]\big((u_1 * v) * w_1\big) \\ &\quad + b\big((u * v_1) * w\big) + c\big((b(u * v_1)) * w_1\big) + [b, c]\big((u * v_1) * w_1\big) \\ &\quad + [a, b]\big((u_1 * v_1) * w\big) + c\big(([a, b](u_1 * v_1)) * w_1\big) \\ &\quad + [[a, b], c]\big((u_1 * v_1) * w_1\big).\end{aligned}$$

Taking the sum of the second, fifth and eighth terms and applying Item (b) in Definition 3.1.6, we have

$$\begin{aligned}(u * v) * w = {} & a\big((u_1 * v) * w\big) + c\big((u * v) * w_1\big) + [a, c]\big((u_1 * v) * w_1\big) \\ & + b\big((u * v_1) * w\big) + [b, c]\big((u * v_1) * w_1\big) \\ & + [a, b]\big((u_1 * v_1) * w\big) + [[a, b], c]\big((u_1 * v_1) * w_1\big).\end{aligned}$$

We likewise have

$$\begin{aligned}u * (v * w) &= u * \big(b(v_1 * w) + c(v * w_1) + [b, c](v_1 * w_1)\big) \\ &= a\big(u_1 * (b(v_1 * w))\big) + b\big(u * (v_1 * w)\big) + [a, b]\big(u_1 * (v_1 * w)\big) \\ &\quad + a\big(u_1 * (c(v * w_1))\big) + c\big(u * (v * w_1)\big) + [a, c]\big(u_1 * (v * w_1)\big) \\ &\quad + a\big(u_1 * ([b, c](v_1 * w_1))\big) + [b, c]\big(u * (v_1 * w_1)\big) \\ &\quad + [a, [b, c]](u_1 * (v_1 * w_1)).\end{aligned}$$

Combining the first, fourth and seventh terms, we have

$$\begin{aligned}u * (v * w) = {} & a\big(u_1 * (v * w)\big) + b\big(u * (v_1 * w)\big) + [a, b]\big(u_1 * (v_1 * w)\big) \\ & + c\big(u * (v * w_1)\big) + [a, c]\big(u_1 * (v * w_1)\big) \\ & + [b, c]\big(u * (v_1 * w_1)\big) + [a, [b, c]](u_1 * (v_1 * w_1)).\end{aligned}$$

By the induction hypothesis, the terms on the right hand side of $(u * v) * w$ match with the right hand side of $u * (v * w)$. This completes the induction. □

Hoffman also showed that, together with the coconcatenation comultiplication

$$\Delta : \mathfrak{A} \to \mathfrak{A} \otimes \mathfrak{A}, \quad w \mapsto \sum_{uv=w} u \otimes v,$$

and counit

$$\epsilon : \mathfrak{A} \to \mathbf{k}, \quad w \mapsto \delta_{w,1},$$

$(\mathfrak{A}, *)$ becomes a connected graded bialgebra, therefore a Hopf algebra, by Theorem 2.3.5. It is in fact the Hopf algebra from a free commutative tridendriform algebra that we will introduce in Chapter 5. See also [136].

3.1.2.2 *A generalization of the quasi-shuffle product*

We next give a generalization of the quasi-shuffle product. Let A be a unitary $\mathbf{k}$-algebra. We define the $\mathbf{k}$-module

$$Ш^+(A) := Ш^+_{\mathbf{k},\lambda}(A) := \bigoplus_{k\in\mathbb{N}} A^{\otimes k} = \mathbf{k} \oplus A \oplus A^{\otimes 2} \oplus \cdots . \tag{3.8}$$

It is the underlying $\mathbf{k}$-module of the tensor algebra $T(A)$ of A regarded as a $\mathbf{k}$-module.

We will define another product $*_\lambda$ on $Ш^+(A)$. For this, we only need to define the product of two pure tensors and then to extend by bilinearity. For two pure tensors $\mathfrak{a} = a_1 \otimes \cdots \otimes a_m \in A^{\otimes m}$ and $\mathfrak{b} = b_1 \otimes \cdots \otimes b_n \in A^{\otimes n}$, we define $\mathfrak{a} *_\lambda \mathfrak{b}$ by induction on $m+n$. If $m = n = 0$, then $a, b \in \mathbf{k}$ and we just define $a *_\lambda b = ab \in \mathbf{k}$. Assume the product has been defined for a, b with $m + n \leqslant k$, and let a, b be pure tensors with $m + n = k + 1$. If either m or n is zero, then we define $a *_\lambda b$ by the scalar product. If none of m or n is zero, then we have $\mathfrak{a} = a_1 \otimes \mathfrak{a}', \mathfrak{b} = b_1 \otimes \mathfrak{b}'$ with $a_1, b_1 \in A$ and $\mathfrak{a}' \in A^{\otimes(m-1)}$ and $\mathfrak{b}' \in A^{\otimes(n-1)}$. We then define

$$\mathfrak{a} *_\lambda \mathfrak{b} = a_1 \otimes (\mathfrak{a}' *_\lambda \mathfrak{b}) + b_1 \otimes (\mathfrak{a} *_\lambda \mathfrak{b}') + \lambda(a_1 b_1) \otimes (\mathfrak{a}' *_\lambda \mathfrak{b}'). \tag{3.9}$$

Here the terms on the right hand side are well defined by the induction hypothesis, and $a_1 b_1$ is the product in A.

Example 3.1.9. For example, we have

$$\begin{aligned} a_1 *_\lambda (b_1 \otimes b_2) &= a_1 \otimes (1 *_\lambda (b_1 \otimes b_2)) + b_1 \otimes (a_1 *_\lambda b_2) + \lambda(a_1 b_1) \otimes (1 *_\lambda b_2) \\ &= a_1 \otimes (b_1 \otimes b_2) + b_1 \otimes (a_1 *_\lambda b_2) + \lambda(a_1 b_1) \otimes b_2 \\ &= a_1 \otimes b_1 \otimes b_2 + b_1 \otimes a_1 \otimes b_2 + b_1 \otimes b_2 \otimes a_1 \\ &\quad + b_1 \otimes \lambda(a_1 b_2) + \lambda(a_1 b_1) \otimes b_2. \end{aligned} \tag{3.10}$$

We now relate $*_\lambda$ to the quasi-shuffle product of Hoffman. Let $X = \cup_{n\geqslant 1} X_n$ be a Hoffman set. Then the pairing $[\cdot,\cdot]$ in Eq. (3.6) extends by $\mathbf{k}$-linearity to a binary operation on the free $\mathbf{k}$-module $A = \mathbf{k}X$ on X, making A into a commutative $\mathbf{k}$-algebra without identity, with grading $A_n = \mathbf{k}X_n$. Let $\tilde{A} = \mathbf{k} \oplus A$ be the unitary $\mathbf{k}$-algebra spanned by A. Then $\tilde{A} = \mathbf{k}\tilde{X}$ where $\tilde{X} = \{\mathbf{1}_A\} \cup X$. In this case, the product $*_1$ coincides with $*$.

As in the proof of Theorem 3.1.8, we verify that the product $*_\lambda$ is associative. Therefore we have

Theorem 3.1.10. *The* $\mathbf{k}$*-module* $Ш^+(A)$ *with* $*_\lambda$ *is a unitary commutative* $\mathbf{k}$*-algebra.*

We will also call $*_\lambda$ the *quasi-shuffle product of weight* λ and call $(Ш^+(A), *_\lambda)$ the *quasi-shuffle algebra of weight* λ.

3.1.3 Mixable shuffle product

We next turn to the construction of mixable shuffle algebras [92]. The adjective mixable is to suggest that certain elements in the shuffles can be mixed or merged together. We first give an explicit formula of the product before giving a recursive definition which will be seen to be equivalent to Hoffman's quasi-shuffle product. The same construction has also appeared in the work [81] of Goncharov on motivic shuffle relations and the work [109] of Hazewinkle on overlapping shuffle products.

3.1.3.1 *An intuitive description of mixable shuffles*

The mixable shuffle product of two pure tensors can be described quite easily. But in order to be rigorous, one has to make use of mixable shuffles of permutations which can be counter intuitive. So we will start with an informal discussion here before giving the formal definition.

Let A be a $\mathbf{k}$-algebra. Let two pure tensors $\mathfrak{a} = a_1 \otimes \cdots \otimes a_m \in A^{\otimes m}$ and $\mathfrak{b} = b_1 \otimes \cdots \otimes b_n \in A^{\otimes n}$ be given. Continuing with the intuitive description of the shuffle product of $\mathfrak{a}$ and $\mathfrak{b}$ at the beginning of Section 3.1.1.1, we regard the two tensors as two decks of cards and consider their shuffles. Suppose a shuffle of the two decks (i.e., tensors) is taken and suppose a card (i.e., tensor factor) from the first deck is followed immediately by a card from the second deck. This pair of card is called an admissible pair. Then a proper mixable shuffle is defined to be the result of merging some of the admissible pairs in the shuffle. To get the mixable shuffle product (of weight λ) of the two pure tensors, one takes the linear combination of all the shuffles and proper mixable shuffles with the coefficient of a mixable shuffle to be λ^k if k of the admissible pairs are merged.

Example 3.1.11. Back to the example of $\mathfrak{a} = a_1$ and $\mathfrak{b} = b_1 \otimes b_2$ in Example 3.1.1. In their shuffle $a_1 \otimes b_1 \otimes b_2$, the pair $a_1 \otimes b_1$ is an admissible pair. So $a_1b_1 \otimes b_2$ is a mixable shuffle, where a_1b_1 is the product in the **k**-algebra A. In the shuffle $b_1 \otimes a_1 \otimes b_2$, $a_1 \otimes b_2$ is an admissible pair, so $b_1 \otimes a_1b_2$ is a mixable shuffle. There is no admissible pair in the third shuffle of $\mathfrak{a}$ and $\mathfrak{b}$. Therefore the mixable product (of weight λ) of $\mathfrak{a}$ and $\mathfrak{b}$ is

$$\begin{aligned} & a_1 ш_\lambda (b_1 \otimes b_2) \\ &= a_1 \otimes b_1 \otimes b_2 + b_1 \otimes a_1 \otimes b_2 + b_1 \otimes b_2 \otimes a_1 \quad \text{(shuffles)} \\ &\quad + \lambda a_1 b_1 \otimes b_2 + \lambda b_1 \otimes a_1 b_2 \quad \text{(proper mixable shuffles)}. \end{aligned} \tag{3.11}$$

Example 3.1.12. As another example, consider $\mathfrak{a} = a_1 \otimes a_2$ and $\mathfrak{b} = b_1 \otimes b_2$. Their shuffles are

$$\begin{gathered} a_1 \otimes a_2 \otimes b_1 \otimes b_2, \; a_1 \otimes b_1 \otimes a_2 \otimes b_2, \; a_1 \otimes b_1 \otimes b_2 \otimes a_2, \\ b_1 \otimes a_1 \otimes a_2 \otimes b_2, \; b_1 \otimes a_1 \otimes b_2 \otimes a_2, \; b_1 \otimes b_2 \otimes a_1 \otimes a_2. \end{gathered}$$

The first shuffle yields one proper mixable shuffle $a_1 \otimes a_2b_1 \otimes b_2$. The second shuffle gives three proper mixable shuffles

$$a_1b_1 \otimes a_2 \otimes b_2, \quad a_1 \otimes b_1 \otimes a_2b_2, \quad a_1b_1 \otimes a_2b_2,$$

where the third one comes from merging two admissible pairs. Likewise from the other shuffles we derive the following proper mixable shuffles

$$a_1b_1 \otimes b_2 \otimes a_2, \quad b_1 \otimes a_1 \otimes a_2b_2, \quad b_1 \otimes a_1b_2 \otimes a_2.$$

Note that the last shuffle does not yield any proper mixable shuffles. Thus the mixable shuffle product of $\mathfrak{a}$ and $\mathfrak{b}$ is

$$\begin{aligned} \mathfrak{a} ш_\lambda \mathfrak{b} &= a_1 \otimes a_2 \otimes b_1 \otimes b_2 + a_1 \otimes b_1 \otimes a_2 \otimes b_2 + a_1 \otimes b_1 \otimes b_2 \otimes a_2 \\ &\quad + b_1 \otimes a_1 \otimes a_2 \otimes b_2 + b_1 \otimes a_1 \otimes b_2 \otimes a_2 + b_1 \otimes b_2 \otimes a_1 \otimes a_2 \\ &\quad + \lambda a_1 \otimes a_2 b_1 \otimes b_2 + \lambda a_1 b_1 \otimes a_2 \otimes b_2 + \lambda a_1 \otimes b_1 \otimes a_2 b_2 \\ &\quad + \lambda^2 a_1 b_1 \otimes a_2 b_2 + \lambda a_1 b_1 \otimes b_2 \otimes a_2 + \lambda b_1 \otimes a_1 \otimes a_2 b_2 \\ &\quad + \lambda b_1 \otimes a_1 b_2 \otimes a_2. \end{aligned}$$

The reader is encouraged to work out a couple of more examples to get familiar with this procedure.

3.1.3.2 *Mixable shuffles of permutations*

We now formalize the process of taking mixable shuffle product.

Given an (m, n)-shuffle $\sigma \in S(m, n)$, a pair of indices $(k, k+1)$, $1 \leqslant k < m+n$, is called an *admissible pair* for σ if $\sigma(k) \leqslant m < \sigma(k+1)$. Denote $\mathcal{T}^\sigma$ for the set of admissible pairs for σ. For a subset T of $\mathcal{T}^\sigma$, call the pair (σ, T) a *mixable (m,n)-shuffle*. Let $|T|$ be the cardinality of T. By convention, if $T = \varnothing$, then $(\sigma, T) = \sigma$ is just a shuffle. If $T \neq \varnothing$, then call $(\sigma; T)$ a *proper mixable (m, n)-shuffle*. Denote

$$\bar{S}(m, n) = \{(\sigma, T) \mid \sigma \in S(m, n), \quad T \subset \mathcal{T}^\sigma\} \tag{3.12}$$

for the set of mixable (m, n)-shuffles.

As remarked in Section 3.1.1.1, an (m, n)-shuffle is a permutation σ of $\{1, \ldots, m,\ m+1, \ldots, m+n\}$ such that the natural orders of $\{1, \ldots, m\}$ and $\{m+1, \ldots, m+n\}$ are preserved in

$$\{\sigma(1), \ldots, \sigma(m), \sigma(m+1), \ldots, \sigma(m+n)\}.$$

Further, a mixable (m, n)-shuffle (σ, T) is an (m, n)-shuffle σ in which pairs of indices from T represent positions where "merging" will occur.

Example 3.1.13. By Example 3.1.2, the $(1, 2)$-shuffles are

$$\sigma_1 = \begin{pmatrix} 1 & 2 & 3 \\ 1 & 2 & 3 \end{pmatrix}, \quad \sigma_2 = \begin{pmatrix} 1 & 2 & 3 \\ 2 & 1 & 3 \end{pmatrix}, \quad \sigma_3 = \begin{pmatrix} 1 & 2 & 3 \\ 2 & 3 & 1 \end{pmatrix}.$$

For σ_1, the pair $(1, 2)$ is an admissible pair. For σ_2, the pair $(2, 3)$ is an admissible pair. There are no admissible pairs for σ_3.

3.1.3.3 *Mixable shuffles of tensors*

Let A be a commutative nonunitary $\mathbf{k}$-algebra. For $\mathfrak{a} = a_1 \otimes \cdots \otimes a_m \in A^{\otimes m}$, $\mathfrak{b} = b_1 \otimes \cdots \otimes b_n \in A^{\otimes n}$ and $(\sigma, T) \in \bar{S}(m, n)$, define

$$\sigma(\mathfrak{a} \otimes \mathfrak{b}; T) = u_{\sigma(1)} \hat{\diamond} u_{\sigma(2)} \hat{\diamond} \cdots \hat{\diamond} u_{\sigma(m+n)} \in A^{\otimes(m+n-|T|)},$$

where for each pair $(k, k+1)$, $1 \leqslant k < m+n$,

$$u_{\sigma(k)} \hat{\diamond} u_{\sigma(k+1)} = \begin{cases} u_{\sigma(k)} u_{\sigma(k+1)}, & (k, k+1) \in T \\ u_{\sigma(k)} \otimes u_{\sigma(k+1)}, & (k, k+1) \notin T. \end{cases}$$

$\sigma(\mathfrak{a} \otimes \mathfrak{b}; T)$ is called a *mixable shuffle* of the pure tensors $\mathfrak{a}$ and $\mathfrak{b}$ (with respect to T). When $T = \varnothing$, a mixable shuffle is just a shuffle and when $T \neq \varnothing$, we call $\sigma(\mathfrak{a} \otimes \mathfrak{b}; T)$ a proper mixable shuffle of $\mathfrak{a}$ and $\mathfrak{b}$.

Example 3.1.14. Back to the case when $\mathfrak{a} = a_1$ and $\mathfrak{b} = b_1 \otimes b_2$ considered in Example 3.1.11. Then we have

$$\sigma_1(\mathfrak{a} \otimes \mathfrak{b}; (1, 2)) = a_1 b_1 \otimes b_2, \quad \sigma_2(\mathfrak{a} \otimes \mathfrak{b}; (2, 3)) = b_1 \otimes a_1 b_2,$$

recovering the two proper mixed shuffles found there.

Now fix $\lambda \in \mathbf{k}$. Define, for $\mathfrak{a}$ and $\mathfrak{b}$ as above, the *mixable shuffle product*

$$\mathfrak{a} ш_\lambda \mathfrak{b} := \sum_{(\sigma,T)\in \bar{S}(m,n)} \lambda^{|T|}\sigma(\mathfrak{a}\otimes\mathfrak{b};T) \in \bigoplus_{k\leqslant m+n} A^{\otimes k}. \tag{3.13}$$

We then extend the operation $ш_\lambda$ to a binary operation on

$$\bigoplus_{k\geqslant 1} A^{\otimes k} = A \oplus A^{\otimes 2} \oplus \ldots$$

by biadditivity. Further extend $ш_\lambda$ to

$$Ш^+(A) := Ш^+_{\mathbf{k},\lambda}(A) := \bigoplus_{k\in\mathbb{N}} A^{\otimes k} = \mathbf{k} \oplus A \oplus A^{\otimes 2} \oplus \ldots, \tag{3.14}$$

by letting $\mathbf{1} \in \mathbf{k}$ multiply as the identity.

Example 3.1.15. Continuing with Example 3.1.14, we have

$$\begin{aligned} & a_1 ш_\lambda (b_1 \otimes b_2) \\ = {} & a_1 \otimes b_1 \otimes b_2 + b_1 \otimes a_1 \otimes b_2 + b_1 \otimes b_2 \otimes a_1 \text{ (shuffles)} \\ & +\lambda a_1 b_1 \otimes b_2 + \lambda b_1 \otimes a_1 b_2 \text{ (improper mixed shuffles).} \end{aligned} \tag{3.15}$$

Theorem 3.1.16. *The* $\mathbf{k}$*-module* $Ш^+(A)$ *with the mixable shuffle product* $ш_\lambda$ *is a commutative unitary* $\mathbf{k}$*-algebra.*

Proof. We only need to verify the associativity. For this we can give two proofs. One proof works directly with the mixable shuffles. This is presented in [92]. We will give the other proof by showing, in Section 3.1.3.4, that the mixable shuffle can be defined by the same recursion as the generalized quasi-shuffle product. Then the theorem simply follows from Theorem 3.1.10. □

3.1.3.4 *Mixable shuffles and quasi-shuffles*

We now establish a relationship between the quasi-shuffle product and the mixable shuffle product.

To prove Theorem 3.1.16, we only need to prove the following statement.

Proposition 3.1.17. *The two products* $ш_\lambda$ *in Eq.* (3.13) *and* $*_\lambda$ *in Eq.* (3.9) *are the same.*

Proof. We just need to show that both products satisfy the same recursive relations. We first note that the recursive relation of $*_\lambda$ in Eq. (3.9) can be rewritten as follows. For any $m, n \geqslant 1$ and $\mathfrak{a} := a_1 \otimes \cdots \otimes a_m$, $\mathfrak{b} := b_1 \otimes \cdots \otimes b_n$

with $a_i, b_j \in X,\ 1 \leqslant i \leqslant m,\ 1 \leqslant j \leqslant n$, then

$$\begin{aligned}
&(1)\ \mathfrak{a} *_\lambda \mathfrak{b} = a_1 \otimes b_1 + b_1 \otimes a_1 + \lambda a_1 b_1, \quad \text{when } m, n = 1,\\
&(2)\ \mathfrak{a} *_\lambda \mathfrak{b} = a_1 \otimes b_1 \otimes \cdots \otimes b_n + b_1 \otimes \big(a_1 *_\lambda (b_2 \otimes \cdots \otimes b_n)\big)\\
&\qquad\qquad + \lambda(a_1 b_1) \otimes b_2 \otimes \cdots \otimes b_n, \ \text{when } m = 1, n \geqslant 2,\\
&(3)\ \mathfrak{a} *_\lambda \mathfrak{b} = a_1 \otimes \big((a_2 \otimes \cdots \otimes a_m) *_\lambda b_1\big) + b_1 \otimes a_1 \otimes \cdots \otimes a_m\\
&\qquad\qquad + \lambda(a_1 b_1) \otimes a_2 \otimes \cdots \otimes a_m, \ \text{when } m \geqslant 2, n = 1,\\
&(4)\ \mathfrak{a} *_\lambda \mathfrak{b} = a_1 \otimes \big((a_2 \otimes \cdots \otimes a_m) *_\lambda (b_1 \otimes \cdots \otimes b_n)\big)\\
&\qquad\qquad + b_1 \otimes \big((a_1 \otimes \cdots \otimes a_m) *_\lambda (b_2 \otimes \cdots \otimes b_n)\big)\\
&\qquad\qquad + \lambda(a_1 b_1) \otimes \big((a_2 \otimes \cdots \otimes a_m) *_\lambda (b_2 \otimes \cdots \otimes b_n)\big),\\
&\quad \text{when } m, n \geqslant 2.
\end{aligned} \tag{3.16}$$

On the other hand, consider the set $\bar{S}(m,n)$ of mixable (m,n)-shuffles defined in Eq. (3.12). Noting that either $\sigma^{-1}(1) = 1$ or $\sigma^{-1}(m+1) = 1$, we have the disjoint union (see [92, Eq. (4)])

$$\bar{S}(m,n) = \bar{S}_{1,0}(m,n) \bigsqcup \bar{S}_{0,1}(m,n) \bigsqcup \bar{S}_{1,1}(m,n),$$

where

$$\begin{aligned}
\bar{S}_{1,0}(m,n) &= \{(\sigma,T) \in \bar{S}(m,n) \mid (1,2) \notin T, \sigma^{-1}(1) = 1\},\\
\bar{S}_{0,1}(m,n) &= \{(\sigma,T) \in \bar{S}(m,n) \mid (1,2) \notin T, \sigma^{-1}(m+1) = 1\},\\
\bar{S}_{1,1}(m,n) &= \{(\sigma,T) \in \bar{S}(m,n) \mid (1,2) \in T\}.
\end{aligned}$$

Therefore (See [92, p. 137]),

$$\begin{aligned}
\sum_{(\sigma,T)\in\bar{S}_{1,0}(m,n)} \lambda^{|T|}\sigma(\mathfrak{a}\otimes\mathfrak{b};T) &= a_1 \otimes \sum_{(\sigma,T)\in\bar{S}(m-1,n)} \lambda^{|T|}\sigma((a_2\otimes\cdots\otimes a_m)\otimes\mathfrak{b};T)\\
&= \begin{cases} a_1 \otimes \big((a_2\otimes\cdots\otimes a_m) ш_\lambda \mathfrak{b}\big), & m \geqslant 2,\\ a_1 \otimes \mathfrak{b}, & m = 1.\end{cases}
\end{aligned}$$

Similarly,

$$\begin{aligned}
\sum_{(\sigma,T)\in\bar{S}_{0,1}(m,n)} \lambda^{|T|}\sigma(\mathfrak{a}\otimes\mathfrak{b};T) &= b_1 \otimes \sum_{(\sigma,T)\in\bar{S}(m,n-1)} \lambda^{|T|}\sigma(\mathfrak{a}\otimes(b_2\otimes\cdots\otimes b_n);T)\\
&= \begin{cases} b_1 \otimes \big(\mathfrak{a} ш_\lambda (b_2\otimes\cdots\otimes b_n)\big), & n \geqslant 2,\\ b_1 \otimes \mathfrak{a}, & n = 1;\end{cases}
\end{aligned}$$

and

$$\begin{aligned}
&\sum_{(\sigma,T)\in\bar{S}_{1,1}(m,n)} \lambda^{|T|}\sigma(\mathfrak{a}\otimes\mathfrak{b};T)\\
= &\sum_{(\sigma,T)\in\bar{S}(m-1,n-1)} \lambda(a_1 b_1) \otimes \lambda^{|T|}\sigma((a_2\otimes\cdots\otimes a_m)\otimes(b_2\otimes\cdots\otimes b_n);T)
\end{aligned}$$

$$= \begin{cases} \lambda(a_1b_1) \otimes \big((a_2 \otimes \cdots \otimes a_m) *_\lambda (b_2 \otimes \cdots \otimes b_n)\big), & m \geqslant 2, n \geqslant 2, \\ \lambda(a_1b_1) \otimes (b_2 \otimes \cdots \otimes b_n), & m = 1, n \geqslant 2, \\ \lambda(a_1b_1) \otimes (a_2 \otimes \cdots \otimes a_m), & m \geqslant 2, n = 1, \\ \lambda(a_1b_1), & m = n = 1. \end{cases}$$

Since

$$\begin{aligned} \mathfrak{a}Ш_\lambda \mathfrak{b} :&= \sum_{(\sigma,T)\in\bar{S}(m,n)} \lambda^{|T|}\sigma(\mathfrak{a}\otimes\mathfrak{b};T) \\ &= \sum_{(\sigma,T)\in\bar{S}_{1,0}(m,n)} \lambda^{|T|}\sigma(\mathfrak{a}\otimes\mathfrak{b};T) + \sum_{(\sigma,T)\in\bar{S}_{0,1}(m,n)} \lambda^{|T|}\sigma(\mathfrak{a}\otimes\mathfrak{b};T) \\ &\quad + \sum_{(\sigma,T)\in\bar{S}_{1,1}(m,n)} \lambda^{|T|}\sigma(\mathfrak{a}\otimes\mathfrak{b};T), \end{aligned}$$

we get

$$\begin{aligned} &(1)\ \mathfrak{a}Ш_\lambda\mathfrak{b} = a_1\otimes b_1 + b_1\otimes a_1 + \lambda(a_1b_1), \quad \text{when } m,n=1, \\ &(2)\ \mathfrak{a}Ш_\lambda\mathfrak{b} = a_1\otimes b_1\otimes\cdots\otimes b_n + b_1\otimes\big(a_1Ш_\lambda(b_2\otimes\cdots\otimes b_n)\big) \\ &\qquad\qquad +\lambda(a_1b_1)\otimes b_2\otimes\cdots\otimes b_n, \quad \text{when } m=1, n\geqslant 2, \\ &(3)\ \mathfrak{a}Ш_\lambda\mathfrak{b} = a_1\otimes\big((a_2\otimes\cdots\otimes a_m)Ш_\lambda b_1\big) + a_1\otimes\cdots\otimes a_m\otimes b_1 \\ &\qquad\qquad +\lambda(a_1b_1)\otimes a_2\otimes\cdots\otimes a_m, \quad \text{when } m\geqslant 2, n=1, \\ &(4)\ \mathfrak{a}Ш_\lambda\mathfrak{b} = a_1\otimes\big((a_2\otimes\cdots\otimes a_m)Ш_\lambda(b_1\otimes\cdots\otimes b_n)\big) \\ &\qquad\qquad +b_1\otimes\big((a_1\otimes\cdots\otimes a_m)Ш_\lambda(b_2\otimes\cdots\otimes b_n)\big) \\ &\qquad\qquad +\lambda(a_1b_1)\otimes\big((a_2\otimes\cdots\otimes a_m)Ш_\lambda(b_2\otimes\cdots\otimes b_n)\big), \\ &\quad \text{when } m,n\geqslant 2. \end{aligned} \tag{3.17}$$

Comparing Eq. (3.16) with Eq. (3.17), we conclude that $*_\lambda = Ш_\lambda$. □

As remarked above, this completes the proof of Theorem 3.1.16.

3.1.4 The stuffle product

Yet another interpretation of the mixable shuffle or quasi-shuffle product can be given in terms of order preserving maps that are called stuffle in the study of multiple zeta values but could be traced back to Cartier's work [33] on free commutative Rota-Baxter algebras. See also the work [74] of Ehrenborg on monomial quasi-symmetric functions and the work [29] of Bradley on q-multiple zeta values.

3.1.4.1 *Stuffles as order preserving maps*

For positive integers m and n, denote $[m] = \{1, \ldots, m\}$ and $[m+1, m+n] = \{m+1, \ldots, m+n\}$. Define

$$\mathcal{J}_{m,n} = \left\{ (\varphi, \psi) \,\middle|\, \begin{array}{l} \varphi : [m] \to [m+n], \psi : [n] \to [m+n] \text{ are order} \\ \text{preserving injective maps and } \operatorname{im}(\varphi) \cup \operatorname{im}(\psi) \\ = [m+n] \end{array} \right\}. \quad (3.18)$$

Let $\mathfrak{a} \in A^{\otimes m}$, $\mathfrak{b} \in A^{\otimes n}$ and $(\varphi, \psi) \in \mathcal{J}_{m,n}$. We define $\mathfrak{a} ш_{(\varphi,\psi)} \mathfrak{b}$ to be the tensor whose i-th factor is

$$(\mathfrak{a} ш_{(\varphi,\psi)} \mathfrak{b})_i = \begin{cases} a_j, & \text{if } i = \varphi(j) \\ b_j, & \text{if } i = \psi(j) \end{cases} = a_{\varphi^{-1}(i)} b_{\psi^{-1}(i)}, \quad 1 \leqslant i \leqslant m+n, \quad (3.19)$$

with the convention that $a_\varnothing = b_\varnothing = 1$. Then we have

$$\mathfrak{a} ш \mathfrak{b} = \sum_{(\varphi,\psi) \in \mathcal{J}_{m,n}} \mathfrak{a} ш_{(\varphi,\psi)} \mathfrak{b}. \quad (3.20)$$

More generally, for $0 \leqslant r \leqslant \min(m, n)$, define

$$\mathcal{J}_{m,n,r} = \left\{ (\varphi, \psi) \,\middle|\, \begin{array}{l} \varphi : [m] \to [m+n-r], \psi : [n] \to [m+n-r] \\ \text{are order preserving injective maps and} \\ \operatorname{im}(\varphi) \cup \operatorname{im}(\psi) = [m+n-r] \end{array} \right\}. \quad (3.21)$$

Clearly, $\mathcal{J}_{m,n,0} = \mathcal{J}_{m,n}$. Denote

$$\overline{\mathcal{J}}_{m,n} = \bigcup_{r=0}^{\min(m,n)} \mathcal{J}_{m,n,r}. \quad (3.22)$$

Let $\mathfrak{a} \in A^{\otimes m}$, $\mathfrak{b} \in A^{\otimes n}$ and $(\varphi, \psi) \in \mathcal{J}_{m,n,r}$. We define $\mathfrak{a} ш_{(\varphi,\psi)} \mathfrak{b}$ to be the pure tensor of length $m+n-r$ whose i-th factor is

$$(\mathfrak{a} ш_{(\varphi,\psi)} \mathfrak{b})_i = \begin{cases} a_j, & \text{if } i = \varphi(j), i \notin \operatorname{im}\psi, \\ b_j, & \text{if } i = \psi(j), i \notin \operatorname{im}\varphi, \\ a_j b_{j'}, & \text{if } i = \varphi(j), i = \psi(j'). \end{cases}$$

In short,

$$\mathfrak{a} ш_{(\varphi,\psi)} \mathfrak{b} = a_{\varphi^{-1}(1)} b_{\psi^{-1}(1)} \otimes \cdots \otimes a_{\varphi^{-1}(m+n-r)} b_{\psi^{-1}(m+n-r)},$$

with the convention that $a_\varnothing = b_\varnothing = 1$. $\mathfrak{a} ш_{(\varphi,\psi)} \mathfrak{b}$ is called a *stuffle of* $\mathfrak{a}$ *and* $\mathfrak{b}$ since it involves "stuffing" two tensor factors of a shuffle into one tensor factor. Define the *stuffle product* of $\mathfrak{a}$ and $\mathfrak{b}$ by

$$\mathfrak{a} ш_{st,\lambda} \mathfrak{b} = \sum_{r=0}^{\min(m,n)} \lambda^r \Big(\sum_{(\varphi,\psi) \in \mathcal{J}_{m,n,r}} \mathfrak{a} ш_{(\varphi,\psi)} \mathfrak{b} \Big). \quad (3.23)$$

Example 3.1.18. When $m = 1$ and $n = 2$, we have $\mathfrak{a} = a_1$ and $\mathfrak{b} = b_1 \otimes b_2$. Then we have

$m+n-r$	φ	ψ	$\mathfrak{a}ш_{(\varphi,\psi)}\mathfrak{b}$
3	$\varphi(1) = 1$	$\psi(1) = 2, \psi(2) = 3$	$a_1 \otimes b_1 \otimes b_2$
3	$\varphi(1) = 2$	$\psi(1) = 1, \psi(2) = 3$	$b_1 \otimes a_1 \otimes b_2$
3	$\varphi(1) = 3$	$\psi(1) = 1, \psi(2) = 2$	$b_1 \otimes b_2 \otimes a_1$
2	$\varphi(1) = 1$	$\psi(1) = 1, \psi(2) = 2$	$a_1 b_1 \otimes b_2$
2	$\varphi(1) = 2$	$\psi(1) = 1, \psi(2) = 2$	$b_1 \otimes a_1 b_2$

Thus we have

$$a_1 ш_{st,\lambda}(b_1 \otimes b_2) = a_1 \otimes b_1 \otimes b_2 + b_1 \otimes a_1 \otimes b_2 + b_1 \otimes b_2 \otimes a_1 + \lambda a_1 b_1 \otimes b_2 + \lambda b_1 \otimes a_1 b_2.$$

This agrees with Eq. (3.10).

The next theorem shows that this is not a coincidence.

Theorem 3.1.19. *The multiplications* $ш_{st,\lambda}$ *and* $*_\lambda$ *are the same. Particularly,*

$$\mathfrak{a} * \mathfrak{b} = \mathfrak{a} *_1 \mathfrak{b} = \sum_{r=0}^{\min(m,n)} \Big(\sum_{(\varphi,\psi)\in \mathcal{J}_{m,n,r}} \mathfrak{a}ш_{(\varphi,\psi)}\mathfrak{b} \Big) = \sum_{(\varphi,\psi)\in \bar{\mathcal{J}}_{m,n}} \mathfrak{a}ш_{(\varphi,\psi)}\mathfrak{b}.$$

Proof. For simplicity, we take $\lambda = 1$. We will prove that the product $ш_{st} := ш_1$ satisfies the same recursion that defines $*_1$ in Eq. (3.9).

We first introduce some notations. For positive integers $i < j$, denote $[i, j] := \{i, i+1, \dots, j\}$. For positive integers $i < j$, $i < k$, we define

$$\overline{\mathcal{J}}_{[i,j],[i,k]} = \left\{ (\varphi, \psi) \;\middle|\; \begin{array}{l} \varphi : [i,j] \to [i,\ell], \psi : [i,k] \to [i,\ell] \\ \text{are order preserving injective maps and} \\ \mathrm{im}(\varphi) \cup \mathrm{im}(\psi) = [i,\ell] \text{ for some } \ell \end{array} \right\},$$

$$\overline{\mathcal{J}}_{[i,i],[i,k]} = \{\mathrm{id}_{[i,k]}\},$$
$$\overline{\mathcal{J}}_{[i,j],[i,i]} = \{\mathrm{id}_{[i,j]}\}.$$

Thus $\overline{\mathcal{J}}_{m,n} = \overline{\mathcal{J}}_{[m],[n]}$.

With these notations, we have the disjoint union

$$\overline{\mathcal{J}}_{m,n} = \overline{\mathcal{J}}'_{m,n} \bigsqcup \overline{\mathcal{J}}''_{m,n} \bigsqcup \overline{\mathcal{J}}'''_{m,n}, \tag{3.24}$$

where

$$\overline{\mathcal{J}}'_{m,n} := \{(\varphi,\psi) \in \overline{\mathcal{J}}_{m,n} \mid \varphi(1) = 1, \psi(1) \neq 1\},$$
$$\overline{\mathcal{J}}''_{m,n} := \{(\varphi,\psi) \in \overline{\mathcal{J}}_{m,n} \mid \varphi(1) \neq 1, \psi(1) = 1\},$$
$$\overline{\mathcal{J}}'''_{m,n} := \{(\varphi,\psi) \in \overline{\mathcal{J}}_{m,n} \mid \varphi(1) = 1, \psi(1) = 1\}.$$

Further,

$$\overline{\mathcal{J}}'_{m,n} = \left\{(\varphi,\psi) \mid \varphi(1)=1, \quad \left(\varphi|_{[2,m]},\psi\right) \in \overline{\mathcal{J}}_{[2,m],[2,n]}\right\},$$
$$\overline{\mathcal{J}}''_{m,n} = \left\{(\varphi,\psi) \mid \psi(1)=1, \quad \left(\varphi,\psi|_{[2,n]}\right) \in \overline{\mathcal{J}}_{[2,m],[2,n]}\right\},$$
$$\overline{\mathcal{J}}'''_{m,n} = \left\{(\varphi,\psi) \mid \varphi(1)=1, \quad \psi(1)=1, \left(\varphi|_{[2,m]},\psi|_{[2,n]}\right) \in \overline{\mathcal{J}}_{[2,m],[2,n]}\right\}.$$

Therefore we have

$$\begin{aligned}
\mathfrak{a}\,\text{Ш}_{st}\,\mathfrak{b} &= \sum_{(\varphi,\psi)\in\overline{\mathcal{J}}_{m,n}} \mathfrak{a}\,\text{Ш}_{(\varphi,\psi)}\,\mathfrak{b} \\
&= \sum_{(\varphi,\psi)\in\overline{\mathcal{J}}'_{m,n}} \mathfrak{a}\,\text{Ш}_{(\varphi,\psi)}\,\mathfrak{b} + \sum_{(\varphi,\psi)\in\overline{\mathcal{J}}''_{m,n}} \mathfrak{a}\,\text{Ш}_{(\varphi,\psi)}\,\mathfrak{b} + \sum_{(\varphi,\psi)\in\overline{\mathcal{J}}'''_{m,n}} \mathfrak{a}\,\text{Ш}_{(\varphi,\psi)}\,\mathfrak{b} \\
&= \sum_{(\varphi',\psi)\in\overline{\mathcal{J}}_{m-1,n}} a_1 \otimes (\mathfrak{a}'\,\text{Ш}_{(\varphi',\psi)}\,\mathfrak{b}) + \sum_{(\varphi,\psi')\in\overline{\mathcal{J}}_{m,n-1}} b_1 \otimes (\mathfrak{a}\,\text{Ш}_{(\varphi,\psi')}\,\mathfrak{b}') \\
&\quad + \sum_{(\varphi',\psi')\in\overline{\mathcal{J}}_{m-1,n-1}} a_1 b_1 \otimes (\mathfrak{a}'\,\text{Ш}_{(\varphi',\psi')}\,\mathfrak{b}') \\
&= a_1 \otimes (\mathfrak{a}'\,\text{Ш}_{st}\,\mathfrak{b}) + b_1 \otimes (\mathfrak{a}\,\text{Ш}_{st}\,\mathfrak{b}') + a_1 b_1 \otimes (\mathfrak{a}'\,\text{Ш}_{st}\,\mathfrak{b}').
\end{aligned}$$

This is the same recursion for the quasi-shuffle product. We also check directly that

$$a_1\,\text{Ш}_{st}\,b_1 = a_1 \otimes b_1 + b_1 \otimes a_1 + a_1 b_1.$$

Thus the products have the same initial condition. Thus Ш_{st} coincides with $*_1$.

□

Here is an equivalent description of the stuffle product. Let $\mathrm{stfl}(m,n,r)$ denote the set of surjective maps from $[m+n]$ to $[m+n-r]$ that preserve the natural orders of $[m]$ and $[m+1,m+n]$. Let

$$\mathrm{stfl}(m,n) = \bigcup_{r=0}^{\min(m,n)} \mathrm{stfl}(m,n,r).$$

Then

$$(a_1 \otimes \cdots \otimes a_m) * (a_{m+1} \otimes \cdots \otimes a_{m+n}) = \sum_{\pi\in\mathrm{stfl}(m,n)} c_1^\pi \otimes \cdots \otimes c_{m+n-r}^\pi, \quad (3.25)$$

where $c_i^\pi = \prod_{j\in\pi^{-1}(i)} a_j$.

3.1.4.2 *Cartier's construction of stuffles as ordered subsets*

A variation of the stuffle product can already be found in the construction of free commutative Rota-Baxter algebras by Cartier back in the 1970s. See

[33, 92]. The original construction of Cartier is for free objects on a set, in the category of commutative algebras with no identity and with weight -1. We will extend his construction to free objects on a **k**-algebra A with arbitrary weight, using the tensor product notation.

Let a **k**-algebra A and λ in **k** be given. For $m, n \geqslant 1$, let $\overline{S}_c(m,n)$ denote the set of triples (k, P, Q) in which k is an integer in $[m+n]$, P and Q are ordered subsets of $[k]$ with the natural ordering such that $P \cup Q = [k]$, $|P| = m$ and $|Q| = n$. It then follows that $\max\{m,n\} \leqslant k \leqslant m+n$. We also have the following numerical formula for $\overline{S}_c(m,n)$.

Proposition 3.1.20. *For a given k with $\max\{m,n\} \leqslant k \leqslant m+n$, the cardinality of the subset*

$$\overline{S}_c(m,n,k) := \{(k,P,Q)\}$$

of $\overline{S}_c(m,n)$ is

$$\binom{k}{m}\binom{m}{m+n-k}.$$

Proof. For the given k, any triple (k, P, Q) can be uniquely determined by first taking a subset of $[k]$ of cardinality m for the subset P, and then taking a subset S of P of cardinality $m+n-k$ for $P \cap Q$. These determine $Q = S \cup ([k] \backslash P)$. Thus there are $\binom{k}{m}\binom{m}{m+n-k}$ such triples. □

For $\mathfrak{a} = a_1 \otimes \cdots \otimes a_m \in A^{\otimes m}$, $\mathfrak{b} = b_1 \otimes \cdots \otimes b_n \in A^{\otimes n}$ and $(k, P, Q) \in \overline{S}_c(m,n)$, define

$$\Phi_{k,P,Q}(\mathfrak{a},\mathfrak{b}) := c_1 \otimes \cdots \otimes c_k \in A^{\otimes k},$$

where

$$c_j = \begin{cases} a_\alpha, & \text{if } j \text{ is the } \alpha\text{-th element in } P \text{ and } j \notin Q; \\ b_\beta, & \text{if } j \text{ is the } \beta\text{-th element in } Q \text{ and } j \notin P; \\ a_\alpha b_\beta, & \text{if } j \text{ is the } \alpha\text{-th element in } P \text{ and the } \beta\text{-th element in } Q. \end{cases}$$

Define a **k**-bilinear multiplication $ш_c$ on $Ш^+(A)$ by taking $1 \in A^{\otimes 0} = \mathbf{k}$ to be the identity and

$$\mathfrak{a} ш_c \mathfrak{b} := \mathfrak{a} ш_{c,\lambda} \mathfrak{b} := \sum_{(k,P,Q) \in \overline{S}_c(m,n)} \lambda^{m+n-k} \Phi_{k,P,Q}(\mathfrak{a},\mathfrak{b}). \tag{3.26}$$

Proposition 3.1.21. *The product $ш_c$ coincides with the stuffle product $ш_{st}$.*

Proof. We first give a bijection

$$f_{m,n} : \overline{\mathcal{J}}_{m,n} \to \overline{S}_c(m,n), \quad m,n \geqslant 1,$$

between the set $\overline{\mathcal{J}}_{m,n}$ of stuffles in Eq. (3.22) and the set $\overline{S}_c(m,n)$ defined after Cartier. It is simply given by

$$f_{m,n}(\varphi,\psi) = (|\mathrm{im}(\varphi)\cup \mathrm{im}(\psi)|, \mathrm{im}\varphi, \mathrm{im}\psi), \quad (\varphi,\psi)\in \mathcal{J}(m,n).$$

On the other hand, for $(k,P,Q)\in \overline{S}_c(m,n)$, define

$$g_{m,n}(k,P,Q) = (\varphi,\psi),$$

where $\varphi : [m] \to P$ and $\psi : [n] \to Q$ are the unique order preserving bijections. Then $f_{m,n}$ and $g_{m,n}$ are the inverse of each other. We further check

$$\mathfrak{a} ш_{(\varphi,\psi)} \mathfrak{b} = \Phi_{f_{m,n}(\varphi,\psi)}(\mathfrak{a},\mathfrak{b}), \quad \mathfrak{a}\in A^{\otimes m}, \mathfrak{b}\in A^{\otimes n}.$$

Then the proposition follows from Theorem 3.1.19. □

Combining Proposition 3.1.17 and Proposition 3.1.21 we obtain the following relation.

Proposition 3.1.22. *The pair* $(Ш^+(A), ш_{c,\lambda})$ *is an algebra isomorphic to the mixable shuffle algebra* $(Ш^+(A), ш_\lambda)$.

3.1.5 Delannoy paths

We end this section with a combinatorial interpretation of mixable shuffles in terms of Delannoy paths. It has appeared in the papers of Fares on coalgebras, of Aguiar-Hsiao [5] on quasi-symmetric functions and of Loday [136] on the Zinbiel operad.

Let $D(m,n)$ be the set of lattice paths from $(0,0)$ to (m,n) consisting of steps either to the right, to the above, or to the above-right. Such a path is called a *Delannoy path*. For $d\in D(m,n)$, define $d(\mathfrak{a},\mathfrak{b})$ to be the path d with $\mathfrak{a} = (a_1,\dots,a_m)$ (resp. $\mathfrak{b} = (b_1,\dots,b_n)$) sequentially labeling the horizontal (resp. vertical) and diagonal segments of d.

Define

$$\mathfrak{a} ш_{d,\lambda} \mathfrak{b} = \sum_{d\in D(m,n)} \lambda^r d(\mathfrak{a},\mathfrak{b}), \tag{3.27}$$

where r is the number of diagonal steps in d.

Example 3.1.23. When $\mathfrak{a} = a_1$ and $\mathfrak{b} = b_1\otimes b_2$, we have the following Delannoy paths (in dash dotted line) and corresponding tensor lists.

Thus we have

$$\begin{aligned} & a_1 ш_{d,\lambda}(b_1\otimes b_2) \\ &= a_1\otimes b_1\otimes b_2 + b_1\otimes a_1\otimes b_2 + b_1\otimes b_2\otimes a_1 + \lambda a_1b_1\otimes b_2 + \lambda b_1\otimes a_1b_2. \end{aligned}$$

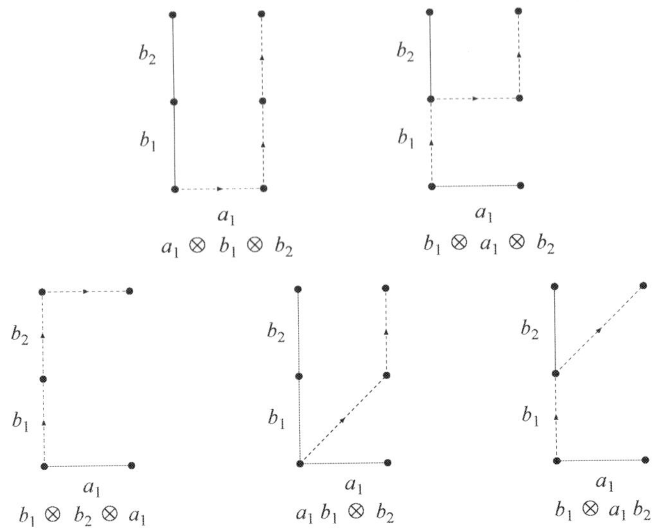

In general, we have the following relation whose proof is similar to the proof of Proposition 3.1.21 and is left as an exercise.

Proposition 3.1.24. *The product* $Ш_{d,\lambda}$ *coincides with the stuffle product* $Ш_{st,\lambda}$.

We summarize the relationship among the products that we have introduced in this section.

Theorem 3.1.25. *The following products on* $Ш^+(A) = \bigoplus_{k\geqslant} A^{\otimes k}$ *are the same: the quasi-shuffle products* $*_\lambda$ *of weight* λ *defined in Eq.* (3.9), *the mixable shuffle product* $Ш_\lambda$ *of weight* λ *defined in Eq.* (3.13), *the stuffle product* $Ш_{st,\lambda}$ *of weight* λ *defined in Eq.* (3.23), *the Cartier product* $Ш_{c,\lambda}$ *of weight* λ *defined in Eq.* (3.26), *the product* $Ш_{d,\lambda}$ *of weight* λ *defined by Delannoy paths in Eq.* (3.27).

3.2 Free commutative Rota-Baxter algebras

We continue with the notations in the last section.

3.2.1 Mixable shuffle Rota-Baxter algebras

Let A be a $\mathbf{k}$-algebra A. Let

$$Ш^+_{\mathbf{k}}(A) = \bigoplus_{k\geqslant 0} A^{\otimes k}$$

be the algebra equipped with the mixable shuffle product $ш_\lambda$ of weight λ defined in Eq. (3.13), or equivalently by Theorem 3.1.25, equipped with either the quasi-shuffle products $*_\lambda$ of weight λ in Eq. (3.9), or the stuffle product $ш_{st,\lambda}$ of weight λ in Eq. (3.23), or the Cartier product $ш_{c,\lambda}$ of weight λ in Eq. (3.26), or the product $ш_{d,\lambda}$ of weight λ defined by Delannoy paths in Eq. (3.27).

Define

$$Ш(A) := Ш_{\mathbf{k},\lambda}(A) := A \otimes Ш^+_{\mathbf{k}}(A) = \oplus_{k\geqslant 1} A^{\otimes k} \tag{3.28}$$

to be the tensor product algebra whose product is denoted $\diamond$ and is called the *augmented mixable shuffle product of weight* λ. More precisely, the product $\diamond := \diamond_\lambda$ on $Ш(A)$ is defined by:

$$(a_0 \otimes \mathfrak{a}) \diamond (b_0 \otimes \mathfrak{b}) := (a_0 b_0) \otimes (\mathfrak{a} ш_\lambda \mathfrak{b}), \quad a_0,\ b_0 \in A,\ \mathfrak{a},\ \mathfrak{b} \in Ш^+(A). \tag{3.29}$$

Thus we have the algebra isomorphism (embedding to the second tensor factor)

$$\alpha : (Ш^+(A), ш_\lambda) \to (\mathbf{1}_A \otimes Ш^+(A), \diamond). \tag{3.30}$$

Define the $\mathbf{k}$-linear endomorphism P_A on $Ш_{\mathbf{k}}(A)$ by assigning

$$P_A(a_0 \otimes \mathfrak{a}) = \mathbf{1}_A \otimes a_0 \otimes \mathfrak{a}, \quad \mathfrak{a} \in A^{\otimes n},\ n \geqslant 1,$$
$$P_A(a_0 \otimes c) = \mathbf{1}_A \otimes c a_0, \quad c \in A^{\otimes 0} = \mathbf{k}$$

and extending by additivity. Let $j_A : A \to Ш_{\mathbf{k}}(A)$ be the canonical inclusion map. Call $(Ш_{\mathbf{k}}(A), P_A)$ the *(mixable) shuffle Rota-Baxter* $\mathbf{k}$*-algebra of weight* λ *on* A. The term is justified by the following theorem [92].

Theorem 3.2.1. *The shuffle Rota-Baxter algebra* $(Ш_{\mathbf{k}}(A), P_A)$, *together with the natural embedding* j_A, *is a free commutative Rota-Baxter* $\mathbf{k}$*-algebra on* A *of weight* λ. *More precisely, for any commutative Rota-Baxter* $\mathbf{k}$*-algebra* (R, P) *of weight* λ *and algebra homomorphism* $f : A \to R$, *there is a unique Rota-Baxter* $\mathbf{k}$*-algebra homomorphism* $\bar{f} : (Ш_{\mathbf{k}}(A), P_A) \to (R, P)$ *such that* $f = \bar{f} \circ j_A$.

We will suppress $\mathbf{k}$ and λ from $Ш_{\mathbf{k},\lambda}(A)$ when there is no danger of confusion.

Proof. We will always assume that a Rota-Baxter algebra has weight λ in this proof. We first prove that $(Ш(A), P_A)$ is a commutative Rota-Baxter algebra. For this we use Eq. (3.17) and check, for pure tensors $\mathfrak{a} = a_1 \otimes \mathfrak{a}'$ and $\mathfrak{b} = b_1 \otimes \mathfrak{b}'$,

$$\begin{aligned}
P_A(a) \diamond P_A(b) &= (1 \otimes \mathfrak{a}) \diamond (1 \otimes \mathfrak{b}) \\
&= 1 \otimes (\mathfrak{a} ш_\lambda \mathfrak{b}) \\
&= 1 \otimes (a_1 \otimes (\mathfrak{a}' ш_\lambda \mathfrak{b})) + 1 \otimes (b_1 \otimes (\mathfrak{a} ш_\lambda \mathfrak{b}')) \\
&\quad + 1 \otimes ((\lambda a_1 b_1) \otimes (\mathfrak{a}' ш_\lambda \mathfrak{b}')) \\
&= P_A(\mathfrak{a} \diamond P_A(\mathfrak{b})) + P_A(\mathfrak{b} \diamond P_A(\mathfrak{a})) + \lambda P_A(\mathfrak{a} \diamond \mathfrak{b}).
\end{aligned}$$

To verify the universal property of $(\text{Ш}(A), P_A)$, let (R, P) be a commutative Rota-Baxter algebra and let $f : A \to R$ be an algebra homomorphism. To define $\bar{f} : \text{Ш}(A) \to R$, we need to define $\bar{f}(\mathfrak{a})$ for any pure tensor $\mathfrak{a} = a_1 \otimes \cdots \otimes a_n \in A^{\otimes n}$. For this we use induction on n. When $n = 1$, we have $A = j_A(A)$. So we must have $\bar{f}(\mathfrak{a}) = f(\mathfrak{a})$. Assume that $\bar{f}(\mathfrak{a})$ has been defined for $n \leqslant k$ and let $\mathfrak{a} \in A^{\otimes(k+1)}$. Then $\mathfrak{a} = a_1 \otimes \mathfrak{a}'$ with $\mathfrak{a}' \in A^{\otimes k}$. Note that $\mathfrak{a} = a_1 \diamond P_A(\mathfrak{a}')$. Since $\bar{f}$ is to be a Rota-Baxter algebra homomorphism, we must have

$$\bar{f}(\mathfrak{a}) = \bar{f}(a_1)\bar{f}(P_A(\mathfrak{a}')) = f(a_1)P(\bar{f}(\mathfrak{a}')). \tag{3.31}$$

In fact, if $\mathfrak{a} = a_1 \otimes a_2 \otimes \cdots \otimes a_{k+1}$, then we must have

$$\bar{f}(\mathfrak{a}) = f(a_1)P\big(f(a_2)P(f(a_3)\cdots)\big).$$

So the uniqueness of $\bar{f}$ is proved.

Now we just need to prove that the $\bar{f}$ obtained this way, i.e., by Eq. (3.31), is a Rota-Baxter algebra homomorphism. From the construction we obviously have $\bar{f} \circ P_A = P \circ \bar{f}$. So it remains to prove the multiplicativity. Let $\mathfrak{a} \in A^{\otimes m}$ and $\mathfrak{b} \in A^{\otimes n}$, we will verify

$$\bar{f}(\mathfrak{a} \diamond \mathfrak{b}) = \bar{f}(\mathfrak{a})\bar{f}(\mathfrak{b}) \tag{3.32}$$

by induction on $m + n$. Then $m + n \geqslant 2$. When $m + n = 2$, then $m = n = 1$ and so $a, b \in A$. Then $a \diamond b = ab$, so Eq. (3.32) follows since f is an algebra homomorphism. Assume that the equation is proved for $m + n \leqslant k$ and let $\mathfrak{a} = a_1 \otimes \mathfrak{a}' = a_1 \diamond P_A(\mathfrak{a}') \in A^{\otimes m}$, $\mathfrak{b} = b_1 \otimes \mathfrak{b}' = b_1 \diamond P_A(\mathfrak{b}') \in A^{\otimes n}$ with $m + n = k + 1$. Then by Eq. (3.31) and the induction hypothesis, we have

$$\begin{aligned}
\bar{f}(\mathfrak{a} \diamond \mathfrak{b}) &= \bar{f}\big((a_1b_1) \diamond P_A(\mathfrak{a}') \diamond P_A(\mathfrak{b}')\big) \\
&= \bar{f}\big((a_1b_1) \diamond P_A\big(\mathfrak{a}' \diamond P_A(\mathfrak{b}') + P_A(\mathfrak{a}') \diamond \mathfrak{b}' + \lambda \mathfrak{a}' \diamond \mathfrak{b}'\big)\big) \\
&= \bar{f}(a_1b_1)(\bar{f} \circ P_A)\big(\mathfrak{a}' \diamond (1 \otimes \mathfrak{b}') + (1 \otimes \mathfrak{a}') \diamond \mathfrak{b}' + \lambda(\mathfrak{a}' \diamond \mathfrak{b}')\big) \\
&= f(a_1b_1)(P \circ \bar{f})\big(\mathfrak{a}' \diamond (1 \otimes \mathfrak{b}') + (1 \otimes \mathfrak{a}') \diamond \mathfrak{b}' + \lambda(\mathfrak{a}' \diamond \mathfrak{b}')\big) \\
&= f(a_1b_1)P\big(\bar{f}(\mathfrak{a}')\bar{f}(1 \otimes \mathfrak{b}') + \bar{f}(1 \otimes \mathfrak{a}')\bar{f}(\mathfrak{b}') + \lambda \bar{f}(\mathfrak{a}')\bar{f}(\mathfrak{b}')\big) \\
&= f(a_1)f(b_1)P\big(\bar{f}(\mathfrak{a}')P(\bar{f}(\mathfrak{b}')) + P(\bar{f}(\mathfrak{a}'))\bar{f}(\mathfrak{b}') + \lambda \bar{f}(\mathfrak{a}')\bar{f}(\mathfrak{b}')\big) \\
&= f(a_1)f(b_1)P\big(\bar{f}(\mathfrak{a}')\big)P\big(\bar{f}(\mathfrak{b}')\big) \\
&= f(a_1)P\big(\bar{f}(\mathfrak{a}')\big)\, f(b_1)P\big(\bar{f}(\mathfrak{b}')\big) \\
&= \bar{f}(\mathfrak{a})\bar{f}(\mathfrak{b}).
\end{aligned}$$

This completes the induction and the proof of Theorem 3.2.1. □

We next specialize to free objects in the classical sense. For a given set X, let $\mathbf{k}[X]$ be the polynomial $\mathbf{k}$-algebra on X with the natural embedding $X \hookrightarrow \mathbf{k}[X]$. Let $(\text{Ш}_{\mathbf{k}}(X), P_X)$ be the Rota-Baxter $\mathbf{k}$-algebra $(\text{Ш}_{\mathbf{k}}(\mathbf{k}[X]), P_{\mathbf{k}[X]})$.

Proposition 3.2.2. *The Rota-Baxter algebra* $(\text{Ш}_{\mathbf{k}}(X), P_X)$*, together with the set embedding*

$$j_X : X \hookrightarrow \mathbf{k}[X] \stackrel{j_{\mathbf{k}[X]}}{\longrightarrow} \text{Ш}_{\mathbf{k}}(\mathbf{k}[X]),$$

is a free commutative Rota-Baxter $\mathbf{k}$*-algebra on the set* X*, described by the following universal property: for any commutative Rota-Baxter* $\mathbf{k}$*-algebra* (R, P) *and any set map* $\varphi : X \to R$*, there exists a unique Rota-Baxter* $\mathbf{k}$*-algebra homomorphism* $\bar{\varphi} : (\text{Ш}_{\mathbf{k}}(X), P_X) \to (R, P)$ *such that the diagram*

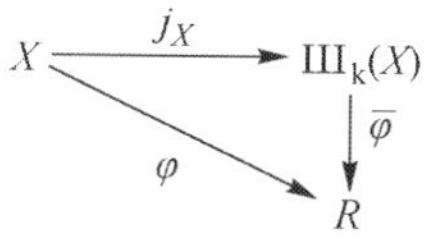

commutes.

Proof. We only need to verify that $(\text{Ш}_{\mathbf{k}}(X), P_X)$ satisfies the universal property of a free commutative Rota-Baxter $\mathbf{k}$-algebra on X. Fix a given commutative Rota-Baxter $\mathbf{k}$-algebra (R, P), and a set map $\varphi : X \to R$. By the universal property of the free commutative $\mathbf{k}$-algebra $\mathbf{k}[X]$ on X, φ extends uniquely to a $\mathbf{k}$-algebra homomorphism $\bar{\varphi} : \mathbf{k}[X] \to R$. Then by Theorem 3.2.1, $\bar{\varphi}$ extends uniquely to a homomorphism of Rota-Baxter $\mathbf{k}$-algebras

$$\tilde{\varphi} : (\text{Ш}_{\mathbf{k}}(X), P_X) \cong (\text{Ш}_{\mathbf{k}}(\mathbf{k}[X]), P_{\mathbf{k}[X]}) \to (R, P).$$

This proves the proposition. □

3.2.2 Special cases of free commutative Rota-Baxter algebras

We now consider special cases of free commutative Rota-Baxter algebras.

3.2.2.1 *Free commutative Rota-Baxter algebra on* k

As a particular example, we consider $\text{Ш}_{\mathbf{k}}(\mathbf{k})$, the free commutative Rota-Baxter $\mathbf{k}$-algebra on $\mathbf{k}$. $\text{Ш}_{\mathbf{k}}(\mathbf{k})$ is also $\text{Ш}_{\mathbf{k}}(\varnothing)$, the free commutative Rota-Baxter $\mathbf{k}$-algebra on the empty set $\varnothing$, defined in Proposition 3.2.2. This free commutative Rota-Baxter algebra not only provides the simplest example of a free commutative Rota-Baxter algebra but also establishes the connection between Rota-Baxter algebras and some well-known concepts such as divided powers and Stirling numbers (see Section 3.3).

If we choose $A = \mathbf{k}$ in the construction of free commutative Rota-Baxter $\mathbf{k}$-algebras, then we get

$$\text{Ш}_{\mathbf{k}}(\mathbf{k}) = \bigoplus_{n=0}^{\infty} \mathbf{k}^{\otimes(n+1)}.$$

Since the tensor product is over $\mathbf{k}$, we have $\mathbf{k}^{\otimes(n+1)} = \mathbf{k}\mathbf{1}^{\otimes(n+1)}$, where $\mathbf{1}^{\otimes(n+1)} = \underbrace{\mathbf{1}\otimes\cdots\otimes\mathbf{1}}_{(n+1)-\text{factors}}$. Particularly, $\mathbf{1}^{\otimes 1} = \mathbf{1}_{\mathbf{k}}$ is the unit of $\mathbf{k}$. Thus $\text{Ш}_{\mathbf{k}}(\mathbf{k})$ is a free $\mathbf{k}$-module on the basis $\mathbf{1}^{\otimes n}, n \geqslant 1$. The following result gives the multiplication table of $\text{Ш}_{\mathbf{k}}(\mathbf{k})$.

Proposition 3.2.3. *For any* $m, n \in \mathbb{N}$,

$$\mathbf{1}^{\otimes(m+1)} \diamond_\lambda \mathbf{1}^{\otimes(n+1)} = \sum_{r=0}^{\min(m,n)} \binom{m+n-r}{m}\binom{m}{r} \lambda^r \mathbf{1}^{\otimes(m+n+1-r)}.$$

Proof. First applying the definitions of the products $\diamond_\lambda$, $ш_\lambda$, $ш_{c,\lambda}$, then Theorem 3.1.25, and then Proposition 3.1.20, we obtain

$$\begin{aligned}
\mathbf{1}^{\otimes(m+1)} \diamond_\lambda \mathbf{1}^{\otimes(n+1)} &= \mathbf{1}\otimes(\mathbf{1}^{\otimes m} ш_\lambda \mathbf{1}^{\otimes n}) \\
&= \mathbf{1}\otimes(\mathbf{1}^{\otimes m} ш_{c,\lambda} \mathbf{1}^{\otimes n}) \\
&= \mathbf{1}\otimes\left(\sum_{(k,P,Q)\in\overline{S}_c(m,n)} \lambda^{m+n-k}\Phi_{k,P,Q}(\mathbf{1}^{\otimes m}, \mathbf{1}^{\otimes n})\right) \\
&= \mathbf{1}\otimes\left(\sum_{k=\max\{m,n\}}^{m+n} \lambda^{m+n-k} \sum_{(k,P,Q)\in\overline{S}_c(m,n,k)} \mathbf{1}^{\otimes k}\right) \\
&= \sum_{k=\max\{m,n\}}^{m+n} \binom{k}{m}\binom{m}{m+n-k} \lambda^{m+n-k}\mathbf{1}^{\otimes(k+1)} \\
&= \sum_{r=0}^{\min\{m,n\}} \binom{m+n-r}{m}\binom{m}{r} \lambda^r \mathbf{1}^{\otimes(m+n+1-r)}.
\end{aligned}$$

Thus we have the conclusion. □

We also have the following consequence of Theorem 3.2.1.

Corollary 3.2.4. *The Rota-Baxter algebra* $\text{Ш}_{\mathbf{k}}(\mathbf{k})$ *is the initial object in the category of commutative Rota-Baxter* $\mathbf{k}$*-algebras. In other words, for any commutative Rota-Baxter* $\mathbf{k}$*-algebra* (R, P)*, there is a unique Rota-Baxter* $\mathbf{k}$*-algebra homomorphism* $(\text{Ш}_{\mathbf{k}}(\mathbf{k}), P_{\mathbf{k}}) \to (R, P)$.

Proof. This follows since for any commutative Rota-Baxter algebra (R, P), there is a unique $\mathbf{k}$-algebra homomorphism $f : \mathbf{k} \to R$, namely, the structure map of the $\mathbf{k}$-algebra R. □

Now we consider the special case when $\lambda = 0$.

Proposition 3.2.5. *Let* $\lambda = 0$ *and consider the free commutative Rota-Baxter algebra* $\text{Ш}_0(k)$ *on* $\mathbf{k}$.

(a) *For any* $m,\ n \in \mathbb{N}$, $\mathbf{1}^{\otimes(m+1)} \diamond \mathbf{1}^{\otimes(n+1)} = \binom{m+n}{n} \mathbf{1}^{\otimes(m+n+1)}$. *Thus* $Ш_0(\mathbf{k})$ *is the divided power algebra.*

(b) *The free commutative Rota-Baxter* $\mathbf{k}$*-algebra* $Ш_0(\mathbf{k})$ *is a Hopf algebra.*

Proof. (a). This follows from Proposition 3.2.3.

(b). This follows from Item (a) and Section 2.2.3.1. □

3.2.2.2 *Free commutative Rota-Baxter algebra on one generator*

The free commutative Rota-Baxter algebra on one generator x is the same as the free commutative Rota-Baxter algebra on the polynomial algebra $\mathbf{k}[x]$. Thus

$$Ш_{\mathbf{k},\lambda}(\mathbf{k}[x]) = \bigoplus_{k \geqslant 0} \mathbf{k}[x]^{\otimes(k+1)}.$$

Since a $\mathbf{k}$-linear basis of $\mathbf{k}[x]^{\otimes(k+1)}$ is the set

$$\{x^{i_0} \otimes x^{i_1} \otimes \cdots \otimes x^{i_k} \mid i_j \geqslant 0, \quad 0 \leqslant j \leqslant k\},$$

with the notation

$$\mathfrak{I} := \bigsqcup_{k \geqslant 0} \mathbb{N}_{\geqslant 0}^{k+1},$$

we have

$$Ш_{\mathbf{k},\lambda}(\mathbf{k}[x]) = \bigoplus_{I \in \mathfrak{I}} \mathbf{k} x^{\otimes I},$$

where, for $I = (i_0, \ldots, i_k)$, we denote $x^{\otimes I} = x^{i_0} \otimes \cdots \otimes x^{i_k}$. With these notations, we also have

$$Ш^{+}_{\mathbf{k},\lambda}(\mathbf{k}[x]) = \bigoplus_{I \in \mathfrak{I} \cup \{0\}} x^{\otimes I}.$$

For positive integers m, n, let $I = (i_1, \ldots, i_m)$ and $J = (j_1, \ldots, j_n)$ be in $\mathfrak{I}$, The mixable shuffle $x^{\otimes I} ш_1 x^{\otimes J}$, or equivalently the stuffle $x^{\otimes I} ш_{st,1} x^{\otimes J}$, can be expressed as a product of the vectors I and J. More precisely, define an additive version of the stuffle product of I and J as follows. Let $(\varphi, \psi) \in \mathcal{J}_{m,n,r}$ as in Eq. (3.21). Thus $\varphi : [m] \to [m+n-r]$ and $\psi : [n] \to [m+n-r]$ and injective order preserving maps with $\operatorname{im}(\varphi) \cup \operatorname{im}(\psi) = [m+n-r]$. Define the *stuffle of* I *and* J (along (φ, ψ)) by

$$I\, ш_{(\varphi,\psi)} J = (i_{\varphi^{-1}(1)} + j_{\psi^{-1}(1)}, \ldots, i_{\varphi^{-1}(m+n-r)} + j_{\psi^{-1}(m+n-r)}), \tag{3.33}$$

with the convention that $\varphi^{-1}(\varnothing) = 0 = \psi^{-1}(\varnothing)$. Thus we have

$$x^{\otimes I} ш_{(\varphi,\psi)} x^{\otimes J} = x^{\otimes (I\, ш_{(\varphi,\psi)} J)}$$

and therefore

$$x^{\otimes I} ш_{st,1} x^{\otimes J} = \sum_{(\varphi,\psi)\in \overline{\mathcal{J}}_{m,n}} x^{\otimes (I\, ш_{(\varphi,\psi)} J)}.$$

We call

$$\sum_{(\varphi,\psi)\in \overline{\mathcal{J}}_{m,n}} I\, ш_{(\varphi,\psi)} J$$

the *stuffle product* of I and J. This is in fact the original meaning of the stuffle product, coming from the study of multiple zeta values in number theory [28, 29].

For I as above with $i_1 \geqslant 2$ and $i_2, \ldots, i_m \geqslant 1$, define the *multiple zeta value*

$$\zeta(I) := \zeta(i_1, \ldots, i_m) := \sum_{1\leqslant n_1 \leqslant \cdots \leqslant n_m} \frac{1}{n_1^{i_1} \cdots n_m^{i_m}}.$$

This is a well-defined real number, specializing on the well-known Riemann zeta values $\zeta(i_1)$ when $m = 1$. For J as above with $j_1 \geqslant 2$, $j_2, \ldots, j_n \geqslant 1$, there is the following fundamental identity among multiple zeta values [110, 174]

$$\zeta(I)\zeta(J) = \sum_{(\varphi,\psi)\in \overline{\mathcal{J}}_{m,n}} \zeta(I\, ш_{\varphi,\psi} J),$$

or formally

$$\zeta(I)\zeta(J) = \zeta(I\, ш_{st,1} J).$$

For example, for $I = (i_1)$ and $J = (j_1, j_2)$ with $i_1, j_1 \geqslant 2$ and $j_2 \geqslant 1$, we have

$$\zeta(i_1)\zeta(j_1,j_2) = \zeta(i_1,j_1,j_2)+\zeta(j_1,i_1,j_2)+\zeta(j_1,j_2,i_1)+\zeta(i_1+j_1,j_2)+\zeta(j_1,i_1+j_2).$$

See Section 3.3.3 for multiplication formulas of special elements in this form in terms of partitions.

3.2.3 Free commutative Rota-Baxter algebras of Cartier

We first recall the construction of free commutative Rota-Baxter algebras by Cartier [33, 92]. The original construction of Cartier is for free objects on a set, in the category of algebras with no identity and with weight -1. We will extend his construction slightly to free objects in the category of algebras with identity and with arbitrary weight. We will still consider free objects on sets. In order to consider free objects on a **k**-algebra A, it will be more convenient to use the tensor product notation introduced in Section 3.1, Section 3.1.4.2 in particular.

Let X be a set. Let λ be a fixed element in $\mathbf{k}$. Let M be the free commutative monoid generated by X. Let $\mathcal{X}$ denote the set of symbols of the form

$$u_0 \cdot [\,], \quad u_0 \in M,$$

and

$$u_0 \cdot [u_1, \ldots, u_m], \quad m \geqslant 1,\ u_0, u_1, \ldots, u_m \in M.$$

Let $\mathfrak{B}(X)$ be the free $\mathbf{k}$-module on $\mathcal{X}$. Cartier gave a $\mathbf{k}$-bilinear multiplication $\diamond_c$ on $\mathfrak{B}(X)$ by defining

$$\begin{aligned}(u_0 \cdot [\,]) \diamond_c (v_0 \cdot [\,]) &= u_0 v_0 \cdot [\,],\\ (u_0 \cdot [\,]) \diamond_c (v_0 \cdot [v_1, \ldots, v_n]) &= (v_0 \cdot [v_1, \ldots, v_n]) \diamond_c (u_0 \cdot [\,])\\ &= u_0 v_0 \cdot [v_1, \ldots, v_n],\end{aligned}$$

and

$$\begin{aligned}&(u_0 \cdot [u_1, \ldots, u_m]) \diamond_c (v_0 \cdot [v_1, \ldots, v_n])\\ =& \sum_{(k,P,Q)\in \overline{S}_c(m,n)} \lambda^{m+n-k} u_0 v_0 \cdot \Phi_{k,P,Q}([u_1, \ldots, u_m], [v_1 \ldots, v_n]). \quad (3.34)\end{aligned}$$

Here $\overline{S}_c(m, n)$, as defined in Section 3.1.4.2, is the set of triples (k, P, Q) in which k is an integer between 1 and $m + n$, P and Q are ordered subsets of $[k]$ such that $|P| = m, |Q| = n$ and $P \cup Q = [k]$. The latter conditions imply that $\max\{m, n\} \leqslant k \leqslant m + n$. For each triple $(k, P, Q) \in \overline{S}_c(m, n)$, define

$$\Phi_{k,P,Q}([u_1, \ldots, u_m], [v_1, \ldots, v_n]) := [w_1, \ldots, w_k] \in \mathcal{X},$$

where

$$w_j = \begin{cases} u_\alpha, & \text{if } j \text{ is the } \alpha\text{-th element in } P \text{ and } j \notin Q;\\ v_\beta, & \text{if } j \text{ is the } \beta\text{-th element in } Q \text{ and } j \notin P;\\ u_\alpha v_\beta, & \text{if } j \text{ is the } \alpha\text{-th element in } P \text{ and the } \beta\text{-th element in } Q.\end{cases}$$

Define a $\mathbf{k}$-linear operator P_X^c on $\mathfrak{B}(X)$ by

$$\begin{aligned}P_X^c(u_0 \cdot [\,]) &= 1 \cdot [u_0],\\ P_X^c(u_0 \cdot [u_1, \ldots, u_m]) &= 1 \cdot [u_0, u_1, \ldots, u_m].\end{aligned}$$

The following theorem is a modification of Theorem 1 in [33] and can be proved in the same way. We will prove it by relating the product of Cartier to the mixable shuffle product.

Theorem 3.2.6. *The triple* $(\mathfrak{B}(X), \diamond_c, P_X^c)$ *is a free commutative Rota-Baxter algebra on* X *of weight* λ.

Proof. We only need to show that $(\mathfrak{B}(X), \diamond_c, P_X^c)$ is canonically isomorphic to the mixable shuffle product construction $\text{Ш}_{\mathbf{k}}(X)$ in Proposition 3.2.2 of free commutative Rota-Baxter algebras.

From the definition of $\diamond_c$, we see that $\mathfrak{B}(X)$ is the tensor product algebra $\mathbf{k}[X] \otimes \mathfrak{B}^+(X)$. As a $\mathbf{k}$-module, we have

$$\mathfrak{B}^+(X) = \bigoplus_{k \geqslant 0} (\mathbf{k}M^k) = \bigoplus_{k \geqslant 0} (\mathbf{k}M)^{\otimes k} = \bigoplus_{k \geqslant 0} \mathbf{k}[X]^{\otimes k} = \text{Ш}^+(\mathbf{k}[X]),$$

by identifying $[u_1, \ldots, u_m]$ with $u_1 \otimes \cdots \otimes u_m$. Further, by Eq. (3.34), the product of $[u_1, \ldots, u_m]$ and $[v_1, \ldots, v_n]$ in $\mathfrak{B}^+(X)$ is given by

$$\sum_{(k,P,Q) \in \overline{S}_c(m,n)} \lambda^{m+n-k} \Phi_{k,P,Q}([u_1, \ldots, u_m], [v_1, \ldots, v_n]).$$

This coincides with the product $\text{ш}_{c,\lambda}$. Thus the product $\diamond_c$ coincides with $\diamond$ by Theorem 3.1.25 and Eq. (3.28). Since the Rota-Baxter operators also correspond to each other, the theorem follows. □

3.2.4 Standard Rota-Baxter algebras, Spitzer's Identity and Waring's Formula

The standard Rota-Baxter algebra constructed by Rota in [156] is a free object in the category of commutative Rota-Baxter algebras not necessarily having an identity. It is described as a Rota-Baxter subalgebra of another commutative Rota-Baxter algebra whose construction goes back to Baxter [24]. In Rota's construction, there are further restrictions that $\mathbf{k}$ can be a field of characteristic zero, the free commutative Rota-Baxter algebra obtained be on a finite set X, and the weight λ be 1. By making use of shuffle Rota-Baxter algebras, we will show that Rota's description can be modified to yield a free commutative Rota-Baxter algebra on an algebra in the category of commutative Rota-Baxter algebras with an identity, with a mild restriction on the weight λ. We can also provide a similar construction for algebras not necessarily having an identity, and for complete Rota-Baxter algebras, but we will not explain it here. See [93].

We first present Rota's construction, modified to give free objects in the category of commutative Rota-Baxter algebras with identity. We then give the general construction.

3.2.4.1 *The standard Rota-Baxter algebra of Rota*

For further details, see [156, 160].

As before, let $\mathbf{k}$ be a commutative ring with an identity, and fix a λ in $\mathbf{k}$. Let X be a given set. For each $x \in X$, let $t^{(x)}$ denote a sequence $(t_1^{(x)}, \ldots, t_n^{(x)}, \ldots)$ of distinct symbols $t_n^{(x)}, n \geqslant 1$. We also require that the sets $\{t_n^{(x_1)}\}_n$ and $\{t_n^{(x_2)}\}_n$ be disjoint for $x_1 \neq x_2$ in X. Denote

$$\overline{X} = \bigcup_{x \in X} \{t_n^{(x)} \mid n \geqslant 1\}$$

and denote by $\mathfrak{A}(X)$ the ring of sequences with entries in $\mathbf{k}[\overline{X}]$, the $\mathbf{k}$-algebra of polynomials with variables in $\overline{X}$. Thus the addition, multiplication and scalar multiplication by $\mathbf{k}[\overline{X}]$ in $\mathfrak{A}(X)$ are defined componentwise. Alternatively, for $k \in \mathbb{N}_{>0}$, let γ_k denote the sequence $(\delta_{n,k})_n$ with $\delta_{n,k}$ being the Kronecker delta. Then we can identify a sequence $(a_n)_n$ in $\mathfrak{A}(X)$ with a series

$$\sum_{n=1}^{\infty} a_n \gamma_n = a_1\gamma_1 + a_2\gamma_2 + \cdots .$$

Then the addition, multiplication and scalar multiplication by $\mathbf{k}[\overline{X}]$ are given termwise.

Define

$$P_X^r = P_{X,\lambda}^r : \mathfrak{A}(X) \to \mathfrak{A}(X)$$

by

$$P_X^r(a_1, a_2, a_3, \ldots) = \lambda(0, a_1, a_1 + a_2, a_1 + a_2 + a_3, \ldots).$$

In other words, each entry of $P_X^r(a)$, $a = (a_1, a_2, \ldots)$, is λ times the sum of the previous entries of a. If elements in $\mathfrak{A}(X)$ are described by series $\sum_{n=1}^{\infty} a_n \gamma_n$ given above, then we simply have

$$P_X^r\left(\sum_{n=1}^{\infty} a_n \gamma_n\right) = \lambda \sum_{n=1}^{\infty}\left(\sum_{i=1}^{n-1} a_i\right)\gamma_n.$$

For $\lambda = 1$, P_X^r is the Rota-Baxter operator of weight 1 defined in Exercise 1.1.6. It follows that, for any $\lambda \in \mathbf{k}$, P_X^r defines a Rota-Baxter operator of weight λ on $\mathfrak{A}(X)$. Hence $(\mathfrak{A}(X), P_X^r)$ is in $\mathrm{CRB}_{\mathbf{k}}$.

Definition 3.2.7. The *standard Rota-Baxter algebra* on X is the Rota-Baxter subalgebra $\mathfrak{S}(X)$ of $\mathfrak{A}(X)$ generated by the sequences

$$t^{(x)} = (t_1^{(x)}, \ldots, t_n^{(x)}, \ldots), \; x \in X.$$

An important result of Rota [156, 160] is

Theorem 3.2.8. *$(\mathfrak{S}(X), P_X^r)$ is a free commutative Rota-Baxter algebra on X.*

We will obtain this as a special case of the more general construction in the next subsection.

3.2.4.2 *The standard Rota-Baxter algebra in general*

Let A be a commutative $\mathbf{k}$-algebra. We give an alternative construction of the free commutative Rota-Baxter algebra on A.

For each $n \in \mathbb{N}_+$, we now let $A^{\otimes n}$ denote the n-th tensor power algebra where the tensor product is taken over $\mathbf{k}$. Note that the multiplication on $A^{\otimes n}$ here is different from the multiplication on $A^{\otimes n}$ when it is regarded as a $\mathbf{k}$-submodule of $Ш_{\mathbf{k}}(A)$.

Consider the direct limit algebra

$$\overline{A} = \varinjlim A^{\otimes n},$$

where the transition map is given by

$$A^{\otimes n} \to A^{\otimes(n+1)}, \quad \mathfrak{a} \mapsto \mathfrak{a} \otimes \mathbf{1}_A.$$

Let $\mathfrak{A}(A)$ be the set of sequences with entries in $\overline{A}$. Thus we have

$$\mathfrak{A}(A) = \prod_{n=1}^{\infty} \overline{A}\gamma_n = \left\{ \sum_{n=1}^{\infty} a_n \gamma_n \,\middle|\, a_n \in \overline{A} \right\}.$$

Define addition, multiplication and scalar multiplication on $\mathfrak{A}(A)$ componentwise, making $\mathfrak{A}(A)$ into a $\overline{A}$-algebra, with the sequence $(1, 1, \ldots)$ as the identity. Define

$$P_A^r = P_{A,\lambda}^r : \mathfrak{A}(A) \to \mathfrak{A}(A)$$

by

$$P_A^r(a_1, a_2, a_3, \ldots) = \lambda(0, a_1, a_1 + a_2, a_1 + a_2 + a_3, \ldots).$$

Then $(\mathfrak{A}(A), P_A^r)$ is a commutative Rota-Baxter algebra of weight 1. For each $a \in A$, define $t^{(a)} = (t_k^{(a)})_k$ in $\mathfrak{A}(A)$ by

$$t_k^{(a)} = \otimes_{i=1}^k a_i (= \otimes_{i=1}^\infty a_i), \quad a_i = \begin{cases} a, \, i = k\,, \\ 1, \, i \neq k\,. \end{cases}$$

Definition 3.2.9. The *standard Rota-Baxter algebra* on A is the Rota-Baxter subalgebra $\mathfrak{S}(A)$ of $\mathfrak{A}(A)$ generated by the sequences

$$t^{(a)} = (t_1^{(a)}, \ldots, t_n^{(a)}, \ldots), \quad a \in A.$$

Since $Ш_{\mathbf{k}}(A)$ is a free commutative Rota-Baxter algebra on A, the commutative algebra homomorphism

$$A \to \mathfrak{A}(A), \quad a \mapsto \mathfrak{t}^{(a)}$$

extends uniquely to a homomorphism of commutative Rota-Baxter algebras

$$\Phi : Ш_{\mathbf{k}}(A) \to \mathfrak{A}(A).$$

Theorem 3.2.10. *Assume that $\lambda \in \mathbf{k}$ is not a zero divisor in $\overline{A}$. The homomorphism of commutative Rota-Baxter algebras*

$$\Phi : Ш_{\mathbf{k}}(A) \to \mathfrak{S}(A)$$

induced by sending $a \in A$ to $\mathfrak{t}^{(a)} = (t_1^{(a)}, \ldots, t_n^{(a)}, \ldots)$ is an isomorphism.

The proof of this theorem will be given in the next subsection. As a result of this theorem, we have the following construction of Rota.

Corollary 3.2.11. *Assume that $\lambda \in \mathbf{k}$ is not a zero divisor in $\overline{A}$. Then $(\mathfrak{S}(A), P_A^r)$ is a free commutative Rota-Baxter algebra on A. Particularly, the standard commutative Rota-Baxter algebra on X is a free commutative Rota-Baxter on X.*

3.2.4.3 *Shuffle Rota-Baxter algebras and standard Rota-Baxter algebras*

We will prove Theorem 3.2.10 in this section by an inductive argument. For this purpose, we introduce some concepts and preliminary results that give filtration on $Ш_{\mathbf{k}}(A)$ and $\mathfrak{A}(A)$.

For $k \in \mathbb{N}_+$, let $Ш^k(A)$ denote the $\mathbf{k}$-submodule $A^{\otimes(k+1)}$ of $Ш_{\mathbf{k}}(A)$. Denote

$$\mathrm{Fil}^k(A) = \bigoplus_{n \geqslant k} Ш_{\mathbf{k}}^n(A).$$

It is a Rota-Baxter ideal of $Ш_{\mathbf{k}}(A)$ generated by pure tensors. Next define

$$F^k\mathfrak{A}(A) = \{(a_i) \in \mathfrak{A}(A) \mid a_i = 0,\ i \leqslant k\} = \left\{ \sum_{n=k+1}^{\infty} a_n \gamma_n \mid a_n \in \overline{A} \right\}.$$

Also denote $F^0\mathfrak{A}(A) = \mathfrak{A}(A)$. Then each $F^k\mathfrak{A}(A)$ is an ideal of $\mathfrak{A}(A)$. Define

$$F^k\mathfrak{S}(A) = F^k\mathfrak{A}(A) \cap \mathfrak{S}(A).$$

Then we have

$$F^k\mathfrak{S}(A) = \{(a_i) \in \mathfrak{S}(A) \mid a_i = 0, \quad i \leqslant k\}.$$

Each $F^k\mathfrak{S}(A)$ is an ideal of $\mathfrak{S}(A)$.

Lemma 3.2.12. *For any $k \in \mathbb{N}_{\geqslant 0}$, we have*
(a) $\mathfrak{A}(A)P_A^r(F^k\mathfrak{A}(A)) \subseteq F^{k+1}\mathfrak{A}(A)$.
(b) $\Phi(\mathrm{Fil}^k Ш_{\mathbf{k}}(A)) \subseteq F^k\mathfrak{A}(A)$.
Similar inclusions hold for $\mathfrak{S}(A)$.

Proof. We only need to verify the inclusions for $\mathfrak{A}(A)$. The inclusions for $\mathfrak{S}(A)$ follow immediately. By the definition of $P_A^r(F^k\mathfrak{A}(A))$ we have $P_A^r(F^k\mathfrak{A}(A)) \subseteq$

$F^{k+1}\mathfrak{A}(A)$. Since $\mathfrak{A}(A)P_A^r(F^k\mathfrak{A}(A))$ equals the ideal of $\mathfrak{A}(A)$ generated by $P_A^r(F^k\mathfrak{A}(A))$, we get the first inclusion.

The second inclusion is clear for $k = 0$. By induction on k, for $k > 0$, we have

$$\begin{aligned}\Phi(\mathrm{Fil}^k Ш_{\mathbf{k}}(A)) &\subseteq \Phi(Ш_{\mathbf{k}}(A)P_A(\mathrm{Fil}^{k-1}Ш_{\mathbf{k}}(A))) \\ &\subseteq \Phi(Ш_{\mathbf{k}}(A))P_A^r(\Phi(\mathrm{Fil}^{k-1}Ш_{\mathbf{k}}(A))) \\ &\subseteq \mathfrak{A}(A)P_A^r(F^{k-1}\mathfrak{A}(A)) \\ &\subseteq F^k\mathfrak{A}(A).\end{aligned}$$

□

Lemma 3.2.13. *For $n \in \mathbb{N}_{>0}$ and $a_1 \otimes \cdots \otimes a_n \in Ш_{\mathbf{k}}^n(A)$, we have*

$$\Phi(a_1 \otimes \cdots \otimes a_n) = \lambda^{n-1}(a_n \otimes \cdots \otimes a_1)\gamma_n + \text{ a term in } F^n\mathfrak{S}(A).$$

Proof. By definition, for $a_1 \in A \subseteq Ш_{\mathbf{k}}(A)$, we have

$$\Phi(a_1) = \sum_{k=1}^{\infty} t_k^{(a_1)}\gamma_k = a_1\gamma_1 + (1 \otimes a_1)\gamma_2 + \cdots .$$

So the lemma is proved for $n = 1$. Assume that the lemma is proved for $n \geqslant 1$, and consider $a_1 \otimes \cdots \otimes a_{n+1} \in Ш_{\mathbf{k}}^n(A)$. Applying Lemma 3.2.12, we have

$$\begin{aligned}\Phi(a_1 \otimes a_2 \otimes \cdots \otimes a_{n+1}) &= \Phi(a_1 P_A(a_2 \otimes \cdots \otimes a_{n+1})) \\ &= \Phi(a_1)\Phi(P_A(a_{n+1} \otimes \cdots \otimes a_2)) \\ &= \Phi(a_1)P_A^r(\Phi(a_{n+1} \otimes \cdots \otimes a_2)) \\ &= \left(\sum_{k=1}^{\infty} t_k^{(a_1)}\gamma_k\right) P_A^r(\lambda^{n-1}(a_{n+1} \otimes \cdots \otimes a_2)\gamma_n \\ &\quad + \text{a term in } F^n\mathfrak{S}(A)) \\ &= \left(\sum_{k=1}^{\infty} t_k^{(a_1)}\gamma_k\right) (\lambda^n(a_{n+1} \otimes \cdots \otimes a_2)\gamma_{n+1} \\ &\quad + \text{a term in } F^{n+1}\mathfrak{S}(A)) \\ &= \lambda^n(a_{n+1} \otimes a_n \otimes \cdots \otimes a_1)\gamma_{n+1} \\ &\quad + \text{a term in } F^{n+1}\mathfrak{S}(A).\end{aligned}$$

This completes the induction. □

We are now ready to prove Theorem 3.2.10. Since $(Ш_{\mathbf{k}}(A), P_A)$ is a free commutative Rota-Baxter algebra on A, the assignment

$$\Phi : A \to \mathfrak{S}(A), \quad a \mapsto \mathfrak{t}^{(a)} = (t_1^{(a)}, \ldots, t_n^{(a)}, \ldots)$$

induces a homomorphism $\Phi : Ш_{\mathbf{k}}(A) \to \mathfrak{S}(A)$ of Rota-Baxter algebras. Since $\mathfrak{S}(A)$ is the Rota-Baxter subalgebra of $\mathfrak{A}(A)$ generated by A, the homomorphism Φ is onto. So we only need to verify that Φ is injective.

Since $Ш_{\mathbf{k}}(X) = \bigoplus_{k\geqslant 0} Ш_{\mathbf{k}}^k(X)$, for any $G \in Ш_{\mathbf{k}}(X)$, we can uniquely write $G = \sum_{n\in\mathbb{N}_{\geqslant 0}} G_n$ with $G_n \in Ш_{\mathbf{k}}^n(A)$. Suppose $\Phi(G) = 0$; we will show by induction on $n \geqslant 0$ that $G_n = 0$. For $n = 0$ we have $Ш_{\mathbf{k}}^0(A) = A$. So G_0 is in A. By Lemma 3.2.12 we find that $\Phi(\sum_{k\geqslant 1} G_k)$ is in $F^1\mathfrak{A}(A)$. Thus the first component of $\Phi(F)$ in $\mathfrak{A}(A)$ is from

$$\Phi(G_0) = G_0\gamma_1 + \text{ a term in } F^1\mathfrak{A}(A).$$

Thus $\Phi(G) = 0$ implies $G_0\gamma_1 = 0$. Therefore, $G_0 = 0$.

Now assume that $G_k = 0$ for $k \leqslant n$ where $n \geqslant 0$ and consider $G_{n+1} \in Ш_{\mathbf{k}}^{n+1}(A) = A^{\otimes(n+2)}$. Thus G_{n+1} can be expressed as

$$G_{n+1} = \sum_{i=1}^{k} a_1^{(i)} \otimes \cdots \otimes a_{n+2}^{(i)}, \quad k \geqslant 1,\ a_j^{(i)} \in A.$$

Since $G_k = 0$ for $k \leqslant n$, and by Lemma 3.2.12, for $k \geqslant n+2$,

$$\begin{aligned}\Phi(G_k) &\in \Phi(Ш_{\mathbf{k}}^k(A))\\ &\subseteq \Phi(\mathrm{Fil}^k Ш_{\mathbf{k}}(A))\\ &\subseteq \Phi(\mathrm{Fil}^{n+2} Ш_{\mathbf{k}}(A))\\ &\subseteq F^{n+2}\mathfrak{S}(A),\end{aligned}$$

the only contribution of $\Phi(G)$ to the coefficient of γ_{n+2} is from $\Phi(G_{n+1})$. By Lemma 3.2.13, this coefficient is

$$\lambda^{n+1} \sum_{i=1}^{k} a_{n+2}^{(i)} \otimes \cdots \otimes a_1^{(i)}.$$

Thus $\Phi(F) = 0$ implies that

$$\lambda^{n+1} \sum_{i=1}^{k} a_{n+2}^{(i)} \otimes \cdots \otimes a_1^{(i)} = 0.$$

Since λ is not a zero divisor in $\bar{A}$, we further have

$$\sum_{i=1}^{k} a_{n+2}^{(i)} \otimes \cdots \otimes a_1^{(i)} = 0$$

as an element in the tensor power algebra $A^{\otimes(n+2)}$. But this element is zero or does not depend only on the $\mathbf{k}$-module structure of $A^{\otimes(n+2)}$ and the $\mathbf{k}$-module map

$$A^{\otimes(n+2)} \to A^{\otimes(n+2)}, \quad a_{n+2} \otimes \cdots \otimes a_1 \mapsto a_1 \otimes \cdots \otimes a_{n+2}$$

is an isomorphism. Thus we also have

$$G_{n+1} = \sum_{i=1}^{k} a_1^{(i)} \otimes \cdots \otimes a_{n+2}^{(i)} = 0.$$

This completes the induction. Thus Φ is injective and hence an isomorphism, completing the proof of Theorem 3.2.10.

3.2.4.4 *Spitzer's identity revisited*

In Section 1.3.1, we derive Waring's formula as an instance of (the classical case of) Spitzer's identity. As an application of the standard Rota-Baxter algebra, we now show, following [156, 160], that conversely, the classical case of Spitzer's identity can be derived from Warning's formula.

We first recall from Section 1.3 the algebraic formulation in the classical commutative case.

Theorem 3.2.14. (Spitzer [163]) *Let (A, P) be a commutative Rota-Baxter $\mathbb{Q}$-algebra of weight $\lambda = -1$. Then for $a \in A$, we have*

$$\exp\left(-P(\log(1 - at))\right) = \sum_{n=0}^{\infty} t^n \underbrace{P(aP(aP(a \cdots P(aP(a)) \cdots)))}_{n\text{-times}} \tag{3.35}$$

in the ring of power series $A[[t]]$.

Proof. We first prove for the weight 1 case using standard Rota-Baxter algebras and then translate it to the weight -1 case.

Proposition 3.2.15. [160] *Let (A, P) be a commutative Rota-Baxter $\mathbf{k}$-algebra of weight 1. Then for $b \in A$, we have*

$$\exp\left(P(\log(1 + tb))\right) = \sum_{n=0}^{\infty} t^n (Pb)^{[n]} \tag{3.36}$$

in the ring of power series $A[[t]]$. Here

$$(Pb)^{[n]} = \underbrace{P(b(P(b \cdots (P(b)) \cdots)))}_{n \text{ iteration}}$$

with the convention that $(Pb)^{[1]} = P(b)$ and $(Pb)^{[0]} = 1$.

Proof. First let $x = (x_1, x_2, \ldots)$ where x_i, $i \geqslant 1$, are symbols and let $X = \{x\}$. Consider the standard Rota-Baxter algebra $\mathfrak{S}(X)$. As we checked in Section 1.3.1,

$$(P_X^r x)^{[n]} = (0, e_n(x_1), e_n(x_1, x_2), e_n(x_1, x_2, x_3), \ldots),$$

where $e_n(x_1,\ldots,x_m)$ is the elementary symmetric function of degree n in the variables $x_1,\ldots,x_m$ with the convention that $e_0(x_1,\ldots,x_m) = 1$ and $e_n(x_1,\ldots,x_m)=0$ if $m<n$. By definition,

$$P_X^r(x^k) = (0, x_1^k, x_1^k + x_2^k, x_1^k + x_2^k + x_3^k, \ldots, p_k(x_1,\ldots,x_m),\ldots),$$

where $p_k(x_1,\ldots,x_m) = x_1^k + x_2^k + \cdots + x_m^k$ is the power sum symmetric function of degree k in the variables $x_1,\ldots,x_m$. These two classes of symmetric functions are related by the Waring's formula [160]

$$\exp\left(-\sum_{k=1}^{\infty}(-1)^k t^k p_k(x_1,\ldots,x_m)/k\right) = \sum_{n=0}^{\infty} e_n(x_1,\ldots,x_m)t^n, \quad \forall\, m \geqslant 1.$$

This proves

$$\exp\left(P_X^r(\log(1+tx))\right) = \sum_{n=0}^{\infty} t^n (P_X^r x)^{[n]}. \tag{3.37}$$

Next let (A,P) be any commutative Rota-Baxter **k**-algebra and let b be any element in A. By the universal property of the free commutative Rota-Baxter algebra $(\mathfrak{S}, P_X^r)$, there is a unique Rota-Baxter algebra homomorphism $\tilde{\varphi}: \mathfrak{S} \to A$ such that $\tilde{\varphi}(x) = b$. Since all the coefficients in the expansion of $\log(1+u)$ and $\exp(u)$ are rational, and $\tilde{\varphi}\circ P_X^r = P\circ\tilde{\varphi}$, applying $\tilde{\varphi}$ to Eq. (3.37) gives the desired equation. □

We can now complete the proof of Theorem 3.2.14 as follows. Let (A,P) be a commutative Rota-Baxter algebra of weight -1. Then $(A,-P)$ is a commutative Rota-Baxter algebra of weight 1. Applying Proposition 3.2.15, we have

$$\exp\left(-P(\log(1+tb))\right) = \sum_{n=0}^{\infty} t^n ((-P)b)^{[n]} = \sum_{n=0}^{\infty} t^n (P(-b))^{[n]}.$$

Replacing b by $-a$ in this equation gives Eq. (3.35). □

3.3 Rota-Baxter algebras, Stirling numbers and partitions

As another application of shuffle Rota-Baxter algebras, we study Stirling numbers, partitions and multinomial coefficients in the context of commutative Rota-Baxter algebras. This allows us to apply congruences from number theory to obtain congruences in commutative Rota-Baxter algebras and obtain an analog of Fermat's little theorem for commutative Rota-Baxter algebras.

3.3.1 Stirling numbers

Stirling numbers of the first kind and the second kind were introduced by J. Stirling [165] in 1730 and attracted interests of well-known mathematicians like Euler, Lagrange, Laplace and Cauchy. These numbers have played important roles in combinatorics, number theory, probability and statistics. There is a large literature on these numbers. See for example the survey article [36].

For each positive integer n, let $(t)_n = t(t-1)\cdots(t-n+1) \in \mathbb{Q}[t]$ be the descending (falling) factorial. Also define $(t)_0 = 1$. *Stirling numbers of first kind*, denoted by $s(n,k)$, are defined to be the coefficients in the identity

$$(t)_n = \sum_{k=0}^{n} s(n,k)t^k.$$

Stirling numbers of the second kind, denoted by $S(n,k)$ with $n,\ k \in \mathbb{N}_{\geqslant 0}$, are defined to be the coefficients in the identity

$$t^n = \sum_{k=0}^{n} S(n,k)(t)_k.$$

Stirling numbers of the second kind can also be recursively defined by

$$S(n+1,k+1) = S(n,k) + (k+1)S(n,k+1), \quad n,\ k \geqslant 0 \tag{3.38}$$

with $S(0,0)=1$, $S(n,0)=S(0,k)=0$ for $n,\ k \geqslant 1$. It follows that

$$S(n,1) = S(n,n) = 1$$

for $n \geqslant 1$.

Stirling numbers of the second kind have the following additional equivalent descriptions:

(a) (*Differential*) $S(n,k) = \frac{1}{k!}[\Delta^k t^n]_{t=0}$, where Δ is the forward difference operator.

(b) (*Combinatorial*) $S(n,k)$ is the number of partitions of n objects into k non-empty cells.

(c) (*Function*) $k!S(n,k)$ is the number of surjective maps from a set with n elements to a set with k elements.

(d) (*Generating function*) $\exp[t(e^u-1)] = \sum_{n=0}^{\infty}\sum_{k=0}^{n} S(n,k)t^k u^n/n!$.

Stirling numbers of the first and second kinds are related by the duality

$$\sum_{k=0}^{\infty} s(n,k)S(k,m) = \sum_{k=0}^{\infty} S(n,k)s(k,m) = \delta_{n,m}, \quad n,\ m \in \mathbb{Z}, \tag{3.39}$$

where $\delta_{n,m}$ is the Kronecker delta.

3.3.2 Stirling numbers in Rota-Baxter algebras

We give an interpretation of Stirling numbers of the first and second kinds in terms of commutative Rota-Baxter algebras.

Theorem 3.3.1. *Let (R, P) be a commutative Rota-Baxter* $\mathbf{k}$*-algebra. For* $n \geqslant 1$,

$$P(1)^n = \sum_{k=1}^{n} k!S(n,k)\lambda^{n-k}P^k(1), \qquad n!P^n(1) = \sum_{k=1}^{n} s(n,k)\lambda^{n-k}P(1)^k. \quad (3.40)$$

From the theorem with the weight $\lambda = 1$, we formally have

$$e^{P(1)u} = \sum_{n=0}^{\infty} P(1)^n u^n/n! = \sum_{n=0}^{\infty}\sum_{k=0}^{n} S(n,k)k!P^k(1)u^n/n! \in R[[u]].$$

So $e^{P(1)u}$ serves as a generating function of $k!S(n,k)$. We likewise obtain a generating function for multinomial coefficients (see Theorem 3.3.4).

Proof. The first equation holds for $n = 1$ since $S(1,1) = 1$. Assume that it holds for an $n \geqslant 1$. By [92, Prop. 6.1],

$$P^k(1)P(1) = (k+1)P^{k+1}(1) + k\lambda P^k(1). \quad (3.41)$$

Then we have

$$\begin{aligned} P(1)^{n+1} &= P(1)^n P(1) \\ &= \left(\sum_{k=1}^{n} k!S(n,k)\lambda^{n-k}P^k(1)\right) P(1) \\ &= \sum_{k=1}^{n} k!S(n,k)\lambda^{n-k}\big((k+1)P^{k+1}(1) + \lambda k P^k(1)\big) \\ &= \sum_{k=1}^{n}(k+1)!S(n,k)\lambda^{n-k}P^{k+1}(1) + \sum_{k=1}^{n} k!S(n,k)\lambda^{n+1-k}kP^k(1). \end{aligned}$$

Replacing k by $k-1$ in the first sum, we have

$$\begin{aligned} P(1)^{n+1} &= \sum_{k=2}^{n+1} k!S(n,k-1)\lambda^{n+1-k}P^k + \sum_{k=1}^{n} k!S(n,k)\lambda^{n+1-k}kP^k \\ &= S(n,1)\lambda^n P(1) + \sum_{k=2}^{n} k!\lambda^{n+1-k}\big(S(n,k-1) + kS(n,k)\big)P^k(1) \\ &\quad +(n+1)!S(n,n)P^{n+1}(1). \end{aligned}$$

So applying $S(k,1) = S(k,k) = 1,\ k \in \mathbb{N}$ and Eq. (3.38), we obtain

$$\begin{aligned} P(1)^{n+1} &= S(n+1,1)\lambda^n P(1) + \sum_{k=2}^{n} k!\lambda^{n+1-k} S(n+1,k) P^k(1) \\ &\quad + (n+1)! S(n+1,n+1) P^{(n+1)}(1) \\ &= \sum_{k=1}^{n+1} k!\lambda^{n+1-k} S(n+1,k) P^k(1). \end{aligned}$$

This proves the first equation. The second equation then follows from the duality in Eq. (3.39) between Stirling numbers of the first and second kinds. □

Stirling numbers of the second kind enjoy many congruences [31, 32, 40]. The most well-known one is the congruence

$$S(p,k) \equiv 0 \pmod{p},\ 0 < k < p, \tag{3.42}$$

for any prime number p. There are several proofs of this property [40, 116]. We will derive a generalization of this formula by computations in commutative Rota-Baxter algebras.

From Theorem 3.3.1 and the congruence in Eq. (3.42) we obtain the following consequence.

Corollary 3.3.2. *Let (R, P) be a commutative Rota-Baxter* **k***-algebra and let p be a prime number.*

$$P(1)^p \equiv \lambda^{p-1} P(1) \pmod{p} \tag{3.43}$$

The more general question of whether Eq. (3.43) holds when 1 is replaced by other elements of R will be addressed in Theorem 3.3.6.

3.3.3 Multinomial coefficients and partitions in Rota-Baxter algebras

We extend the relation in Eq. (3.40) further here and show that multinomial coefficients

$$\binom{n}{n_1, \ldots, n_k} := \frac{n!}{(n_1!)\cdots(n_k!)}$$

and partitions also arise naturally in the products of commutative Rota-Baxter algebras. This allows us to recover the congruence in Eq. (3.42) using multinomial coefficients.

Let $X = \{x\}$. Let

$$\mathcal{J} = \{I = (i_1, \ldots, i_k) | i_s \geqslant 0, 1 \leqslant s \leqslant k, k \geqslant 1\}.$$

For $I \in \mathcal{J}$, define the norm of I to be $|I| = \sum_{s=1}^{k} i_s$ and define the length of I to be $\ell(I) = k$. Define $I > 0$ if $i_s > 0$ for $1 \leqslant s \leqslant k$.

Denote

$$x^{\otimes I} = x^{i_1} \otimes \cdots \otimes x^{i_k}$$

and

$$\mathcal{X} = \{x^{\otimes I} | I \in \mathcal{J}\}.$$

Recall from Section 3.2.2.2 that the free commutative Rota-Baxter algebra on $\mathbf{k}[x]$ is $\bigoplus_{n \geqslant 1} \mathbf{k}[x]^{\otimes n}$ as a $\mathbf{k}$-module, so is a free $\mathbf{k}$-module on $\mathcal{X}$ together with the mixed shuffle product defined in Eq. (3.13) [83, 92]. The Rota-Baxter operator P_X on $Ш_{\mathbf{k}}(X)$ is defined by $P_X(x^{\otimes I}) = 1 \otimes x^{\otimes I}$. Thus $P_X(Ш_{\mathbf{k}}(X)) = 1 \otimes Ш_{\mathbf{k}}(X)$ is a free $\mathbf{k}$-module on $1 \otimes \mathcal{X} := \{1 \otimes x^{\otimes I} \mid I \in \mathcal{J}\}$.

By the Rota-Baxter identity in Eq. (1.1), $P_X(Ш_{\mathbf{k}}(X))$ is closed under multiplication. Therefore, for each $n \geqslant 1$,

$$(1 \otimes x)^n = \sum_{|I|=n} S(n, I) \otimes x^{\otimes I} \tag{3.44}$$

for elements $S(n, I) \in \mathbf{k}$. The first few terms of $S(n, I)$ are given by

$$\begin{aligned}
(1 \otimes x) &= 1 \otimes x, \\
(1 \otimes x)^2 &= 2(1 \otimes x \otimes x) + \lambda(1 \otimes x^2), \\
(1 \otimes x)^3 &= 6(1 \otimes x \otimes x \otimes x) + 3\lambda(1 \otimes x \otimes x^2) + 3\lambda(1 \otimes x^2 \otimes x) \\
&\quad + \lambda^2(1 \otimes x^3), \\
(1 \otimes x)^4 &= 24(1 \otimes x \otimes x \otimes x \otimes x) + 12\lambda(1 \otimes x \otimes x \otimes x^2) \\
&\quad + 12\lambda(1 \otimes x \otimes x^2 \otimes x) + 12\lambda(1 \otimes x^2 \otimes x \otimes x) + 6\lambda^2(1 \otimes x^2 \otimes x^2) \\
&\quad + 4\lambda^2(1 \otimes x \otimes x^3) + 4\lambda^2(1 \otimes x^3 \otimes x) + \lambda^3(1 \otimes x^4).
\end{aligned}$$

Lemma 3.3.3. *$S(n, I) \neq 0$ only if $I > 0$.*

Proof. We prove by induction on n with $n = 1$ being obvious. Assuming the lemma holds for n. So

$$(1 \otimes x)^n = \sum_{|I|=n, I>0} S(n, I) \otimes x^{\otimes I}.$$

For $x^{\otimes I} = x^{i_1} \otimes \cdots \otimes x^{i_k}$, the mixable shuffle product (3.13) gives

$$\begin{aligned}
(1 \otimes x^{\otimes I})(1 \otimes x) = 1 \otimes \big(& x^{\otimes(i_1, \ldots, i_k, 1)} + x^{\otimes(i_1, \ldots, i_{k-1}, 1, i_k)} + \cdots + x^{\otimes(1, i_1, \ldots, i_k)} \\
& + \lambda x^{\otimes(i_1, \ldots, i_{k-1}, i_k+1)} + \cdots + \lambda x^{\otimes(i_1+1, i_2, \ldots, i_k)}\big).
\end{aligned}$$

This shows that the lemma holds for $n + 1$. □

By Lemma 3.3.3, the sum in Eq. (3.44) is over the set of *ordered* partitions of n. For example, when $n = 3$, the terms on the right of $(1 \otimes x)^3$ correspond to the ordered partitions

$$n = 1 + 1 + 1 = 1 + 2 = 2 + 1 = 3.$$

For the coefficients in the sum, we have

Theorem 3.3.4. (a) *For $I > 0$, we have* $S(n, I) = \binom{n}{i_1, \ldots, i_k} \lambda^{n-\ell(I)}$.

(b) $\sum\limits_{|I|=n, \ell(I)=k} S(n, I) = k! S(n, k) \lambda^{n-k}$.

Recall that the multinomial coefficient $\binom{n}{i_1, \ldots, i_k}$ is the number of ways to distribute n distinct objects into k distinct sets $S_1, \ldots, S_k$ with $i_1, \ldots, i_k$ objects.

By Eq. (3.44) and the above theorem, we formally have

$$\frac{1}{1 - (1 \otimes x)} = \sum_{n=0}^{\infty} (1 \otimes x)^n = \sum_{n=0}^{\infty} \sum_{|I|=n} \binom{n}{I} \otimes x^I.$$

Here $\binom{n}{I} = \binom{n}{i_1, \ldots, i_k}$ for $I = (i_1, \ldots, i_k)$ by abbreviation and $(1 \otimes x)^0 = 1$ by convention. Thus $1/(1 - 1 \otimes x)$ can be regarded as a generating function for the multinomial coefficients $\binom{n}{I}$.

Proof. (a). We prove by induction on $\ell(I)$. It is obviously true for $\ell(I) = 1$. Assume that it is true for all $\ell(I)$ with $\ell(I) = n$. Then

$$\begin{aligned} \sum_{|J|=n+1} S(n+1, J)(1 \otimes x^{\otimes J}) &= (1 \otimes x)^{n+1} \\ &= \sum_{|I|=n} S(n, I)(1 \otimes x^{\otimes I})(1 \otimes x). \end{aligned} \tag{3.45}$$

Let $I = (i_1, \ldots, i_k)$. Using the notation $x^{\otimes I} = x^{i_1} \otimes \cdots \otimes x^{i_k}$, the mixable shuffle product in Eq. (3.13) gives

$$\begin{aligned} (1 \otimes x^{\otimes I})(1 \otimes x) = 1 \otimes \big(& x^{\otimes (i_1, \ldots, i_k, 1)} + x^{\otimes (i_1, \ldots, i_{k-1}, 1, i_k)} + \cdots + x^{\otimes (1, i_1, \ldots, i_k)} \\ & + \lambda x^{\otimes (i_1, \ldots, i_{k-1}, i_k + 1)} + \ldots + \lambda x^{\otimes (i_1 + 1, i_2, \ldots, i_k)} \big). \end{aligned}$$

Applying this to the right hand side of Eq. (3.45), we see that the coefficient $S(n+1, J)$ with $J=(j_1,\ldots,j_k)$ on the left hand side of Eq. (3.45) has contributions from precisely those $S(n, I_t)$ in the sum on the right hand side where I_t is defined by $I_t=(j_1,\ldots,j_{t-1}, j_t-1, j_{t+1},\ldots,j_k)$, $1\leqslant t\leqslant k$, with the convention that $i_t=j_t-1$ is deleted from I_t in the event that $j_t=1$. Further $\ell(I_t)=\ell(J)$ if $j_t>1$ and $\ell(I_t)=\ell(J)-1$ if $j_t=1$. For example, when $J=(2,1,4)$, we have $I_1=(1,1,4)$, $I_2=(2,4)$, and $I_3=(2,1,3)$. We then have

$$S(n+1,J)=\sum_{1\leqslant t\leqslant k, j_t=1} S(n,I_t)+\sum_{1\leqslant t\leqslant k, j_t>1}\lambda S(n,I_t).$$

So by the induction hypothesis, we have

$$\begin{aligned}
S(n+1,J) &= \sum_{j_t=1}\binom{n}{j_1,\ldots,j_{t-1},j_t-1,j_{t+1},\ldots,j_k}\lambda^{n-\ell(I_t)}\\
&\quad+\sum_{j_t>1}\lambda\binom{n}{j_1,\ldots,j_{t-1},j_t-1,j_{t+1},\ldots,j_k}\lambda^{n-\ell(I_t)}\\
&=\sum_{t=1}^{k}\binom{n}{j_1,\ldots,j_{t-1},j_t-1,j_{t+1},\ldots,j_k}\lambda^{n+1-\ell(J)}\\
&=\binom{n+1}{j_1,\ldots,j_k}\lambda^{n+1-\ell(J)},
\end{aligned}$$

where the last equation is the generalized Pascal identity [151, p.24].

(b) Once Item (a) is proved, Item (b) is a standard property on Stirling numbers and multinomial coefficients. We give a direct proof using properties of free commutative Rota-Baxter algebras instead of Item (a). Since $\text{Ш}(\{x\})$ is the free commutative Rota-Baxter algebra on $\{x\}$, there is a unique Rota-Baxter algebra homomorphism

$$f:\text{Ш}(\{x\})\to\text{Ш}(\mathbf{k})$$

induced by $f(x)=1$. We have

$$\begin{aligned}
(1\otimes 1)^n &= f((1\otimes x)^n)\\
&=\sum_{|I|=n} f(S(n,I)\otimes x^{\otimes I})\\
&=\sum_{|I|=n} S(n,I)\otimes f(x^{\otimes I})\\
&=\sum_{|I|=n} S(n,I)\otimes f(x)^{\otimes I}
\end{aligned}$$

$$= \sum_{|I|=n} S(n,I) \otimes 1^{\otimes I}$$
$$= \sum_{|I|=n} S(n,I) \otimes 1^{\otimes \ell(I)}$$
$$= \sum_{|I|=n} \sum_{\ell(I)=k} S(n,I) \otimes 1^{\otimes k}.$$

On the other hand, by Theorem 3.3.1,

$$(1 \otimes 1)^n = P_X(1)^n = \sum_{k=1}^{n} k! S(n,k) \lambda^{n-k} (1 \otimes 1)^k.$$

Since $\{1 \otimes 1^{\otimes k} |\ k \geqslant 1\}$ is linearly independent over $\mathbf{k}$, we obtain

$$\sum_{|I|=n, \ell(I)=k} S(n,I) = k!\, S(n,k) \lambda^{n-k},$$

as needed. □

We now deduce congruences of $S(n,I)$. As a result, we obtain a proof of the classical congruence in Eq. (3.42).

Proposition 3.3.5. *Let p be a prime number.*

(a) *$p|S(p,I)$ for any $I \in \mathcal{J}$ with $|I| = p$ and $I \neq (p)$.*

(b) *$p|S(p,k)$* for *$1 < k < p$.*

(c) *$(1 \otimes x)^p \equiv \lambda^{p-1} \otimes x^p \pmod p$.*

(d) *For any commutative Rota-Baxter algebra (R,P) and $a \in R$, we have $P(a)^p \equiv \lambda^{p-1} P(a^p) \pmod p$.*

(e) *Let $\lambda^{p-1} = 1$ and let $Ш_{\mathbb{F}_p}(\mathbb{F}_p)$ be the free commutative Rota-Baxter algebra over $\mathbb{F}_p$ of weight λ. For $a \in Ш_{\mathbb{F}_p}(\mathbb{F}_p)$, we have $a^p = \lambda^{p-1} a$.*

Proof. (a) By Theorem 3.3.4, $S(p,I) = \dfrac{p!}{n_1! \cdots n_k!} \lambda^{p-k}$. For $I \neq (p)$, we have $p \nmid n_1! \cdots n_k!$ since $n_i < p$. The claim follows.

(b) From Item (a) and Theorem 3.3.4. (b), we have $p|k! S(p,k) \lambda^{p-k}$, $1 < k < p$. So we just need to take $\lambda = 1$.

(c) This follows from

$$(1 \otimes x)^p = \sum_{|I|=p, \ell(I)=k} S(p,I) \lambda^{p-\ell(I)} \otimes x^{\otimes I} \equiv \lambda^{p-1} \otimes x^{\otimes (p)} = \lambda^{p-1} \otimes x^p \mod p.$$

(d) Since $Ш(\{x\})$ is the free commutative Rota-Baxter algebra on $\{x\}$, the assignment $x \mapsto a$ induces a Rota-Baxter algebra homomorphism $f : Ш(\{x\}) \to R$. Applying f to the congruence in Item (c) gives Item (d).

(e) We first prove $(1^{\otimes n})^p = 1^{\otimes n}$ by induction on $n \geqslant 1$. The case when $n = 1$ is clear. Using Item (d) and the induction hypothesis, we have

$$\begin{aligned}(1^{\otimes(n+1)})^p &= P(1^{\otimes n})^p \\ &= \lambda^{p-1} P((1^{\otimes n})^p) \\ &= P(1^{\otimes n}) \\ &= 1^{\otimes(n+1)}.\end{aligned}$$

Since $\{1^{\otimes n} \big| n \geqslant 1\}$ is a basis of $Ш_{\mathbb{F}_p}(\mathbb{F}_p)$ and $Ш_{\mathbb{F}_p}(\mathbb{F}_p)$ has characteristic p, Item (e) follows. □

3.3.4 Fermat's little theorem for Rota-Baxter algebras

Note that the congruence in Eq. (3.42) is equivalent to Fermat's little theorem

$$a^p \equiv a \pmod p, \; a \in \mathbb{Z}.$$

We apply Eqs. (3.40) and (3.42) to generalize this theorem for elements in any commutative Rota-Baxter algebra.

We consider free commutative Rota-Baxter algebras in general. Let A be a commutative **k**-algebra. Recall that the free commutative Rota-Baxter algebra $Ш(A)$ on A is

$$Ш(A) = \bigoplus_{n \geqslant 1} A^{\otimes n}$$

with multiplication given by the augmented mixable shuffle product and with Rota-Baxter operator

$$P_A(a_1 \otimes \cdots \otimes a_n) = 1 \otimes a_1 \otimes \cdots \otimes a_n$$

for $a_1 \otimes \cdots \otimes a_n \in A^{\otimes n}$.

Theorem 3.3.6. *Let p be a prime number. Assume that the weight λ of the commutative Rota-Baxter algebra is invertible.*

(a) (**Tensor form of freshman's dream**) *For $a_1 \otimes \cdots \otimes a_n \in A^{\otimes n}$,*

$$(a_1 \otimes \cdots \otimes a_n)^p \equiv \lambda^{(n-1)(p-1)} a_1^p \otimes \cdots \otimes a_n^p \mod p.$$

Particularly, if $\lambda = 1$, then

$$(a_1 \otimes \cdots \otimes a_n)^p \equiv a_1^p \otimes \cdots \otimes a_n^p \mod p.$$

(b) *Let I be the Rota-Baxter ideal of $Ш_{\mathbf{k}}(A)$ generated by p and $\{a^p - a \big| a \in A\}$. Then for any $\mathfrak{a} \in Ш_{\mathbf{k}}(A)$, we have $\mathfrak{a}^p \equiv \mathfrak{a} \mod I$.*

(c) *Let A be a commutative* $\mathbf{k}$*-algebra of characteristic p such that $a^p = a$ for all $a \in A$. Then $\mathfrak{a}^p = \mathfrak{a}$ for all $\mathfrak{a} \in Ш_{\mathbf{k}}(A)$.*

Remark 3.3.1. (a) The operations of taking a tensor product and taking a power cannot be interchanged in general in $Ш_{\mathbf{k}}(A)$. This is similar to the operations of taking a sum and taking a power. This is why Item (a) is called a tensor form of Freshman's Dream $(a+b)^p \equiv a^p + b^p \mod p$.

(b) Let A be the $\mathbb{F}_p$-algebra $\mathbb{F}_p[x]/(x^p - x)$, then $a^p = a$ for all $a \in A$. This follows from Fermat's little theorem and the Freshman's Dream. The free commutative Rota-Baxter algebra $Ш_{\mathbb{F}_p}(A)$ is a quite large algebra over A,

$$Ш_{\mathbb{F}_p}(A) = \bigoplus_{n \geqslant 1} A^{\otimes n}.$$

But we still have $\mathfrak{a}^p = \mathfrak{a}$ for all $\mathfrak{a} \in Ш_{\mathbb{F}_p}(A)$ by the theorem.

Proof. (a) By Proposition 3.3.5. (d), we have

$$\begin{aligned}(a_1 \otimes \cdots \otimes a_n)^p &\equiv (a_1 P_A(a_2 \otimes \cdots \otimes a_n))^p \\ &\equiv a_1^p P_A(a_2 \otimes \cdots \otimes a_n)^p \\ &\equiv a_1^p \lambda^{p-1} P_A((a_2 \otimes \cdots \otimes a_n)^p) \\ &\equiv \lambda^{p-1} a_1^p \otimes (a_2 \otimes \cdots \otimes a_n)^p \mod p.\end{aligned}$$

So by induction, we obtain

$$(a_1 \otimes \cdots \otimes a_n)^p \equiv \lambda^{(p-1)(n-1)} a_1^p \otimes a_2^p \otimes \cdots \otimes a_n^p \mod I.$$

(b) By Item (a) we have

$$(a_1 \otimes \cdots \otimes a_n)^p \equiv \lambda^{(n-1)(p-1)} a_1^p \otimes \cdots \otimes a_n^p \equiv \lambda^{(n-1)(p-1)} a_1 \otimes \cdots \otimes a_n \mod I.$$

Since λ is invertible, from $\lambda^p \equiv \lambda \mod I$ we have $\lambda^{p-1} \equiv 1 \mod I$. Then Item (b) follows.

(c) This follows from Item (b), since the ideal I is trivial here. □

3.4 Some related literature and problems

This chapter mainly follows [87, 92, 93]. The first construction of free Rota-Baxter algebras in the commutative case was the standard Rota-Baxter algebra [156] obtained by Rota as an internal construction, in the sense that it is obtained as a Rota-Baxter subalgebra of another commutative Rota-Baxter algebra, in opposite to the external construction of most other free objects, by starting from stretch so to speak. Thus this method is not that standard despite its name. But his construction motivated the first construction in the "standard" way given by Cartier [33]. The key ingredient in Cartier's construction was "stuffing" two finite sequences into one by merging some of the terms together. It is interesting to notice that this construction was ignored by the mathematical community including its author, just to be rediscovered twenty years later under the name of stuffle product in the study of multiple zeta values. See for example [28, 29, 34].

The construction of free commutative Rota-Baxter algebras by mixable shuffles [92] was followed by further studies of these algebras, including their completions [93], basic commutative algebra properties of such zero divisors and radicals [83], Hopf algebra structures [11, 53] and localization [38], as well as applications to umbral calculus in combinatorics [84], multiple zeta values in number theory [57, 103, 104, 105, 106, 107, 108, 146], Stirling numbers and partitions [87]. The ring theoretic structure of mixable shuffle algebras and free commutative Rota-Baxter algebras are studied in [102]. Relationship between Rota-Baxter algebras and differential algebras was considered in [86]. Then the combined structure of Rota-Baxter algebras and differential algebras were obtained in two variations, one under the name of differential Rota-Baxter algebras [94] and the other under the name of integro-differential algebras [155].

Finding a suitable class of ideals in commutative Rota-Baxter algebras that satisfies the ascending chain condition is essential in order to generalize the rich theories of commutative algebra and commutative differential algebra to commutative Rota-Baxter algebra. The Hilbert's basis theorem shows that the ideals of a polynomial algebra with finitely many generators satisfy the ascending chain condition. The Ritt-Raudenbush basis theorem [120, 152] states that the radical differential ideals of a differential polynomial algebra satisfy the ascending chain condition. Finding a suitable class of ideals in Rota-Baxter polynomial algebras (i.e., free commutative Rota-Baxter algebras) that satisfies the ascending chain condition is a fundamental problem in Rota-Baxter algebra that remains unsolved. A preliminary discussion can be found in [85].

For Rota-Baxter algebra-related structures, see [64] for free commutative Nijenhuis algebras and TD-algebra using shuffle type products.

Chapter 4

Free Noncommutative Rota-Baxter Algebras and Rooted Trees

Free commutative Rota-Baxter algebras have been constructed in Chapter 3. We consider the noncommutative case in this chapter. The study of Rota-Baxter algebras in the noncommutative case has its motivation from two quite recent developments. As we see in Chapters 1 and 2, noncommutative Rota-Baxter algebras played a fundamental role in the theory of Connes and Kreimer [42, 43, 61, 62] on renormalization of perturbative quantum field theory. This moves the constructions of the corresponding free objects to the forefront.

Another motivation came from the connection [2, 47] between Rota-Baxter algebras and dendriform algebras of Loday and Ronco [131] that we will study in Chapter 5, their Hopf algebra of planar trees in particular.

We will in fact consider several variations of free Rota-Baxter algebras, including the free Rota-Baxter algebra on a set and the free Rota-Baxter algebra on another algebra, a concept introduced in [92] in the commutative case. We also consider free objects in both the category of unitary Rota-Baxter algebras and that of nonunitary Rota-Baxter algebras. For each of these contexts, we give explicit constructions in terms of words, and in terms of planar rooted trees, with the latter being related to planar trees of Loday-Ronco [131, 137] and rooted trees of Connes-Kreimer [42].

In order to give a uniform treatment of these various constructions of free Rota-Baxter algebras, we derive them from a Rota-Baxter algebra structure on planar rooted trees. This Rota-Baxter algebra of rooted trees not only dictates the construction of free Rota-Baxter algebras, but also gives a natural tree interpretations of elements of free Rota-Baxter algebras as angularly decorated planar rooted trees. These Rota-Baxter algebras of rooted trees will be considered in Section 4.1.

Then free unitary and nonunitary Rota-Baxter algebras on a set are constructed in Section 4.2 in both the contexts of words and of trees as mentioned above.

As an application of these free Rota-Baxter algebras, the unitarization of Rota-Baxter algebras is studied in Section 4.3. To prepare for the application of free Rota-Baxter algebras to dendriform algebras in Chapter 5, we construct the free Rota-Baxter algebra over another algebra in Section 4.4. This process is similar to the one in Section 4.2 for the more special free Rota-Baxter algebras.

4.1 The Rota-Baxter algebra of trees

We first obtain a Rota-Baxter algebra structure on planar rooted trees and their various subsets. This allows us to give a uniform construction of free RBAs in different settings in later sections.

4.1.1 Planar rooted trees

For the convenience of the reader and for fixing notations, we start by recalling concepts and facts of planar rooted trees. We will only consider undecorated trees. But all results of this section can be easily generated to decorated trees. With these generalizations, we obtain free Rota-Baxter algebras with multiple Rota-Baxter operators.

We refer the reader to Section 2.3.4 for the concept of a rooted tree. A *plane rooted tree* is a rooted tree that can be embedded into the plane and in which a vertex has been distinguished as the root, so that there is only one incoming edge at each vertex except the root which has no incoming edges. A *planar rooted tree* is a plane rooted tree with a fixed embedding into the plane. There are two drawings of planar rooted trees. In one drawing all vertices are represented by a dot and the root is usually at the top of the tree. The following list shows the first few of them.

Note that we distinguish the branches of the trees, so the trees are non-symmetric, unlike the nonplanar rooted trees in Section 2.3.4.

In the second drawing the leave vertices are removed with only the edges leading to them left, and the root, placed at the bottom in opposite to the first drawing, gets an extra leave pointing down. The following list shows the first few of them.

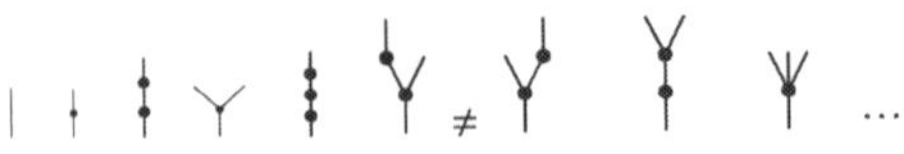

In the following, we will mostly follow the first set of notations though the second set of notations will also be used to relate to the Hopf algebra of planar rooted trees of Loday and Ronco.

Let $\mathcal{T}$ be the set of planar rooted trees and let $\mathfrak{T}$ be the free semigroup generated by $\mathcal{T}$ in which the product is denoted by $\sqcup$. Thus each element in $\mathfrak{T}$ is a noncommutative product of trees in $\mathcal{T}$, hence called a *forest*. So a forest is of the form $T_1 \sqcup \cdots \sqcup T_n$ consisting of trees $T_1, \ldots, T_n$. Here $\sqcup$ means putting two trees next to each other, i.e., the concatenation. We use $\lfloor T_1 \sqcup \cdots \sqcup T_n \rfloor$ to denote the tree with a new root and an edge from the new root to each of the trees $T_1, \ldots, T_n$. In the new tree, the trees $T_1, \ldots, T_n$, now the branches, are in the same order as in $T_1 \sqcup \cdots \sqcup T_n$. This is the grafting operator B^+ [42] introduced in Section 2.3.4. The operation is also denoted by $T_1 \vee \cdots \vee T_n$ in some other literature, such as [131, 137]. Our operation $\sqcup$ is different from $\vee$. Their relation is

$$\lfloor T_1 \sqcup \cdots \sqcup T_n \rfloor = T_1 \vee \cdots \vee T_n.$$

So $\sqcup$ just put trees next to each other to form a forest while $\lfloor\ \rfloor$ graft a forest into a new tree. Note that parentheses for $\sqcup$ is associative and so nonconsequential:

$$(T_1 \sqcup T_2) \sqcup T_3 = T_1 \sqcup T_2 \sqcup T_3 = T_1 \sqcup (T_2 \sqcup T_3)$$

while brackets for $\sqcup$ are highly nonassociative: $\lfloor T_1 \sqcup T_2 \rfloor \sqcup T_3$ and $T_1 \sqcup \lfloor T_2 \sqcup T_3 \rfloor$ are different forests.

A path is a linearly ordered chain from one vertex of the tree to the other. The length of a path is the number of edges it contains. The *depth* (or *height*) $d(T)$ of a tree T is the maximal length of paths from the root to the leaves of the tree. The depth of a forest is the maximal length of trees in the forest. It is clear that the depth of a forest T is increased by one in $\lfloor T \rfloor$.

Furthermore, for a forest $T = T_1 \sqcup \cdots \sqcup T_b$ with trees $T_1, \ldots, T_b$, we define b to be the *breadth* $b(T)$ of T, define the *head index* $h(T)$ of T to be 0 if T_1 is $\bullet$ and to be 1 otherwise. So $h(T) = 1$ if and only if $T_1 = \lfloor \overline{T} \rfloor$ for a forest $\overline{T}$. Similarly define the *tail index* $t(T)$ of T to be 0 (resp. 1) if T_b is $\bullet$ (resp. is not $\bullet$).

Example 4.1.1. The forest $\lfloor \lfloor \bullet \rfloor \bullet \rfloor \bullet$ has head index 1, tail index 0, breadth 2 and depth 2.

We now show that planar rooted forests have a recursive structure. For any subset X of $\mathfrak{T}$, let $\langle X \rangle$ be the sub-semigroup of $\mathfrak{T}$ generated by X. Let $\mathfrak{T}_0 = \langle \bullet \rangle$, consisting of forests generated by $\bullet$. These are also the forests of depth zero. Then recursively define

$$\mathfrak{T}_n = \langle \{\bullet\} \cup \lfloor \mathfrak{T}_{n-1} \rfloor \rangle. \tag{4.1}$$

Exercise 4.1.2. Prove by induction on the depth of trees that $\mathfrak{T}_n$ is the set of trees with depth less or equal to n.

As a result, the subsets $\mathfrak{T}_n, n \geqslant 0$, form a linear direct system: $\mathfrak{T}_n \supseteq \mathfrak{T}_{n-1}, n \geqslant 1$, and $\mathfrak{T}$ is the direct limit

$$\mathfrak{T} = \cup_{n\geqslant 0}\mathfrak{T}_n = \lim_{\rightarrow} \mathfrak{T}_n. \tag{4.2}$$

Then we have

$$\lfloor \mathfrak{T} \rfloor \subseteq \mathfrak{T}.$$

We next show that $\mathfrak{T}$ satisfies a universal property with respect to the product $\sqcup$ and the operator $\lfloor\ \rfloor$. Recall that a pointed set is a set together with a particular element of the set. We similarly define

Definition 4.1.3. ([88]) An *operated semigroup* is a semigroup U together with a map $\alpha : U \to U$. A morphism from an operated semigroup (U, α) to an operated semigroup (V, β) is a semigroup homomorphism $f : U \to V$ such that $f \circ \alpha = \beta \circ f$.

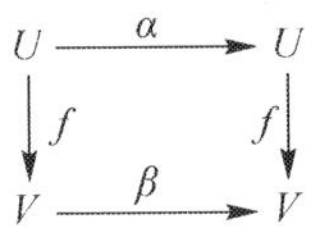

When the term semigroup is replaced by monoid, **k**-algebra or nonunitary **k**-algebra in the above definition, we obtain the concept of an *operated monoid*, *operated* **k**-*algebras* or *operated nonunitary* **k**-*algebra*. A morphism between operated (nonunitary) **k**-algebras is assumed to be **k**-linear. For example, the semigroup $\mathfrak{T}$ of forests with the map $\lfloor\ \rfloor$ and the product $\sqcup$ is an operated semigroup. The free **k**-module $\mathbf{k}\mathfrak{T}$ generated by $\mathfrak{T}$ is an operated **k**-algebra.

Definition 4.1.4. A *free operated semigroup on a set* X is an operated semigroup (U_X, α_X) together with a map $j_X : X \to U_X$ with the property that, for any operated semigroup (V, β) together with a map $f : X \to V$, there is a unique morphism $\overline{f} : (U_X, \alpha_X) \to (V, \beta)$ of operated semigroups such that $f = \overline{f} \circ j_X$.

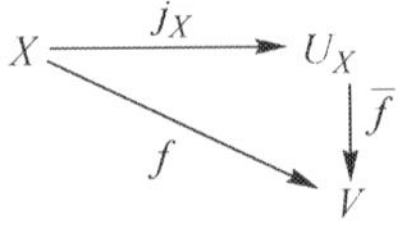

We similarly define the concept of free operated **k**-algebras.

Proposition 4.1.5. *Let* $j_\bullet : \{\bullet\} \to \mathfrak{T}$ *be the natural embedding* $j_\bullet(\bullet) = \bullet \in \mathcal{T}$.

(a) *The triple* $(\mathfrak{T}, (\lfloor\ \rfloor), j_\bullet)$ *is the free operated semigroup on one generator* $\bullet$.
(b) *The triple* $(\mathbf{k}\mathfrak{T}, (\lfloor\ \rfloor), j_\bullet)$ *is the free operated nonunitary algebra on one generator* $\bullet$.

We will see later that this free property dictates the construction of free RBAs .

Proof. (a) Let (V, β) be an operated semigroup with product denoted by $*$, and let $f : \{\bullet\} \to V$ be a set map. Recursively define semigroup homomorphisms $\overline{f}_n : \mathfrak{T}_n \to V, n \geqslant 0$, as follows.

First define $\overline{f}_0 : \mathfrak{T}_0 = \langle \bullet \rangle \to V$ by multiplicativity:

$$\overline{f}_0(\bullet \sqcup \cdots \sqcup \bullet) := f(\bullet) * \cdots * f(\bullet).$$

Suppose that $\overline{f}_n : \mathfrak{T}_n \to V$ has been defined. Define

$$\overline{f}_{n+1} : \lfloor \mathfrak{T}_n \rfloor \to V, \quad \overline{f}_{n+1}(\lfloor T \rfloor) = \beta(\overline{f}_n(T)), \quad T \in \mathfrak{T}_n.$$

Then extend the map to

$$\overline{f}_{n+1} : \mathfrak{T}_{n+1} := \langle \mathfrak{T}_0 \cup \lfloor \mathfrak{T}_n \rfloor \rangle \to V$$

by multiplicativity. Then define $\overline{f} = \varinjlim \overline{f}_n : \mathfrak{T} \to V$.

By construction, $\overline{f}$ is a morphism of operated semigroups, and the unique one such that $\overline{f}(\bullet) = f(\bullet)$.

The proof of (b) is similar. $\square$

We note that $\mathbf{k}\mathfrak{T}$ is also the free noncommutative polynomials on $\mathcal{T}$ with coefficients in $\mathbf{k}$ and with the product $\sqcup$. We are going to define another product $\diamond$ on $\mathbf{k}\mathfrak{T}$, making it into an RBA . This is achieved by using the filtration $\mathfrak{T} = \cup_{n \geqslant 0} \mathfrak{T}_n$ to recursively define a set map

$$\diamond : \mathfrak{T} \times \mathfrak{T} \to \mathbf{k}\mathfrak{T}$$

and then extending bilinearly. With the notation

$$T^{\sqcup n} = \underbrace{T \sqcup \cdots \sqcup T}_{n \text{ terms}},$$

we first define

$$\diamond : \mathfrak{T}_0 \times \mathfrak{T}_0 \to \mathbf{k}\mathfrak{T}, \ \bullet^{\sqcup m} \diamond \bullet^{\sqcup n} := \bullet^{\sqcup (m+n-1)}. \tag{4.3}$$

For any $k \geqslant 0$, assume that

$$\diamond : \mathfrak{T}_i \times \mathfrak{T}_j \to \mathbf{k}\mathfrak{T}$$

is defined for $i+j \leqslant k$. Consider trees $T \in \mathfrak{T}_i, T' \in \mathfrak{T}_j$ with $i+j=k+1$. Then $i+j \geqslant 1$. We first note that a tree is $\bullet$ or is of the form $\lfloor \overline{T} \rfloor$ for a forest $\overline{T}$ of smaller depth. From $i+j \geqslant 1$, either $i \geqslant 1$ or $j \geqslant 1$. Thus at least either T or T' is not $\bullet$. Thus we can define

$$T \diamond T' = \begin{cases} \bullet, & \text{if } T = T' = \bullet, \\ T, & \text{if } T' = \bullet, \\ T', & \text{if } T = \bullet, \\ \lfloor T \diamond \overline{T}' \rfloor + \lfloor \overline{T} \diamond T' \rfloor + \lambda \lfloor \overline{T} \diamond \overline{T}' \rfloor, & \text{if } T = \lfloor \overline{T} \rfloor, T' = \lfloor \overline{T}' \rfloor, \end{cases} \tag{4.4}$$

since for the three products on the right hand of the fourth equation, we have

$$\begin{aligned} d(T) + d(\overline{T}') &= d(T) + d(T') - 1, \\ d(\overline{T}) + d(T') &= d(T) + d(T') - 1, \\ d(\overline{T}) + d(\overline{T}') &= d(T) + d(T') - 2, \end{aligned}$$

which are all less than or equal to k. Note that in each case, $T \diamond T'$ is a tree or a linear combination of trees.

Now consider forests $T = T_1 \cdots T_b$ and $T' = T_1' \cdots T_{b'}'$ with $b > 1$ or $b' > 1$. We then define

$$T \diamond T' = T_1 \sqcup \cdots \sqcup T_{b-1} \sqcup (T_b \diamond T_1') \sqcup T_2' \sqcup \cdots \sqcup T_{b'}, \tag{4.5}$$

where $T_b \diamond T_1'$ is defined by Eq. (4.4) and the rest are given by the $\sqcup$-product. This completes the definition of the binary operation $\diamond$ on $\mathbf{k}\mathfrak{T}$.

We record the following simple properties of $\diamond$ for later applications.

Lemma 4.1.6. *For any forests T, T', T'', we have*

$$(T \sqcup T') \diamond T'' = T \sqcup (T' \diamond T''), \quad T'' \diamond (T \sqcup T') = (T'' \diamond T) \sqcup T'.$$

So $\mathfrak{T}$ with the operations $\sqcup$ and $\diamond$ forms a 2-associative algebra [138, 139, 149].

Proof. Let $T = T_1 \sqcup \cdots \sqcup T_b$, $T' = T_1' \sqcup \cdots \sqcup T_{b'}'$ and $T'' = T_1'' \sqcup \cdots \sqcup T_{b''}''$ be the decomposition of the forests into trees. Recall that parentheses for $\sqcup$ are associative. Then by definition,

$$\begin{aligned} &(T \sqcup T') \diamond T'' \\ &= (T_1 \sqcup \cdots \sqcup T_b \sqcup T_1' \sqcup \cdots \sqcup T_{b'}') \diamond (T_1'' \sqcup T_2'' \sqcup \cdots \sqcup T_{b''}'') \\ &= T_1 \sqcup \cdots \sqcup T_b \sqcup T_1' \sqcup \cdots \sqcup T_{b'-1}' \sqcup (T_{b'}' \diamond T_1'') \sqcup T_2'' \sqcup \cdots \sqcup T_{b''}'' \\ &= (T_1 \sqcup \cdots \sqcup T_b) \sqcup (T_1' \sqcup \cdots \sqcup T_{b'-1}' \sqcup (T_{b'}' \diamond T_1'') \sqcup T_2'' \sqcup \cdots \sqcup T_{b''}'') \\ &= T \sqcup (T' \diamond T''). \end{aligned}$$

The proof of the second equation is the same. □

Extending $\diamond$ bilinearly, we obtain a binary operation

$$\diamond : \mathbf{k}\mathfrak{T} \otimes \mathbf{k}\mathfrak{T} \to \mathbf{k}\mathfrak{T}.$$

For $T \in \mathfrak{T}$, define

$$P_{\mathcal{T}}(T) = \lfloor T \rfloor. \tag{4.6}$$

Then $P_{\mathcal{T}}$ defines a linear operator on $\mathbf{k}\mathfrak{T}$.

We can now state our main result on Rota-Baxter algebra on trees. It will be proved in the next subsection.

Theorem 4.1.7. (a) *The pair* $(\mathbf{k}\mathfrak{T}, \diamond)$ *is a unitary associative algebra.*

(b) *The triple* $(\mathbf{k}\mathfrak{T}, \diamond, P_{\mathcal{T}})$ *is a unitary* RBA *of weight* λ.

We will construct a nonunitary sub-RBA in $\mathbf{k}\mathfrak{T}$. Recall that a ladder tree is a tree in which every vertex has valent one or two. In terms of the brackets $\lfloor\ \rfloor$, a ladder tree is a nested sequence of i pairs of brackets, $i \geqslant 0$, with $\bullet$ in the middle

$$\underbrace{\lfloor \cdots \lfloor}_{i\ \text{brackets}} \bullet \underbrace{\rfloor \cdots \rfloor}_{i\ \text{brackets}} .$$

Particularly $\bullet$ is a ladder tree. It is easy to see that a tree or forest contains a ladder subtree if and only if it contains $\lfloor \bullet \rfloor$.

Corollary 4.1.8. *Let* $\mathfrak{T}^0$ *be the subset of* $\mathfrak{T}$ *consisting of forests that does not contain any ladder subtrees, that is, does not contain* $\lfloor \bullet \rfloor$. *The submodule* $\mathbf{k}\mathfrak{T}^0$ *of* $\mathbf{k}\mathfrak{T}$ *is a nonunitary Rota-Baxter subalgebra of* $\mathbf{k}\mathfrak{T}$ *under the product* $\diamond$.

Proof. We only need to check that $\mathbf{k}\mathfrak{T}^0$ is closed under $\diamond$ and $P_{\mathcal{T}} = \lfloor\ \rfloor$.

We remark that a forest T is in $\mathfrak{T}^0$ if and only if $\lfloor T \rfloor$ is in $\mathfrak{T}^0$, as can be easily seen by the definition of $\mathfrak{T}^0$. So $\mathbf{k}\mathfrak{T}^0$ is closed under the Rota-Baxter operator $P_{\mathcal{T}}$.

Note also that a forest is in $\mathfrak{T}^0$ if and only if each of its constituent trees is in $\mathfrak{T}^0$. So by Lemma 4.1.6, we only need to show that if two trees T, T' are in $\mathfrak{T}^0$, then $T \diamond T'$ is in $\mathbf{k}\mathfrak{T}^0$. We prove this by induction on $n := d(T) + d(T')$. Since T and T' are not ladder trees, they cannot be $\bullet$. Then we must have $d(T) \geqslant 1$ and $d(T') \geqslant 1$, and so $n \geqslant 2$. When $n = 2$, we have $T = \lfloor \overline{T} \rfloor$ and $T' = \lfloor \overline{T}' \rfloor$ with $\overline{T}, \overline{T}' \in \langle \bullet \rangle$. Since $\overline{T}$ and $\overline{T}'$ are in $\mathfrak{T}^0$ by the above remark, we must have $\overline{T} = \bullet^{\sqcup i}, \overline{T}' = \bullet^{\sqcup i'}$ with $i, i' \geqslant 2$. Then by Eqs. (4.3), (4.4) and (4.5), we have

$$\begin{aligned} T \diamond T' &= \lfloor \lfloor \bullet^{\sqcup i} \rfloor \diamond \bullet^{\sqcup i'} \rfloor + \lfloor \bullet^{\sqcup i} \diamond \lfloor \bullet^{\sqcup i'} \rfloor \rfloor + \lambda \lfloor \bullet^{\sqcup i} \diamond \bullet^{\sqcup i'} \rfloor \\ &= \lfloor (\lfloor \bullet^{\sqcup i} \rfloor \diamond \bullet) \sqcup \bullet^{\sqcup (i'-1)} \rfloor + \lfloor \bullet^{\sqcup (i-1)} \sqcup (\bullet \diamond \lfloor \bullet^{\sqcup i'} \rfloor) \rfloor + \lambda \lfloor \bullet^{\sqcup (i+i'-1)} \rfloor \\ &= \lfloor \lfloor \bullet^{\sqcup i} \rfloor \sqcup \bullet^{\sqcup (i'-1)} \rfloor + \lfloor \bullet^{\sqcup (i-1)} \sqcup \lfloor \bullet^{\sqcup i'} \rfloor \rfloor + \lambda \lfloor \bullet^{\sqcup (i+i'-1)} \rfloor . \end{aligned}$$

None of the terms contains $\lfloor \bullet \rfloor$, so they are in $\mathbf{k}\mathfrak{T}^0$.

Assume that $T \diamond T'$ is in $\mathbf{k}\mathfrak{T}^0$ for trees T, T' in $\mathfrak{T}^0$ with $n := d(T) + d(T') \geqslant k \geqslant 2$ and suppose $T, T' \in \mathfrak{T}^0$ with $d(T) + d(T') = k + 1$. Then by Eq. (4.4),

$$T \diamond T' = \lfloor \lfloor \overline{T} \rfloor \diamond \overline{T}' \rfloor + \lfloor \overline{T} \diamond \lfloor \overline{T}' \rfloor \rfloor + \lambda \lfloor \overline{T} \diamond \overline{T}' \rfloor.$$

By the induction hypothesis, each term in the brackets on the right hand side is in $\mathbf{k}\mathfrak{T}^0$, so the terms on the right hand side themselves are in $\mathbf{k}\mathfrak{T}^0$. □

4.1.2 The proof of Theorem 4.1.7

Proof. (a) By the definition of $\diamond$ in Eq. (4.4), $\bullet$ is the identity under the product $\diamond$. So we just need to verify the associativity. For this we only need to verify

$$(T' \diamond T'') \diamond T''' = T' \diamond (T'' \diamond T''') \tag{4.7}$$

for forests $T', T'', T''' \in \mathfrak{T}$. We will accomplish this by induction on the sum of the depths $n := d(T') + d(T'') + d(T''')$. If $n = 0$, then all of T', T'', T''' have depth zero and so are in $\mathfrak{T}_0 = \langle \bullet \rangle$, the sub-semigroup of $\mathfrak{T}$ generated by $\bullet$. Then we have $T' = \bullet^{\sqcup i'}$, $T'' = \bullet^{\sqcup i''}$ and $T''' = \bullet^{\sqcup i'''}$, for $i', i'', i''' \geqslant 1$. Then the associativity follows from Eq. (4.3) since both sides of Eq. (4.7) are $\bullet^{i'+i''+i'''-2}$ in this case.

Assume the associativity holds for $n \leqslant k$ and assume that $T', T'', T''' \in \mathfrak{T}$ satisfy $n = d(T') + d(T'') + d(T''') = k + 1$. We next reduce the breadths of the words.

Lemma 4.1.9. *If the associativity*

$$(T' \diamond T'') \diamond T''' = T' \diamond (T'' \diamond T''')$$

holds when T', T'' and T''' are trees, then it holds when they are forests.

Proof. We use induction on the sum of breadths $m := b(T') + b(T'') + b(T''')$. Then $m \geqslant 3$. The case when $m = 3$ is the assumption of the lemma. Assume the associativity holds for $3 \leqslant m \leqslant j$ and take $T', T'', T''' \in \mathfrak{T}$ with $m = j + 1$. Then $j + 1 \geqslant 4$. So at least either T' or T'' or T''' has breadth greater than or equal to 2.

First assume $b(T') \geqslant 2$. Then $T' = T'_1 \sqcup T'_2$ with $T'_1, T'_2 \in \mathfrak{T}$. Thus by Lemma 4.1.6, we have

$$\begin{aligned}(T' \diamond T'') \diamond T''' &= ((T'_1 \sqcup T'_2) \diamond T'') \diamond T''' \\ &= (T'_1 \sqcup (T'_2 \diamond T'')) \diamond T''' \\ &= T'_1 \sqcup ((T'_2 \diamond T'') \diamond T''').\end{aligned}$$

Similarly,

$$\begin{aligned} T' \diamond (T'' \diamond T''') &= (T'_1 \sqcup T'_2) \diamond (T'' \diamond T''') \\ &= T'_1 \sqcup (T'_2 \diamond (T'' \diamond T''')). \end{aligned}$$

Thus

$$(T' \diamond T'') \diamond T''' = T' \diamond (T'' \diamond T''')$$

whenever

$$(T'_2 \diamond T'') \diamond T''' = T'_2 \diamond (T'' \diamond T''')$$

which follows from the induction hypothesis.

A similar proof works if $b(T''') \geqslant 2$.

Finally if $b(T'') \geqslant 2$, then $T'' = T''_1 \sqcup T''_2$ with $T''_1, T''_2 \in \mathfrak{T}$. Using Lemma 4.1.6 repeatedly, we have

$$\begin{aligned} (T' \diamond T'') \diamond T''' &= (T' \diamond (T''_1 \sqcup T''_2)) \diamond T''' \\ &= ((T' \diamond T''_1) \sqcup T''_2) \diamond T''' \\ &= (T' \diamond T''_1) \sqcup (T''_2 \diamond T'''). \end{aligned}$$

In the same way, we have

$$T' \diamond (T'' \diamond T''') = (T' \diamond T''_1) \sqcup (T''_2 \diamond T''').$$

This again proves the associativity. □

To summarize, our proof of the associativity has been reduced to the special case when the forests $T', T'', T''' \in \mathfrak{T}$ are chosen such that

(a) $n := d(T') + d(T'') + d(T''') = k + 1 \geqslant 1$ with the assumption that the associativity holds when $n \leqslant k$, and

(b) the elements are of breadth one.

By (b), all the three elements are trees. If either one of them is $\bullet$, which is the identity under the product $\diamond$, then the associativity is clear.

So it remains to consider the case when T', T'', T''' are all in $\lfloor \mathfrak{T} \rfloor$. Then $T' = \lfloor \overline{T}' \rfloor, T'' = \lfloor \overline{T}'' \rfloor, T''' = \lfloor \overline{T}''' \rfloor$ with $\overline{T}', \overline{T}'', \overline{T}''' \in \mathfrak{T}$. Using Eq. (4.4) and the bilinearity of the product $\diamond$, we have

$$\begin{aligned} &(T' \diamond T'') \diamond T'' = (\lfloor \lfloor \overline{T}' \rfloor \diamond \overline{T}'' \rfloor + \lfloor \overline{T}' \diamond \lfloor \overline{T}'' \rfloor \rfloor + \lambda \lfloor \overline{T}' \diamond \overline{T}'' \rfloor) \diamond \lfloor \overline{T}''' \rfloor \\ &= \lfloor \lfloor \overline{T}' \rfloor \diamond \overline{T}'' \rfloor \diamond \lfloor \overline{T}''' \rfloor + \lfloor \overline{T}' \diamond \lfloor \overline{T}'' \rfloor \rfloor \diamond \lfloor \overline{T}''' \rfloor + \lambda \lfloor \overline{T}' \diamond \overline{T}'' \rfloor \diamond \lfloor \overline{T}''' \rfloor \\ &= \lfloor \lfloor \lfloor \overline{T}' \rfloor \diamond \overline{T}'' \rfloor \diamond \overline{T}''' \rfloor + \lfloor (\lfloor \overline{T}' \rfloor \diamond \overline{T}'') \diamond \lfloor \overline{T}''' \rfloor \rfloor + \lambda \lfloor (\lfloor \overline{T}' \rfloor \diamond \overline{T}'') \diamond \overline{T}''' \rfloor \\ &\quad + \lfloor \lfloor \lfloor \overline{T}' \diamond \lfloor \overline{T}'' \rfloor \rfloor \diamond \overline{T}''' \rfloor + \lfloor (\overline{T}' \diamond \lfloor \overline{T}'' \rfloor) \diamond \lfloor \overline{T}''' \rfloor \rfloor + \lambda \lfloor (\overline{T}' \diamond \lfloor \overline{T}'' \rfloor) \diamond \overline{T}''' \rfloor \\ &\quad + \lambda \lfloor \lfloor \overline{T}' \diamond \overline{T}'' \rfloor \diamond \overline{T}''' \rfloor + \lambda \lfloor (\overline{T}' \diamond \overline{T}'') \diamond \lfloor \overline{T}''' \rfloor \rfloor + \lambda^2 \lfloor (\overline{T}' \diamond \overline{T}'') \diamond \overline{T}''' \rfloor. \end{aligned}$$

Applying the induction hypothesis in n to the fifth term $(\overline{T'}\diamond\lfloor\overline{T''}\rfloor)\diamond\lfloor\overline{T'''}\rfloor$ and then using Eq. (4.4) again, we have

$$\begin{aligned}(T'\diamond T'')\diamond T'' = {} & \lfloor\lfloor\lfloor\overline{T'}\rfloor\diamond\overline{T''}\rfloor\diamond\overline{T'''}\rfloor + \lfloor(\lfloor\overline{T'}\rfloor\diamond\overline{T''})\diamond\lfloor\overline{T'''}\rfloor\rfloor\\ & +\lambda\lfloor(\lfloor\overline{T'}\rfloor\diamond\overline{T''})\diamond\overline{T'''}\rfloor\\ & +\lfloor\lfloor\overline{T'}\diamond\lfloor\overline{T''}\rfloor\rfloor\diamond\overline{T'''}\rfloor + \lfloor\overline{T'}\diamond\lfloor\lfloor\overline{T''}\rfloor\diamond\overline{T'''}\rfloor\rfloor\\ & +\lfloor\overline{T'}\diamond\lfloor\overline{T''}\diamond\lfloor\overline{T'''}\rfloor\rfloor\rfloor\\ & +\lambda\lfloor\overline{T'}\diamond\lfloor\overline{T''}\diamond\overline{T'''}\rfloor\rfloor + \lambda\lfloor(\overline{T'}\diamond\lfloor\overline{T''}\rfloor)\diamond\overline{T'''}\rfloor\\ & +\lambda\lfloor\lfloor\overline{T'}\diamond\overline{T''}\rfloor\diamond\overline{T'''}\rfloor + \lambda\lfloor(\overline{T'}\diamond\overline{T''})\diamond\lfloor\overline{T'''}\rfloor\rfloor\\ & +\lambda^2\lfloor(\overline{T'}\diamond\overline{T''})\diamond\overline{T'''}\rfloor.\end{aligned}$$

Similarly we obtain

$$\begin{aligned} & T'\diamond(T''\diamond T''')\\ & = \lfloor\overline{T'}\rfloor\diamond(\lfloor\lfloor\overline{T''}\rfloor\diamond\overline{T'''}\rfloor + \lfloor\overline{T''}\diamond\lfloor\overline{T'''}\rfloor\rfloor + \lambda\lfloor\overline{T''}\diamond\overline{T'''}\rfloor)\\ & = \lfloor\lfloor\overline{T'}\rfloor\diamond(\lfloor\overline{T''}\rfloor\diamond\overline{T'''})\rfloor + \lfloor\overline{T'}\diamond\lfloor\lfloor\overline{T''}\rfloor\diamond\overline{T'''}\rfloor\rfloor + \lambda\lfloor\overline{T'}\diamond(\lfloor\overline{T''}\rfloor\diamond\overline{T'''})\rfloor\\ & \quad +\lfloor\lfloor\overline{T'}\rfloor\diamond(\overline{T''}\diamond\lfloor\overline{T'''}\rfloor)\rfloor + \lfloor\overline{T'}\diamond\lfloor\overline{T''}\diamond\lfloor\overline{T'''}\rfloor\rfloor\rfloor + \lambda\lfloor\overline{T'}\diamond(\overline{T''}\diamond\lfloor\overline{T'''}\rfloor)\rfloor\\ & \quad +\lambda\lfloor\lfloor\overline{T'}\rfloor\diamond(\overline{T''}\diamond\overline{T'''})\rfloor + \lambda\lfloor\overline{T'}\diamond\lfloor\overline{T''}\diamond\overline{T'''}\rfloor\rfloor + \lambda^2\lfloor\overline{T'}\diamond(\overline{T''}\diamond\overline{T'''})\rfloor\\ & = \lfloor\lfloor\lfloor\overline{T'}\rfloor\diamond\overline{T''}\rfloor\diamond\overline{T'''}\rfloor + \lfloor\lfloor\overline{T'}\diamond\lfloor\overline{T''}\rfloor\rfloor\diamond\overline{T'''}\rfloor + \lambda\lfloor\lfloor\overline{T'}\diamond\overline{T''}\rfloor\diamond\overline{T'''}\rfloor\\ & \quad +\lfloor\overline{T'}\diamond\lfloor\lfloor\overline{T''}\rfloor\diamond\overline{T'''}\rfloor\rfloor + \lambda\lfloor\overline{T'}\diamond(\lfloor\overline{T''}\rfloor\diamond\overline{T'''})\rfloor\\ & \quad +\lfloor\lfloor\overline{T'}\rfloor\diamond(\overline{T''}\diamond\lfloor\overline{T'''}\rfloor)\rfloor + \lfloor\overline{T'}\diamond\lfloor\overline{T''}\diamond\lfloor\overline{T'''}\rfloor\rfloor\rfloor + \lambda\lfloor\overline{T'}\diamond(\overline{T''}\diamond\lfloor\overline{T'''}\rfloor)\rfloor\\ & \quad +\lambda\lfloor\lfloor\overline{T'}\rfloor\diamond(\overline{T''}\diamond\overline{T'''})\rfloor + \lambda\lfloor\overline{T'}\diamond\lfloor\overline{T''}\diamond\overline{T'''}\rfloor\rfloor + \lambda^2\lfloor\overline{T'}\diamond(\overline{T''}\diamond\overline{T'''})\rfloor.\end{aligned}$$

Now by the induction hypothesis, the i-th term in the expansion of $(T'\diamond T'')\diamond T'''$ matches with the $\sigma(i)$-th term in the expansion of $T'\diamond(T''\diamond T''')$. Here the permutation $\sigma\in\Sigma_{11}$ is

$$\begin{pmatrix} i \\ \sigma(i)\end{pmatrix} = \begin{pmatrix} 1 & 2 & 3 & 4 & 5 & 6 & 7 & 8 & 9 & 10 & 11\\ 1 & 6 & 9 & 2 & 4 & 7 & 10 & 5 & 3 & 8 & 11\end{pmatrix}. \tag{4.8}$$

This completes the proof of the first part of Theorem 4.1.7.

(b) We just need to prove that $P_{\mathcal{T}}(T)=\lfloor T\rfloor$ is a Rota-Baxter operator of weight λ. This is immediate from Eq. (4.4). □

4.2 Free Rota-Baxter algebra on a set

We will consider the constructions of the free unitary and nonunitary RBA on a set, generalizing the commutative case considered in [33, 92, 93, 156]. We describe a canonical basis of the algebra in several ways, by words recursively and explicitly, and also by rooted trees.

Variations of free RBAs , such as the more general free nonunitary RBAs and free RBAs on another algebra, will be constructed in Section 4.4. Doing so allows us to extend the constructions and methods for free RBAs on a set to this more general class of RBAs . Indeed, as we will see, these more general RBAs share similar properties of free RBAs on a set.

Free nonunitary RBAs are more convenient in some applications, say, to study universal enveloping RBAs of dendriform algebras and tridendriform algebras, and free dendriform algebras and tridendriform algebras [56]. We also use free RBAs to study the unitarization process of RBAs . This process is well known and simple for algebras, but is much more involved for RBAs . This will be presented in Section 4.3.

4.2.1 Constructions of a basis

We start with the basis definition of free RBAs

Definition 4.2.1. A *free* RBA *on a set* X is an RBA $F(X)$ with a Rota-Baxter operator P_X and a set map $j_X : X \to F(X)$ such that, for any RBA R and any set map $f : X \to R$, there is a unique RBA homomorphism $\overline{f} : F(X) \to R$ such that $\overline{f} \circ j_X = f$.

When the term algebra is replaced by nonunitary algebra in the definition, we obtain the definition of a free nonunitary Rota-Baxter algebra on X. When the set X is replaced to an algebra A, and when the term set map is replaced by algebra homomorphism, we obtain the definition of *free* RBA $F(A)$ on the algebra A. Similarly define the nonunitary variation. It is easy to check to see that $F(X)$ is the special case of $F(A)$ when A is taken to be $\mathbf{k}\langle X\rangle$, the (noncommutative) free algebra on X.

We will start with a recursive construction of a canonical $\mathbf{k}$-basis of the free RBA as certain words on the set X, followed by an explicit description of these words. We then interpret these words in terms of rooted trees. This tree interpretation allows us to use the Rota-Baxter algebra obtained in the last section to construct free Rota-Baxter algebras from these words.

4.2.1.1 *Recursive definition of a basis*

We first slightly extend the universal property of rooted trees in Section 4.1. Let X be a set. The category of operated monoids and free operated monoid on X are defined in the same way as in the case of operated semigroup in Section 4.1. We have the following construction of free operated monoid on X.

For any set Y, let $S(Y)$ denote the free semigroup generated by Y, let $M(Y)$ denote the free monoid generated by Y and let $\lfloor Y \rfloor$ denote the set $\{\lfloor y \rfloor \mid y \in Y\}$. Thus $\lfloor Y \rfloor$ is a set that is indexed by Y but disjoint with Y. We recursively define a direct system

$$\{\mathfrak{S}_n, i_{n,n+1} : \mathfrak{S}_n \to \mathfrak{S}_{n+1}\}$$

of free semigroups and a direct system

$$\{\mathfrak{M}_n, \tilde{i}_{n,n+1} : \mathfrak{M}_n \to \mathfrak{M}_{n+1}\}$$

of free monoids, both with injective transition maps. We do this by first letting

$$\mathfrak{S}_0 := S(X), \quad \mathfrak{M}_0 := M(X),$$

and then define

$$\mathfrak{S}_1 := S(X \cup \lfloor \mathfrak{S}_0 \rfloor) = S(X \cup \lfloor S(X) \rfloor), \quad \mathfrak{M}_1 := M(X \cup \lfloor M(X) \rfloor)$$

with $i_{0,1}$ and $\tilde{i}_{0,1}$ being the natural injections

$$i_{0,1} : \mathfrak{S}_0 = S(X) \hookrightarrow \mathfrak{S}_1 = S(X \cup \lfloor \mathfrak{S}_0 \rfloor),$$
$$\tilde{i}_{0,1} : \mathfrak{M}_0 = M(X) \hookrightarrow \mathfrak{M}_1 = M(X \cup \lfloor \mathfrak{M}_0 \rfloor).$$

We remark that elements in $\lfloor M(X) \rfloor$ are only symbols indexed by elements in $M(X)$. Particularly, $\lfloor \mathbf{1} \rfloor$ is not the identity. We identify $\mathfrak{S}_0$ and $\mathfrak{M}_0$ with their images in $\mathfrak{S}_1$ and $\mathfrak{M}_1$. Particularly, $\mathbf{1} \in \mathfrak{M}_0$ is sent to $\mathbf{1} \in \mathfrak{M}_1$.

Inductively assuming that $\mathfrak{S}_{n-1}$ and $\mathfrak{M}_{n-1}$ have been defined for $n \geqslant 2$, we define

$$\mathfrak{S}_n := S(X \cup \lfloor \mathfrak{S}_{n-1} \rfloor), \quad \mathfrak{M}_n := M(X \cup \lfloor \mathfrak{M}_{n-1} \rfloor). \tag{4.9}$$

Further assume that the embeddings

$$i_{n-2,n-1} : \mathfrak{S}_{n-2} \hookrightarrow \mathfrak{S}_{n-1}, \quad \tilde{i}_{n-2,n-1} : \mathfrak{M}_{n-2} \to \mathfrak{M}_{n-1}$$

have been obtained. Then we have the injections

$$\lfloor \mathfrak{S}_{n-2} \rfloor \hookrightarrow \lfloor \mathfrak{S}_{n-1} \rfloor \quad \text{and} \quad \lfloor \mathfrak{M}_{n-2} \rfloor \hookrightarrow \lfloor \mathfrak{M}_{n-1} \rfloor.$$

Thus by the freeness of $\mathfrak{S}_{n-1} = S(X \cup \lfloor \mathfrak{S}_{n-2} \rfloor)$ and $\mathfrak{M}_{n-1} = M(X \cup \lfloor \mathfrak{M}_{n-2} \rfloor)$, we have

$$\mathfrak{S}_{n-1} = S(X \cup \lfloor \mathfrak{S}_{n-2} \rfloor) \hookrightarrow S(X \cup \lfloor \mathfrak{S}_{n-1} \rfloor) = \mathfrak{S}_n,$$
$$\mathfrak{M}_{n-1} = M(X \cup \lfloor \mathfrak{M}_{n-2} \rfloor) \hookrightarrow M(X \cup \lfloor \mathfrak{M}_{n-1} \rfloor) = \mathfrak{M}_n.$$

We finally define the semigroup

$$\mathfrak{S}(X) := \varinjlim \mathfrak{S}_n$$

and the monoid

$$\mathfrak{M}(X) := \varinjlim \mathfrak{M}_n$$

with identity (the image of) $\mathbf{1}$.

The same proof for Proposition 4.1.5 gives

Proposition 4.2.2. *Let $j_X : X \to \mathfrak{M}(X)$ be the natural embedding.*

(a) *The triple $(\mathfrak{M}(X), (\lfloor\ \rfloor), j_X)$ is the free operated monoid on X.*

(b) *The triple $(\mathbf{k}\,\mathfrak{M}(X), (\lfloor\ \rfloor), j_X)$ is the free operated unitary $\mathbf{k}$-algebra on X.*

(c) *The triple $(\mathfrak{S}(X), (\lfloor\ \rfloor), j_X)$ is the free operated semigroup on X.*

(d) *The triple $(\mathbf{k}\,\mathfrak{S}(X), (\lfloor\ \rfloor), j_X)$ is the free operated nonunitary $\mathbf{k}$-algebra on X.*

Exercise 4.2.3. Give the proof of Proposition 4.2.2. For Item (c), one can prove that it is an operated semigroup isomorphic to the operated semigroup of rooted trees in Proposition 4.1.5.

Let Y and Z be subsets of $\mathfrak{M}(X)$. Define the *alternating products* of Y and Z by

$$\Lambda(Y,Z) := \Big(\bigcup_{r\geqslant 1} (Y\lfloor Z\rfloor)^r\Big) \bigcup \Big(\bigcup_{r\geqslant 0} (Y\lfloor Z\rfloor)^r Y\Big) \bigcup \Big(\bigcup_{r\geqslant 1} (\lfloor Z\rfloor Y)^r\Big) \bigcup \Big(\bigcup_{r\geqslant 0} (\lfloor Z\rfloor Y)^r \lfloor Z\rfloor\Big), \tag{4.10}$$

and

$$\tilde{\Lambda}(Y,Z) = \Big(\bigcup_{r\geqslant 1} (Y\lfloor Z\rfloor)^r\Big) \bigcup \Big(\bigcup_{r\geqslant 0} (Y\lfloor Z\rfloor)^r Y\Big) \bigcup \Big(\bigcup_{r\geqslant 1} (\lfloor Z\rfloor Y)^r\Big) \bigcup \Big(\bigcup_{r\geqslant 0} (\lfloor Z\rfloor Y)^r \lfloor Z\rfloor\Big) \bigcup \{\mathbf{1}\}, \tag{4.11}$$

where $\mathbf{1}$ is the identity in $\mathfrak{M}(X)$. They are again subsets of $\mathfrak{M}(X)$.

Using this notation, we construct a sequence of subsets of $\mathfrak{M}(X)$ by the following recursion. Let

$$\mathfrak{X}_0 = S(X), \quad \tilde{\mathfrak{X}}_0 = \mathfrak{X}_0 \cup \{\mathbf{1}\} = M(X).$$

In general, for $n \geqslant 1$, define

$$\mathfrak{X}_n = \Lambda(\mathfrak{X}_0, \mathfrak{X}_{n-1}), \quad \tilde{\mathfrak{X}}_n = \tilde{\Lambda}(\mathfrak{X}_0, \tilde{\mathfrak{X}}_{n-1}) = \Lambda(\mathfrak{X}_0, \tilde{\mathfrak{X}}_{n-1}) \cup \{\mathbf{1}\}. \tag{4.12}$$

More precisely,

$$\begin{aligned}\mathfrak{X}_n = &\Big(\bigcup_{r\geqslant 1}(\mathfrak{X}_0\lfloor\mathfrak{X}_{n-1}\rfloor)^r\Big)\bigcup\Big(\bigcup_{r\geqslant 0}(\mathfrak{X}_0\lfloor\mathfrak{X}_{n-1}\rfloor)^r\mathfrak{X}_0\Big)\\ &\bigcup\Big(\bigcup_{r\geqslant 1}(\lfloor\mathfrak{X}_{n-1}\rfloor\mathfrak{X}_0)^r\Big)\bigcup\Big(\bigcup_{r\geqslant 0}(\lfloor\mathfrak{X}_{n-1}\rfloor\mathfrak{X}_0)^r\lfloor\mathfrak{X}_{n-1}\rfloor\Big).\end{aligned} \tag{4.13}$$

$$\begin{aligned}\tilde{\mathfrak{X}}_n = &\Big(\bigcup_{r\geqslant 1}(\mathfrak{X}_0\lfloor\tilde{\mathfrak{X}}_{n-1}\rfloor)^r\Big)\bigcup\Big(\bigcup_{r\geqslant 0}(\mathfrak{X}_0\lfloor\tilde{\mathfrak{X}}_{n-1}\rfloor)^r\mathfrak{X}_0\Big)\\ &\bigcup\Big(\bigcup_{r\geqslant 1}(\lfloor\tilde{\mathfrak{X}}_{n-1}\rfloor\mathfrak{X}_0)^r\Big)\bigcup\Big(\bigcup_{r\geqslant 0}(\lfloor\tilde{\mathfrak{X}}_{n-1}\rfloor\mathfrak{X}_0)^r\lfloor\tilde{\mathfrak{X}}_{n-1}\rfloor\Big)\bigcup\{\mathbf{1}\}.\end{aligned} \tag{4.14}$$

So, for instance, $\mathfrak{X}_1 = \Lambda(S(X), S(X))$, $\tilde{\mathfrak{X}}_1 = \Lambda(S(X), M(X)) \cup \{\mathbf{1}\}$. We have

$$\mathfrak{X}_1 \supseteq \mathfrak{X}_0, \quad \tilde{\mathfrak{X}}_1 \supseteq \tilde{\mathfrak{X}}_0.$$

Assuming

$$\mathfrak{X}_n \supseteq \mathfrak{X}_{n-1}, \quad \tilde{\mathfrak{X}}_n \supseteq \tilde{\mathfrak{X}}_{n-1},$$

we get

$$\mathfrak{X}_{n+1} = \Lambda(\mathfrak{X}_0, \mathfrak{X}_n) \supseteq \Lambda(\mathfrak{X}_0, \mathfrak{X}_{n-1}) = \mathfrak{X}_n$$

and

$$\tilde{\mathfrak{X}}_{n+1} = \tilde{\Lambda}(\mathfrak{X}_0, \tilde{\mathfrak{X}}_n) \supseteq \tilde{\Lambda}(\mathfrak{X}_0, \tilde{\mathfrak{X}}_{n-1}) = \tilde{\mathfrak{X}}_n.$$

Thus we can define

$$\mathfrak{X}_\infty = \bigcup_{n\geqslant 0}\mathfrak{X}_n = \varinjlim \mathfrak{X}_n, \tag{4.15}$$

$$\tilde{\mathfrak{X}}_\infty = \bigcup_{n\geqslant 0}\tilde{\mathfrak{X}}_n = \varinjlim \tilde{\mathfrak{X}}_n. \tag{4.16}$$

Definition 4.2.4. Elements in $\tilde{\mathfrak{X}}_\infty$ are called *Rota-Baxter bracketed words (RBWs)*. Elements in $\mathfrak{X}_\infty$ are called *nonunitary Rota-Baxter bracketed words.*

The following properties of RBWs are easily verified.

Lemma 4.2.5. (a) *For each $n \geqslant 1$, the union of $\tilde{\mathfrak{X}}_n = \tilde{\Lambda}(\mathfrak{X}_0, \tilde{\mathfrak{X}}_{n-1})$ expressed in Eq.* (4.14) *is disjoint:*

$$\begin{aligned}\tilde{\mathfrak{X}}_n = &\Big(\bigsqcup_{r\geqslant 1}(\mathfrak{X}_0\lfloor\tilde{\mathfrak{X}}_{n-1}\rfloor)^r\Big)\bigsqcup\Big(\bigsqcup_{r\geqslant 0}(\mathfrak{X}_0\lfloor\tilde{\mathfrak{X}}_{n-1}\rfloor)^r\mathfrak{X}_0\Big)\\ &\bigsqcup\Big(\bigsqcup_{r\geqslant 1}(\lfloor\tilde{\mathfrak{X}}_{n-1}\rfloor\mathfrak{X}_0)^r\Big)\bigsqcup\Big(\bigsqcup_{r\geqslant 0}(\lfloor\tilde{\mathfrak{X}}_{n-1}\rfloor\mathfrak{X}_0)^r\lfloor\tilde{\mathfrak{X}}_{n-1}\rfloor\Big)\bigsqcup\{\mathbf{1}\}.\end{aligned} \tag{4.17}$$

This disjointness of the union is also preserved under the inclusion $\tilde{\mathfrak{X}}_n \to \tilde{\mathfrak{X}}_{n+1}$.

(b) *For each* $n \geqslant 1$, *the union of* $\mathfrak{X}_n = \Lambda(\mathfrak{X}_0, \mathfrak{X}_{n-1})$ *expressed in Eq.* (4.13) *is disjoint:*

$$\mathfrak{X}_n = \Big(\bigsqcup_{r\geqslant 1} (\mathfrak{X}_0 \lfloor \mathfrak{X}_{n-1} \rfloor)^r\Big) \bigsqcup \Big(\bigsqcup_{r\geqslant 0} (\mathfrak{X}_0 \lfloor \mathfrak{X}_{n-1} \rfloor)^r \mathfrak{X}_0\Big) \tag{4.18}$$
$$\bigsqcup \Big(\bigsqcup_{r\geqslant 1} (\lfloor \mathfrak{X}_{n-1} \rfloor \mathfrak{X}_0)^r\Big) \bigsqcup \Big(\bigsqcup_{r\geqslant 0} (\lfloor \mathfrak{X}_{n-1} \rfloor \mathfrak{X}_0)^r \lfloor \mathfrak{X}_{n-1} \rfloor\Big).$$

This disjointness of the union is also preserved under the inclusion $\mathfrak{X}_n \to \mathfrak{X}_{n+1}$.

(c) *Every* RBW $\mathbf{x} \neq \mathbf{1}$ *has a unique decomposition*

$$\mathbf{x} = \mathbf{x}_1 \cdots \mathbf{x}_b, \tag{4.19}$$

where $\mathbf{x}_i$, $1 \leqslant i \leqslant b$, *is alternatively in* $\mathfrak{X}_0$ *or in* $\lfloor \mathfrak{X}_\infty \rfloor$. *This decomposition will be called the standard decomposition of* $\mathbf{x}$.

For an RBW $\mathbf{x}$ in $\tilde{\mathfrak{X}}_n$ with standard decomposition $\mathbf{x}_1 \cdots \mathbf{x}_b$. We define b to be the *breadth* $b(\mathbf{x})$ of $\mathbf{x}$, then we define the *head index* $h(\mathbf{x})$ of $\mathbf{x}$ to be 0 (resp. 1) if $\mathbf{x}_1$ is in $M(X)$ (resp. in $\lfloor \tilde{\mathfrak{X}}_{n-1} \rfloor$). Similarly define the *tail index* $t(\mathbf{x})$ of $\mathbf{x}$ to be 0 (resp. 1) if $\mathbf{x}_b$ is in $M(X)$ (resp. in $\lfloor \tilde{\mathfrak{X}}_{n-1} \rfloor$). In terms of the decomposition of $\mathbf{x}$ in Eq. (4.14), the head index, tail index and breadth of an RBW $\mathbf{x}$ are given in the following table.

$\mathbf{x}$	$(\mathfrak{X}_0 \lfloor \tilde{\mathfrak{X}}_{n-1} \rfloor)^r$	$(\mathfrak{X}_0 \lfloor \tilde{\mathfrak{X}}_{n-1} \rfloor)^r \mathfrak{X}_0$	$(\lfloor \tilde{\mathfrak{X}}_{n-1} \rfloor \mathfrak{X}_0)^r$	$(\lfloor \tilde{\mathfrak{X}}_{n-1} \rfloor \mathfrak{X}_0)^r \lfloor \tilde{\mathfrak{X}}_{n-1} \rfloor$
$h(\mathbf{x})$	0	0	1	1
$t(\mathbf{x})$	1	0	0	1
$b(\mathbf{x})$	$2r$	$2r+1$	$2r$	$2r+1$

Finally, define the *depth* $d(\mathbf{x})$ of $\mathbf{x}$ to be

$$d(\mathbf{x}) = \min\{n \mid \mathbf{x} \in \tilde{\mathfrak{X}}_n\}.$$

So, particularly, the depth of elements in $\tilde{\mathfrak{X}}_0$ is 0 and depth of elements in $\lfloor \tilde{\mathfrak{X}}_0 \rfloor$ is one.

Example 4.2.6. For $t_1, t_2, t_3 \in X$, the word $\lfloor \lfloor t_1 \rfloor t_2 \rfloor t_3$ has head index 1, tail index 0, breadth 2 and depth 2.

4.2.1.2 *The explicit description*

To easily tell whether a word is an RBW, we give an equivalent description RBWs without using a recursion. This part will not be needed in the rest of the section and can be skipped at the first reading.

Definition 4.2.7. Let $X' = X \cup \{\lfloor, \rfloor\}$ and let $\tilde{\mathfrak{X}}'$ be the set of words in $M(X')$ that satisfies the following conditions.

(a) The number of $\lfloor$ in the word equals the number of $\rfloor$ in the word;

(b) Counting from the left to the right, the accumulative number of $\lfloor$ at each location is always greater or equal to that of $\rfloor$;

(c) There is no occurrence of $\rfloor\lfloor$ in the word;

A word $\mathfrak{z}$ in $\tilde{\mathfrak{X}}'$ can be uniquely written as $\mathfrak{z} = z_1 \cdots z_m$ with $z_i \in X', 1 \leqslant i \leqslant m$. We define the *length* $\ell(\mathfrak{z})$ of $\mathfrak{z}$ to be m, define $h'(\mathfrak{z})$ to be 0 (resp. 1) if z_1 is in X (resp. in $\{\lfloor, \rfloor\}$). Similarly define $t'(\mathfrak{z})$ to be 0 (resp. 1) if z_m is in X (resp. in $\{\lfloor, \rfloor\}$). Let $d_i(\mathfrak{z})$ be the number of $\lfloor$ in $z_1, \ldots, z_i$ minus the number of $\rfloor$ in $z_1, \ldots, z_i$. Define $d'(\mathfrak{z}) := \max\{d_i(\mathfrak{z}) \mid 1 \leqslant i \leqslant m\}$. Define $b'(\mathfrak{z})$ of $\mathfrak{z}$ to be the number of zeros of $d_i(\mathfrak{z}), 1 \leqslant i \leqslant m$. We will see that for a word $\mathfrak{z} \in \tilde{\mathfrak{X}}$, the numbers h', t', d', b' of $\mathfrak{z}$ agree with the numbers h, t, d, b defined above, once words in $\tilde{\mathfrak{X}}_\infty$ and in $\tilde{\mathfrak{X}}'$ are identified.

To identify $\tilde{\mathfrak{X}}_\infty$ with $\tilde{\mathfrak{X}}'$, we start with giving a sequence of embeddings

$$\tilde{\phi}_n : \mathfrak{M}_n \to M(X'), \quad n \geqslant 0,$$

for the sets $\mathfrak{M}_n$ defined in Eq. (4.9). First, by the freeness of the monoid $M(X)$ generated by X, the natural embedding

$$\phi_0 : X \to S(X'), \quad \phi_0(x) = x, \ x \in X,$$

extends uniquely to a monoid homomorphism

$$\tilde{\phi}_0 : M(X) \to M(X')$$

identifying $M(X)$ with the free (sub)monoid of $M(X')$ generated by $X = \phi_0(X)$. So the monoid homomorphism ϕ_0 must be injective.

To illustrate this recursive definition of $\tilde{\phi}_n$, we give another step before giving the inductive definition. We define

$$\tilde{\phi}'_0 : \lfloor M(X) \rfloor \to M(X'), \quad \tilde{\phi}'_0(\lfloor \mathbf{x} \rfloor) = \lfloor \tilde{\phi}_0(\mathbf{x}) \rfloor, \quad \mathbf{x} \in S(X), \tilde{\phi}'_0(\lfloor \mathbf{1} \rfloor) = \lfloor\,\rfloor.$$

Here the brackets on the left hand side are just to distinguish $\lfloor \mathbf{x} \rfloor$ as a new letter from $\mathbf{x}$; while on the right hand side, the brackets are themselves letters, giving a new word $\lfloor \phi_0(\mathbf{x}) \rfloor$ from the word $\phi_0(\mathbf{x}) \in S(X')$. We emphasize that $\tilde{\phi}'_0(\lfloor \mathbf{1} \rfloor) = \lfloor\,\rfloor$, and note that $\lfloor \mathbf{1} \rfloor$ is not an element in $M(X')$.

The map $\tilde{\phi}'_0$ is again injective, extended to the injective map

$$\begin{aligned} \phi_1 := \phi_0 \bigsqcup \tilde{\phi}'_0 \colon X \bigsqcup \lfloor M(X) \rfloor &\to \phi_0(X) \bigsqcup \tilde{\phi}'_0(\lfloor M(X) \rfloor) \\ &= X \bigsqcup \lfloor S(X) \rfloor \bigsqcup \{\lfloor\,\rfloor\}. \end{aligned}$$

Then ϕ_1 extended to an injective monoid homomorphism on the free monoid:

$$\tilde{\phi}_1 : M(X \bigsqcup \lfloor M(X) \rfloor) \to M(X').$$

In general, inductively assume that we have defined an injective monoid homomorphism

$$\tilde{\phi}_n : M(X \cup \lfloor \mathfrak{M}_{n-1} \rfloor) \to M(X').$$

We then get the injective map

$$\tilde{\phi}'_n : \lfloor \mathfrak{M}_n \rfloor \to M(X'), \quad \lfloor \mathbf{x} \rfloor \mapsto \lfloor \mathbf{x} \rfloor, \quad \lfloor 1 \rfloor \mapsto \lfloor\, \rfloor,$$

which extends to the injective map

$$\tilde{\phi}'_n : X \bigsqcup \lfloor \mathfrak{M}_n \rfloor \to M(X').$$

We then get the injective monoid homomorphism

$$\tilde{\phi}_{n+1} : M(X \bigsqcup \lfloor \mathfrak{M}_n \rfloor) \to M(X').$$

We thus get a morphism of injective systems

$$\begin{array}{ccc} \mathfrak{M}_n & \xrightarrow{\tilde{\phi}_n} & M(X') \\ \downarrow \subseteq & & \downarrow \mathrm{id} \\ \mathfrak{M}_{n+1} & \xrightarrow{\tilde{\phi}_{n+1}} & M(X') \end{array}$$

with injective transition maps. Here the direct system on the right is $\{M(X') \xrightarrow{\mathrm{id}} M(X')\}$.

We then get an injective map

$$\tilde{\phi}_\infty : \mathfrak{M}(X) := \varinjlim \mathfrak{M}_n \to S(X').$$

It restricts to injective maps

$$\phi_n : \mathfrak{X}_n \to M(X'), \quad \tilde{\phi}_n : \tilde{\mathfrak{X}}_n \to M(X'), \quad 1 \leqslant n \leqslant \infty. \tag{4.20}$$

By restriction we have

$$\begin{aligned} \phi_\infty(\mathfrak{X}_\infty) &= \varinjlim \phi_n(\mathfrak{X}_n), \\ \tilde{\phi}_\infty(\tilde{\mathfrak{X}}_\infty) &= \varinjlim \tilde{\phi}_n(\tilde{\mathfrak{X}}_n). \end{aligned}$$

Proposition 4.2.8. (a) *We have* $\tilde{\mathfrak{X}}' = \tilde{\phi}_\infty(\tilde{\mathfrak{X}}_\infty)$.

(b) *For an* RBW $\mathbf{x} \in \tilde{\mathfrak{X}}_\infty$, *we have*

$$\begin{aligned} h'(\tilde{\phi}_\infty(\mathbf{x})) &= h(\mathbf{x}), \quad t'(\tilde{\phi}_\infty(\mathbf{x})) = t(\mathbf{x}), \\ d'(\tilde{\phi}_\infty(\mathbf{x})) &= d(\mathbf{x}), \quad b'(\tilde{\phi}_\infty(\mathbf{x})) = b(\mathbf{x}). \end{aligned} \tag{4.21}$$

From this proposition, it is immediate that

$$x_1\lfloor x_2\rfloor x_3, \quad \lfloor\lfloor x_1\rfloor x_2\rfloor x_3,\ x_1\lfloor\,\rfloor, \quad \lfloor x_1x_2\rfloor$$

are RBWs while

$$\lfloor x_1\lfloor x_2\rfloor x_3, \quad x_1\rfloor x_2\lfloor, \quad \lfloor x_1\rfloor\lfloor x_2\rfloor$$

are not RBWs.

Proof. (a) Since $\tilde{\phi}_\infty$ is injective as a direct system of injective maps, we only need to prove $\phi_\infty(\tilde{\mathfrak{X}}_\infty) = \tilde{\mathfrak{X}}'$. Let $\tilde{\mathfrak{X}}'_n$ be the set of words $\mathfrak{z}$ in $\tilde{\mathfrak{X}}'$ with $d'(\mathfrak{z}) \leqslant n$. Then we only need to prove

$$\phi_n(\tilde{\mathfrak{X}}_n) = \tilde{\mathfrak{X}}'_n, \quad n \geqslant 0, \tag{4.22}$$

for which we will verify by induction on n.

When $n = 0$, we have $\tilde{\mathfrak{X}}_0 = M(X)$. On the other hand, recall that a word in $M(X')$ is of the form $\mathfrak{z} = z_1 \cdots z_m$, $z_i \in X' := X \cup \{\lfloor, \rfloor\}$, $1 \leqslant i \leqslant m$. The first two conditions in Definition 4.2.7 for $\mathfrak{z}$ in $\tilde{\mathfrak{X}}'$ are equivalent to $d_i(\mathfrak{z}) \geqslant 0$ for $1 \leqslant i \leqslant m$ and $d_m(\mathfrak{z}) = 0$. So words $\mathfrak{z}$ of $\tilde{\mathfrak{X}}'$ with $d'(\mathbf{x}) = 0$ do not have the occurrence of $\lfloor$ or $\rfloor$. So they are precisely $M(X)$. So $\phi_0(\tilde{\mathfrak{X}}_0) = \tilde{\mathfrak{X}}'_0$.

Assume $\phi_n(\tilde{\mathfrak{X}}_n) = \tilde{\mathfrak{X}}'_n$ for $0 \leqslant n \leqslant k$. Let $\mathbf{x}$ be in $\tilde{\mathfrak{X}}_{k+1}$ with the standard decomposition $\mathbf{x} = \mathbf{x}_1 \cdots \mathbf{x}_b$, where $\mathbf{x}_i$, $1 \leqslant i \leqslant b$, is in $\mathfrak{X}_0$ or in $\lfloor \tilde{\mathfrak{X}}_k \rfloor$. If $\mathbf{x}_i$ is in $\mathfrak{X}_0$, then by the above, $\phi_\infty(\mathbf{x}_i) = \phi_0(\mathbf{x}_i)$ is in $\tilde{\mathfrak{X}}'_0 \subseteq \tilde{\mathfrak{X}}'_{k+1}$. If $\mathbf{x}_i$ is in $\lfloor \tilde{\mathfrak{X}}_k \rfloor$, then $\mathbf{x}_i = \lfloor \overline{\mathbf{x}}_i \rfloor$ with $\overline{\mathbf{x}}_i \in \tilde{\mathfrak{X}}_k$. Then by the induction hypothesis, $\phi_k(\overline{\mathbf{x}}_i) \in \tilde{\mathfrak{X}}'_k$. So $d'(\phi_k(\overline{\mathbf{x}}_i)) \leqslant k$. Since

$$\phi_{n+1}(\mathbf{x}_i) = \phi_{n+1}(\lfloor \overline{\mathbf{x}}_i \rfloor) = \lfloor \phi_n(\overline{\mathbf{x}}_i) \rfloor,$$

by the definition of d', we have $d'(\phi_{k+1}(\mathbf{x}_i)) \leqslant k+1$. Hence $\mathbf{x}_i$ is in $\tilde{\mathfrak{X}}'_{k+1}$. Since $\phi_{k+1} : \mathfrak{M}_{k+1} \to M(X')$ is a monoid homomorphism, we have $\phi_{k+1}(\tilde{\mathbf{x}}) \in \tilde{\mathfrak{X}}_{k+1}$.

On the other hand, let $\mathfrak{z}$ be in $\tilde{\mathfrak{X}}'_{k+1}$. We use induction on the length $m := \ell(\mathfrak{z})$ of $\mathfrak{z}$ to prove that $\mathfrak{z}$ is in $\phi_{k+1}(\tilde{\mathfrak{X}}_{k+1})$. When $m = 1$, $\mathfrak{z}$ must be in X and so is in $\phi_{k+1}(\tilde{\mathfrak{X}}_{k+1})$. Assume that this is true for $1 \leqslant m \leqslant j$, and let $\mathfrak{z} = z_1 \cdots z_{j+1}$. Then $j+1 \geqslant 2$. By the conditions on words in $\tilde{\mathfrak{X}}'$, z_1 is in X or is $\lfloor$. If $z_1 \in X$, then $z_2 = \lfloor$ and it is easy to check the conditions in Definition 4.2.7 and see that $z_2 \cdots z_{j+1}$ is still in $\tilde{\mathfrak{X}}'$ with its d'-value at most $k+1$. By the induction hypothesis, $z_2 \cdots z_{j+1}$ is in $\phi_{k+1}(\tilde{\mathfrak{X}}_{k+1})$. Since $z_2 = \lfloor$, using Eq. (4.13), we see that the word $z_2 \cdots z_{j+1}$ must be the image under ϕ_{k+1} of an element in

$$\Big(\bigcup_{r\geqslant 1} (\lfloor \tilde{\mathfrak{X}}_k \rfloor \mathfrak{X}_0)^r\Big) \bigcup \Big(\bigcup_{r\geqslant 0} (\lfloor \tilde{\mathfrak{X}}_k \rfloor \mathfrak{X}_0)^r \lfloor \tilde{\mathfrak{X}}_k \rfloor\Big).$$

Therefore, again by the multiplicativity of ϕ_{k+1}, $\mathbf{x}$ is the image under ϕ_{k+1} of an element in

$$\Big(\bigcup_{r\geqslant 1} (\mathfrak{X}_0 \lfloor \tilde{\mathfrak{X}}_k \rfloor)^r\Big) \bigcup \Big(\bigcup_{r\geqslant 0} (\mathfrak{X}_0 \lfloor \tilde{\mathfrak{X}}_k \rfloor)^r \mathfrak{X}_0\Big)$$

and so is in $\phi_{k+1}(\tilde{\mathfrak{X}}'_{k+1})$.

Now if $z_1 = \lfloor$, then by condition (a) in Definition 4.2.7, there is $1 \leqslant \ell \leqslant j+1$ such that $d_\ell(\mathfrak{z}) = 0$. Let ℓ be the minimum with this condition. Then $z_\ell = \rfloor$. So $\mathfrak{z} = \lfloor z_2 \cdots z_{\ell-1} \rfloor z_{\ell+1} \cdots z_{j+1}$ and $z_2 \cdots z_{\ell-1}$ is still in $\tilde{\mathfrak{X}}'$. Since $d'(\mathfrak{z}) \leqslant k+1$, $d'(z_2 \cdots z_{\ell-1}) \leqslant k$. So by the induction hypothesis on k, $z_2 \cdots z_{\ell-1}$ is in $\phi_k(\tilde{\mathfrak{X}}_k)$ and therefore $z_1 \cdots z_\ell$ is in $\phi_{k+1}(\tilde{\mathfrak{X}}_{k+1})$. If $\mathfrak{z} = z_1 \cdots z_\ell$, then we are done. Otherwise, by condition (c) in Definition 4.2.7, we have $z_{\ell+1} \in X$. Further $z_{\ell+1} \cdots z_{j+1}$ is still in $\tilde{\mathfrak{X}}'$, of length not exceeding j and depth at most $k+1$. So the induction hypotheses imply that $z_{\ell+1} \cdots z_{j+1}$ is in $\phi_{k+1}(\tilde{\mathfrak{X}}_{k+1})$. Then the concatenation $z_1 \cdots z_\ell z_{\ell+1} \cdots z_{j+1}$ is still in $\phi_{k+1}(\tilde{\mathfrak{X}}_{k+1})$ by the multiplicativity of ϕ_{k+1}.

In summary, by induction on m, we obtain $\tilde{\mathfrak{X}}'_{k+1} \subseteq \phi_{k+1}(\tilde{\mathfrak{X}}_{k+1})$ and hence $\phi_{k+1}(\tilde{\mathfrak{X}}_{k+1}) = \tilde{\mathfrak{X}}'_{k+1}$. Then by an induction on n, Equation (4.22) holds for every $n \geqslant 0$. This completes the proof of Item (a).

(b) It is clear that $h(\mathbf{x}) = h'(\mathbf{x})$ and $t(\mathbf{x}) = t'(\mathbf{x})$. The proof of $d(\mathbf{x}) = d'(\mathbf{x})$ follows from Eq. (4.22). The proof of $b(\mathbf{x}) = b'(\mathbf{x})$ also follows from the definitions. □

Exercise 4.2.9. ([88, Theorem 3.4]) Note that the monoid $M(X')$ is an operated monoid with the operator

$$\alpha : M(X') \to M(X'), \mathfrak{m} \mapsto \lfloor \mathfrak{m} \rfloor.$$

Let $\tilde{\mathfrak{M}}'$ be the operated submonoid of $M(X')$ generated by X. Show that $\tilde{\mathfrak{M}}' = \tilde{\mathfrak{X}}'$. Thus we have $\tilde{\phi}_\infty(\tilde{\mathfrak{X}}_\infty) = \tilde{\mathfrak{M}}'$.

For the nonunitary Rota-Baxter bracketed words, we have

Definition 4.2.10. Let $\mathfrak{X}'$ be the set of words in $M(X')$ that satisfies the following conditions.

(a) The number of $\lfloor$ in the word equals the number of $\rfloor$ in the word;
(b) Counting from the left to the right, the accumulative number of $\lfloor$ at each location is always greater than or equal to that of $\rfloor$;
(c) There is no occurrence of $\rfloor \lfloor$, $\lfloor \rfloor$ in the word.

Then the injective maps

$$\tilde{\phi}_n : \tilde{\mathfrak{X}}_n \to M(X'), \quad 0 \leqslant n \leqslant \infty,$$

defined in Eq. (4.20) restrict to injective maps

$$\phi_n : \mathfrak{X}_n \to M(X'), \quad 0 \leqslant n \leqslant \infty.$$

We clearly have

$$\phi_\infty(\mathfrak{X}_\infty) = \varinjlim \phi_n(\mathfrak{X}_n).$$

Then from Proposition 4.2.8 we obtain

Proposition 4.2.11. $\mathfrak{X}' = \phi_\infty(\mathfrak{X}_\infty)$.

From this proposition, it is immediate that

$$x_1\lfloor x_2\rfloor x_3,\ \lfloor\lfloor x_1\rfloor x_2\rfloor x_3,\ \lfloor x_1x_2\rfloor$$

are nonunitary RBWs while

$$\lfloor x_1\lfloor x_2\rfloor x_3,\ x_1\rfloor x_2\lfloor,\ x_1\lfloor\,\rfloor,\ \lfloor x_1\rfloor\lfloor x_2\rfloor$$

are not nonunitary RBWs.

4.2.1.3 *Rooted trees and bracketed products*

We will express elements in $\tilde{\mathfrak{X}}_\infty$ as planar rooted trees that are angularly decorated by elements in X. We first relate every rooted forest to a bracketed product, in analogy to [41, 122].

Let X be a set. Let $\Pi(X)$ denote the power set (the set of subsets) of $\tilde{\mathfrak{X}}_\infty$. The fact that the set of RBWs follows from the same recursion as the one for the set $\mathfrak{T}$ of forests suggests the following tree representation of RBWs.

Theorem 4.2.12. (a) *There is a unique map*

$$h : \mathfrak{T} \to \Pi(X), \quad T \mapsto X^T$$

such that
(i) $X^\bullet = \{\mathbf{1}\}$,
(ii) $X^{T_1 \sqcup T_2} = X^{T_1} X X^{T_2}$,
(iii) $X^{\lfloor T_1\rfloor} = \lfloor X^{T_1}\rfloor$.

(b) $\tilde{\mathfrak{X}}_\infty = \bigsqcup_{T\in\mathfrak{T}} X^T$.

We recall that $\mathbf{1}$ is the identity of $M(X)$, so

$$x\mathbf{1} = \mathbf{1}x = x, \quad x \in X \cup \{\mathbf{1}\}. \tag{4.23}$$

Note that in Item (a)(ii) there is an extra copy of X inserted between X^{T_1} and X^{T_2}. For example,

$$X^{\bullet\ \bullet} = X^{\bullet\sqcup\bullet} = X^\bullet X X^\bullet = \mathbf{1}X\mathbf{1} = X.$$

As another example, we check

$$\begin{aligned} X^{\bullet\,\overset{\bullet}{\bullet\wedge\bullet}} &= X^{\bullet\sqcup\lfloor\bullet\sqcup\bullet\rfloor} \\ &= X^\bullet X X^{\lfloor\bullet\sqcup\bullet\rfloor} \quad \text{by Item (a) (ii)} \\ &= X^\bullet X\lfloor X^{\bullet\sqcup\bullet}\rfloor \quad \text{by Item (a) (iii)} \\ &= X^\bullet X\lfloor X^\bullet X X^\bullet\rfloor \quad \text{by Item (a) (ii)} \\ &= \mathbf{1}X\lfloor \mathbf{1}X\mathbf{1}\rfloor \quad \text{by Item (a) (i)} \end{aligned}$$

$$= X\lfloor X \rfloor \quad \text{by Eq. (4.23).}$$

To simply put, to get X^T from a forest T, express T in terms of $\sqcup, \bullet$ and brackets (called the standard representation of T). Then replace $\sqcup$ by X, keep the brackets and delete $\bullet$.

Proof. (a) Define a product

$$\Pi(X) \times \Pi(X) \to \Pi(X), \quad (Y, Z) \mapsto YXZ.$$

Then $\Pi(X)$ becomes a semigroup, in fact, an operated semigroup with the operator $\lfloor, \rfloor$. Since $\mathfrak{T}$ is the free operated semigroup on $\bullet$, the assignment $\bullet \mapsto \{\mathbf{1}\}$ extends uniquely to a morphism of operated semigroups from $\mathfrak{T}$ to $\Pi(X)$. Therefore the conditions in Item (a) are satisfied.

(b) We will prove by induction on $n \geqslant 0$ that

$$\tilde{\mathfrak{X}}_n = \bigsqcup_{T \in \mathfrak{T}_n} X^T. \tag{4.24}$$

Here $\mathfrak{T}_n$, defined in Eq. (4.1), is the set of planar rooted forests of depth at most n. The disjointness of the union follows from Eq. (4.17).

When $n = 0$, we have $\tilde{\mathfrak{X}}_n = M(X) = \bigsqcup_{m \geqslant 0} X^m$ with the convention that $X^0 = \{\mathbf{1}\}$. On the other hand, $T \in \mathfrak{T}_0$ are forests in $\bullet$, so are of the form $T = \bullet^{\sqcup i}$, $i \geqslant 1$. Since we have $X^{\bullet^{\sqcup i}} = X^{i-1}$ by Item (a), we are done with $n = 0$.

Assume Eq. (4.24) holds for $n \leqslant k$ with $k \geqslant 0$. Any $T \in \mathfrak{T}_{n+1}$ is of the form $T_1 \sqcup \cdots \sqcup T_b$ with $T_i = \bullet$ or $T_i = \lfloor \overline{T}_i \rfloor$ with $\overline{T}_i \in \mathfrak{T}_n$. Thus by the induction hypothesis, either $X^{T_i} = \{\mathbf{1}\}$ or $X^{\overline{T}_i} \subset \tilde{\mathfrak{X}}_n$ and hence

$$X^{T_i} = \lfloor X^{\overline{T}_i} \rfloor \subset \lfloor \tilde{\mathfrak{X}}_n \rfloor \subseteq \tilde{\mathfrak{X}}_{n+1}.$$

This proves $\bigsqcup_{T \in \mathfrak{T}_{n+1}} X^T \subseteq \tilde{\mathfrak{X}}_{n+1}$. The other inclusion is similar, completing the induction. □

For ease of recording a Rota-Baxter word in the tree form, we use the following notation. By Theorem 4.2.12, for any Rota-Baxter word $\mathbf{x} \in \tilde{\mathfrak{X}}_\infty$, there is a unique forest T so that $\mathbf{x} \in X^T$. We call T the *tree* or *skeleton* of $\mathbf{x}$ and denote it by $\mathrm{tr}(\mathbf{x})$. Let $y(T)$ be the number of leaves of the tree T. We also use the shorthand notation $y(\mathbf{x}) := y(\mathrm{tr}(\mathbf{x}))$. Let $|\mathbf{x}|$, called the *fruits* of $\mathbf{x}$, be the *reduced word* in $M(X)$ obtained by ignoring the brackets in $\mathbf{x}$. For example, for $x_1, x_2 \in X$, $|x_1\lfloor x_2 \rfloor| = x_1x_2$ and $|x_1\lfloor \mathbf{1} \rfloor x_2| = x_1x_2$. Note that $|x_1\lfloor \mathbf{1} \rfloor x_2| \neq x_1\mathbf{1}x_2$ since it is not reduced. If $|x| = x_1 \cdots x_w$, then call w the *weight* of $|\mathbf{x}|$. We then denote

$$\mathbf{x} = |\mathbf{x}|^{\mathrm{tr}(\mathbf{x})} = (x_1 \cdots x_w)^{\mathrm{tr}(\mathbf{x})}, \tag{4.25}$$

called the *power form* of $\mathbf{x}$. For example, for $x, y \in X$,

$x\lfloor y\rfloor = (xy)^{\bullet\sqcup\lfloor\bullet\sqcup\bullet\rfloor} = (xy)^{\text{}}$  $= (xy)^{\text{}}$  ,

$x\lfloor \mathbf{1}\rfloor y = (xy)^{\bullet\sqcup\lfloor\bullet\rfloor\sqcup\bullet} = (xy)^{\text{}}$  $= (xy)^{\text{}}$  , (4.26)

$xy\lfloor \mathbf{1}\rfloor = (xy)^{\bullet\sqcup\bullet\sqcup\lfloor\bullet\rfloor} = (xy)^{\text{}}$  $= (xy)^{\text{}}$ 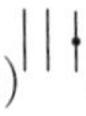 .

We express the exponents in both drawings of the trees in Section 4.1.1. Because of Theorem 4.2.12, there is no ambiguity in the power form notation of Rota-Baxter words. In other words, for any Rota-Baxter word $\mathbf{x}$, there is unique forest $T = \mathrm{tr}(\mathbf{x})$ and unique $(x_1, \ldots, x_w) \in X^w$ such that $\mathbf{x} = (x_1 \cdots x_w)^T$. In fact, there is a close relation between $w(\mathbf{x})$ and the number $y(T)$ of leaves of T.

Proposition 4.2.13. *For any $\mathbf{x} \in \mathfrak{X}_\infty$, we have $w(\mathbf{x}) = y(\mathrm{tr}(\mathbf{x})) - 1$.*

By convention, we define $w(\mathbf{1}) = 0$. Then the proposition is true for all $\mathbf{x} \in \tilde{\mathfrak{X}}_\infty$.

Proof. Fix $\mathbf{x} \in \mathfrak{X}_\infty$. Then there is a unique $T \in \mathfrak{T}\backslash\{\bullet\}$ such that $\mathbf{x} \in X^T$. We use induction on the depth $d(T)$ of T. Recall that $d(T)$ is defined to be the length of a longest path in the forest T from one of its roots to its leaves. When $d(T) = 0$, we have $T = \bullet^{\sqcup n}$, $n \geqslant 2$. Then $\mathbf{x} \in X^T = X^{(n-1)}$ by Theorem 4.2.12 and the proposition is proved.

Assume that the proposition is proved for all words $\mathbf{x}$ in X^T with $d(T) \leqslant k$ and $k \geqslant 0$, and let $\mathbf{x} \in X^T$ with $d(T) = k+1$. Then $k+1 \geqslant 1$. If T is a tree, then $T = \lfloor \overline{T}\rfloor$ with $\overline{T} \in \mathfrak{T}_k$. So $\mathbf{x} = \lfloor \overline{\mathbf{x}}\rfloor$ with $\overline{\mathbf{x}} \in X^{\overline{T}}$. By the induction hypothesis, $w(\overline{\mathbf{x}}) = y(\overline{T}) - 1$. But $w(\lfloor\overline{\mathbf{x}}\rfloor) = w(\overline{\mathbf{x}})$ and $y(\lfloor\overline{T}\rfloor) = y(\overline{T})$. So we get $w(\mathbf{x}) = y(T) - 1$. If T is a forest, then $T = T_1 \sqcup \cdots \sqcup T_b$ with trees $T_1, \ldots, T_b$. So $X^T = X^{T_1} X\, X^{T_2} X \cdots X\, X^{T_b}$. Thus our $\mathbf{x} \in X^T$ is of the form $\mathbf{x}_1 x_1 \mathbf{x}_2 x_2 \cdots x_{b-1}\mathbf{x}_b$ with $\mathbf{x}_i \in X^{T_i}$, $i = 1, \ldots, b$ and $x_i \in X$, $i = 1, \ldots, b-1$. By what have just been proved, we have $w(\mathbf{x}_i) = y(T_i) - 1$. Therefore,

$$\begin{aligned} w(\mathbf{x}) &= w(\mathbf{x}_1) + \cdots + w(\mathbf{x}_b) + b - 1 = y(T_1) - 1 + \cdots + y(T_b) - 1 + b - 1 \\ &= y(T_1) + \cdots + y(T_b) - 1 = y(T_1 \sqcup \cdots \sqcup T_b) - 1 = y(T) - 1. \end{aligned}$$

This completes the induction. □

To get a tree representation of the Rota-Baxter bracketed words themselves we introduce the concept of an angular decorated tree, similar to [81]. An angular decoration of a tree by a set X is the tree with a decoration on the angles of the trees by elements of X, that is, a decoration between each pair of adjacent non-root edged of a vertex. Similarly an angular decoration of a forest by a set X is the forest with a decoration, by elements of X, on the angles of the trees in the forest and between adjacent trees of the forest. A tree or a forest with an angular decoration is called an *angularly decorated tree or forest.* We denote an angularly decorated rooted forest by $D = (T; \vec{x})$, where T is the tree that is decorated and $\vec{x} = (x_1, \ldots, x_n)$ is the decorations $x_i \in X$, listed from the leftmost decoration to the rightmost one. We let $(T; X)$ denote the set of decorated rooted trees of T with the decoration set X. Let

$$(\mathfrak{T}; X) = \bigsqcup_{T \in \mathfrak{T}} (T; X)$$

be the set of angularly decorated rooted forests with decoration set X. Let $D' = (T'; \vec{x}')$ and $D'' = (T''; \vec{x}'')$ be two decorated forests, consisting of forests T' and T'' decorated by vectors $\vec{x}' = (x'_1, \ldots, x'_{n'})$ and $\vec{x}'' = (x''_1, \ldots, x''_{n''})$ respectively. We will use $D' \sqcup_x D''$ to denote the new decorated rooted tree with x decorating in between D' and D''. So we have

$$D' \sqcup_x D'' = (T' \sqcup T''; (x'_1, \ldots, x'_{n'}, x, x''_1, \ldots, x''_{n''})).$$

Here is the first example in Eq. (4.26) expressed in terms of angularly decorated trees and forests. We give the decoration in both drawings.

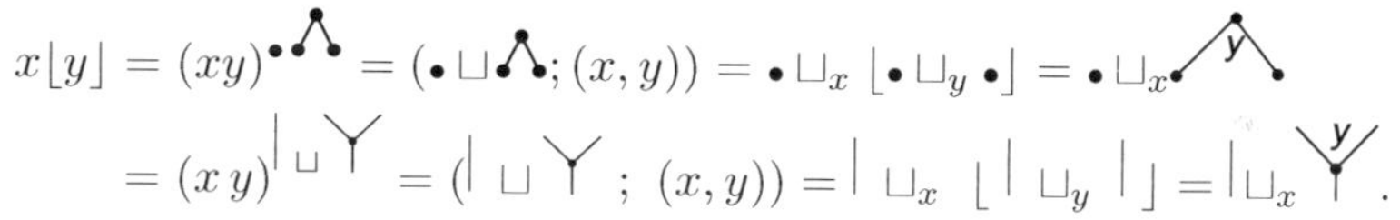

Here are some more examples

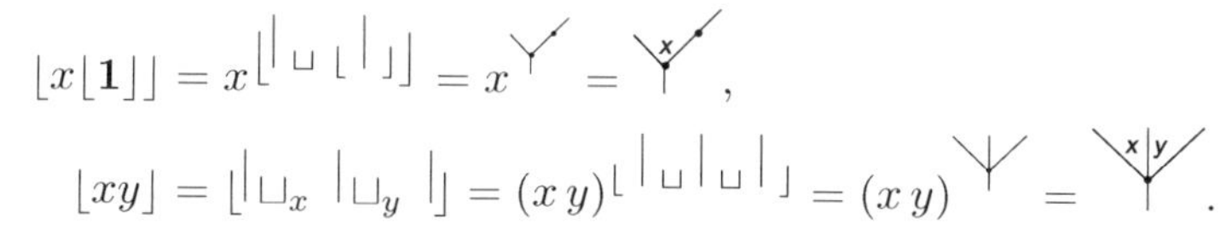

Theorem 4.2.14. *The map*

$$\rho : \tilde{\mathfrak{X}}_\infty \to (\mathfrak{T}; X)$$

sending an RBW in power form $(x_1 \cdots x_w)^T$ *to* $(T; (x_1, \cdots, x_w))$ *is a bijection. Furthermore, it satisfies the following properties.*

(a) $\rho(\mathbf{1}) = \rho(X^{\bullet}) = \{(\bullet; \emptyset)\}$;

(b) *For any two forests* T_1 *and* T_2, $\rho(X^{T_1 \sqcup T_2}) = \bigsqcup_{x \in X} \rho(X^{T_1}) \sqcup_x \rho(X^{T_2})$;

(c) $\rho(X^T) = (T; X)$;

(d) $\rho(X^{\lfloor T \rfloor}) = \lfloor (T; X) \rfloor$.

Therefore, together with Proposition 4.2.8, we have three interpretations of RBWs.

Proof. We first note that by convention, $\rho(\mathbf{1}) = (\bullet; \emptyset)$. Any $\mathbf{1} \neq \mathbf{x} \in \tilde{\mathfrak{X}}_\infty$ has the power form $|\mathbf{x}|^T$ for a unique $T \in \mathfrak{T}$. Here $|\mathbf{x}| = x_1 \cdots x_w$ is the fruit of $\mathbf{x}$ with $w+1$ being the number of leaves of T (Proposition 4.2.13). Therefore, $\mathbf{x}$ gives a unique angularly decoration of T by decorating the w angles between the adjacent leaves of T. This is $\rho(\mathbf{x}) = (T; (x_1, \ldots, x_w))$. Conversely, given an angular decoration $(T; (x_1, \ldots, x_w))$ of T by $x_i \in X$, then by Proposition 4.2.13, $w+1$ is the number of leaves of T. So $(x_1 \cdots x_w)^T$ is a well-defined element of X^T. The two maps are obviously the inverse of each other. □

The map in Theorem 4.2.12

$$h : \mathfrak{T} \to \Pi(X), \quad T \mapsto X^T$$

restricts to a unique map $\mathfrak{T}^0 \to \Pi(X)$ such that
(a) $X^{\bullet \sqcup \bullet} = X$,
(b) $X^{T_1 \sqcup T_2} = X^{T_1} X X^{T_2}$,
(c) $X^{\lfloor T_1 \rfloor} = \lfloor X^{T_1} \rfloor$.
We similarly check

$$\mathfrak{X}_\infty = \bigsqcup_{T \in \mathfrak{T}^0} X^T.$$

As a consequence to Theorem 4.2.14, we have

Corollary 4.2.15. *There is a one-one correspondence between the set $\mathfrak{X}_\infty$ of nonunitary RBWs in X and the set of angularly decorated forests with no ladder subtrees and with decoration set X.*

4.2.2 The product in a free RBA on a set

We define $Ш^{NC}(X)$ to be the free $\mathbf{k}$-module

$$Ш^{NC}(X) := \mathbf{k}\tilde{\mathfrak{X}}_\infty$$

generated by $\tilde{\mathfrak{X}}_\infty$.

We now define a product $\diamond$ on $Ш^{NC}(X)$ by defining $\mathbf{x} \diamond \mathbf{x}'$ for any two words $\mathbf{x}$ and $\mathbf{x}'$ in $\tilde{\mathfrak{X}}_\infty$. and then extending by bilinearity. The construction of the product and the proof of its associativity can be done directly in terms of the words. But it is more concise and intuitive to use the tree representation of the RBWs (Theorem 4.2.12) and the Rota-Baxter algebra structure on trees in Theorem 4.1.7.

4.2.2.1 *Product defined by words*

In this section we define the product on a free RBA in terms of Rota-Baxter words. This product will be given in terms of decorated trees in the next section.

Roughly speaking, the product of such words is defined to be the concatenation product whenever $t(\mathbf{x}) \neq h(\mathbf{x}')$. When $t(\mathbf{x}) = h(\mathbf{x}')$, the product is defined by the product in $M(X)$ or by the relation

$$\lfloor \overline{\mathbf{x}} \rfloor \diamond \lfloor \overline{\mathbf{x}}' \rfloor = \lfloor \lfloor \overline{\mathbf{x}} \rfloor \diamond \overline{\mathbf{x}}' \rfloor + \lfloor \overline{\mathbf{x}} \diamond \lfloor \overline{\mathbf{x}}' \rfloor \rfloor + \lambda \lfloor \overline{\mathbf{x}} \diamond \overline{\mathbf{x}}' \rfloor. \tag{4.27}$$

To be precise, we define the product $\mathbf{x} \diamond \mathbf{x}'$ inductively on the sum $n := d(\mathbf{x}) + d(\mathbf{x}')$ of the depths of $\mathbf{x}$ and $\mathbf{x}'$. Note that we have $n \geqslant 0$. If $n = 0$, then $\mathbf{x}, \mathbf{x}'$ are in $\tilde{\mathfrak{X}}_0 = M(X)$ and we define $\mathbf{x} \diamond \mathbf{x}' = \mathbf{x}\mathbf{x}'$, the concatenation in $M(X)$.

Suppose $\mathbf{x} \diamond \mathbf{x}'$ have been defined for all RBWs $\mathbf{x}, \mathbf{x}' \in \tilde{\mathfrak{X}}_\infty$ with $n \leqslant k$ for a $k \geqslant 0$ and let $\mathbf{x}, \mathbf{x}' \in \tilde{\mathfrak{X}}_\infty$ with $n = k + 1$.

First assume that the breadths of $b(\mathbf{x})$ and $b(\mathbf{x}')$ are both 1. Then $\mathbf{x}$ and $\mathbf{x}'$ are in $\mathfrak{X}_0 = S(X)$ or $\lfloor \tilde{\mathfrak{X}}_\infty \rfloor$ and $\mathbf{x}$ and $\mathbf{x}'$ cannot be both in $S(X)$. We accordingly define

$$\mathbf{x} \diamond \mathbf{x}' = \begin{cases} \mathbf{x}\,\mathbf{x}', & \text{if } \mathbf{x} \in S(X), \mathbf{x} \in \lfloor \tilde{\mathfrak{X}}_\infty \rfloor, \\ \mathbf{x}\,\mathbf{x}', & \text{if } \mathbf{x} \in \lfloor \tilde{\mathfrak{X}}_\infty \rfloor, \mathbf{x}' \in S(X), \\ \lfloor \lfloor \overline{\mathbf{x}} \rfloor \diamond \overline{\mathbf{x}}' \rfloor + \lfloor \overline{\mathbf{x}} \diamond \lfloor \overline{\mathbf{x}}' \rfloor \rfloor + \lambda \lfloor \overline{\mathbf{x}} \diamond \overline{\mathbf{x}}' \rfloor, & \text{if } \mathbf{x} = \lfloor \overline{\mathbf{x}} \rfloor, \mathbf{x}' = \lfloor \overline{\mathbf{x}}' \rfloor \in \lfloor \tilde{\mathfrak{X}}_\infty \rfloor. \end{cases} \tag{4.28}$$

Here the product in the first three cases is by concatenation and in the fourth case is by the induction hypothesis since for the three products on the right hand side we have

$$\begin{aligned} d(\lfloor \overline{\mathbf{x}} \rfloor) + d(\overline{\mathbf{x}}') &= d(\lfloor \overline{\mathbf{x}} \rfloor) + d(\lfloor \overline{\mathbf{x}}' \rfloor) - 1 = d(\mathbf{x}) + d(\mathbf{x}') - 1, \\ d(\overline{\mathbf{x}}) + d(\lfloor \overline{\mathbf{x}}' \rfloor) &= d(\lfloor \overline{\mathbf{x}} \rfloor) + d(\lfloor \overline{\mathbf{x}}' \rfloor) - 1 = d(\mathbf{x}) + d(\mathbf{x}') - 1, \\ d(\overline{\mathbf{x}}) + d(\overline{\mathbf{x}}') &= d(\lfloor \overline{\mathbf{x}} \rfloor) - 1 + d(\lfloor \overline{\mathbf{x}}' \rfloor) - 1 = d(\mathbf{x}) + d(\mathbf{x}') - 2 \end{aligned}$$

which are all less than or equal to k.

Now assume $b(\mathbf{x}) > 1$ or $b(\mathbf{x}') > 1$. Let $\mathbf{x} = \mathbf{x}_1 \cdots \mathbf{x}_b$ and $\mathbf{x}' = \mathbf{x}'_1 \cdots \mathbf{x}'_{b'}$ be the standard decompositions from Lemma 4.2.5. We then define

$$\mathbf{x} \diamond \mathbf{x}' = \mathbf{x}_1 \cdots \mathbf{x}_{b-1} \, (\mathbf{x}_b \diamond \mathbf{x}'_1) \, \mathbf{x}'_2 \cdots \mathbf{x}'_{b'} \tag{4.29}$$

where $\mathbf{x}_b \diamond \mathbf{x}'_1$ is defined by Eq. (4.28) and the rest products are given by the concatenation. The resulting word, or sum of words, is well-defined in $\text{III}^{NC}(X)$ since, by Eq. (4.28), we have $h(\mathbf{x}_b) = h(\mathbf{x}_b \diamond \mathbf{x}'_1)$ and $t(\mathbf{x}'_1) = t(\mathbf{x}_b \diamond \mathbf{x}'_1)$. Therefore, $t(\mathbf{x}_{b-1}) \neq h(\mathbf{x}_b \diamond \mathbf{x}'_1)$ and $h(\mathbf{x}'_2) \neq t(\mathbf{x}_b \diamond \mathbf{x}'_1)$.

4.2.2.2 *Product defined by trees*

By Theorems 4.2.12 and 4.2.14, for any element $\mathbf{x} \in \tilde{\mathfrak{X}}_\infty$, there are unique $(x_1, \ldots, x_w) \in X^w$ and $T \in \mathfrak{T}$ such that $\mathbf{x} = (x_1 \cdots x_w)^T \in X^T$ in the power form of Proposition 4.2.13. Here $w+1$ equals the number $y(T)$ of leaves of T. Similarly, $\mathbf{x}' \in \tilde{\mathfrak{X}}_\infty$ has its power form $(x'_1 \cdots x'_{m'})^{T'} \in X^{T'}$. We then define $\mathbf{x} \diamond \mathbf{x}'$ to have the power form

$$\mathbf{x} \diamond \mathbf{x}' := (x_1 \cdots x_m x'_1 \cdots x'_{m'})^{T_1 \diamond T_2}. \tag{4.30}$$

Here $T_1 \diamond T_2$ is defined in Eq. (4.4). The right hand side is well-defined since by the definition of $\diamond$ for trees, the number $y(T \diamond T')$ of leaves of $T \diamond T'$ is

$$\ell(T \diamond T') = \ell(T) + \ell(T') - 1 = m + 1 + m' + 1 - 1 = m + m' + 1.$$

Comparing the definition of $T_1 \diamond T_2$ in Eq. (4.4) with the product $\diamond$ defined in Eq (4.28) and Eq. (4.29), we see the products $\diamond$ on $\text{Ш}^{\text{NC}}(X)$ defined in terms of RBWs and in terms of trees in Eq. (4.30) are the same.

Extending $\diamond$ bilinearly, we obtain a binary operation

$$\diamond : \text{Ш}^{\text{NC}}(X) \otimes \text{Ш}^{\text{NC}}(X) \to \text{Ш}^{\text{NC}}(X).$$

For $\mathbf{x} \in \tilde{\mathfrak{X}}_\infty$, define

$$P_X(\mathbf{x}) = \lfloor \mathbf{x} \rfloor. \tag{4.31}$$

Then $\lfloor \mathbf{x} \rfloor$ is again in $\tilde{\mathfrak{X}}_\infty$ and hence in $\text{Ш}^{\text{NC}}(X)$. Thus P_X defines a linear operator P_X on $\text{Ш}^{\text{NC}}(X)$. Further, if $\mathbf{x} = (x_1 \cdots x_w)^T$, then we have

$$P_X(\mathbf{x}) = (x_1 \cdots x_w)^{\lfloor T \rfloor} = (x_1 \cdots x_w)^{P_{\mathcal{T}}(T)}. \tag{4.32}$$

Here $P_{\mathcal{T}}$ is the Rota-Baxter operator in Eq. (4.6). Let

$$j_X : X \to \text{Ш}^{\text{NC}}(X) \tag{4.33}$$

be the natural inclusion.

We also define the free $\mathbf{k}$-module

$$\text{Ш}^{\text{NC},0}(X) := \mathbf{k}\tilde{\mathfrak{X}}_\infty.$$

Then it is a submodule of $\text{Ш}^{\text{NC}}(X)$. We define a product $\diamond$ on $\text{Ш}^{\text{NC},0}(X)$ by the restriction of $\diamond$ to $\text{Ш}^{\text{NC}}(X)$, either in terms of words directly or in terms of decorated trees. Also define $\lfloor\ ,\ \rfloor : \text{Ш}^{\text{NC},0}(X) \to \text{Ш}^{\text{NC},0}(X)$ to be the restriction of $\lfloor\ ,\ \rfloor$ on $\text{Ш}^{\text{NC}}(X)$.

4.2.3 The universal property

Theorem 4.2.16. *Let X be a set.*

(a) *The pair $(\text{Ш}^{\text{NC}}(X), \diamond)$ is a $\mathbf{k}$-algebra.*

(b) *The triple $(\text{Ш}^{\text{NC}}(X), \diamond, P_X)$ is an* RBA *of weight λ.*

(c) *The quadruple $(\text{Ш}^{\text{NC}}(X), \diamond, P_X, j_X)$ is the free* RBA *of weight λ on the set X.*

(d) *The pair $(\text{Ш}^{\text{NC},0}(X), \diamond)$ is a nonunitary algebra.*

(e) *The triple $(\text{Ш}^{\text{NC},0}(X), \diamond, P_X)$ is a nonunitary* RBA *of weight λ.*

(f) *The quadruple $(\text{Ш}^{\text{NC},0}(X), \diamond, P_X, j_X)$ is the free nonunitary* RBA *of weight λ on the set X.*

Proof. We just prove Items (a)—(c). The proofs of Items (d)— (f) are similar.

(a) For the associativity of $\diamond$ on $\text{Ш}^{\text{NC}}(X)$ we only need to prove

$$(\mathbf{x} \diamond \mathbf{x}') \diamond \mathbf{x}'' = \mathbf{x} \diamond (\mathbf{x}' \diamond \mathbf{x}'')$$

for any $\mathbf{x}, \mathbf{x}'$ and $\mathbf{x}'' \in \tilde{\mathfrak{X}}_\infty$. Let $(x_1 \cdots x_m)^T$, $(x'_1 \cdots x'_{m'})^{T'}$ and $(x''_1 \cdots x''_{m''})^{T''}$ be their power form with $T, T', T'' \in \mathfrak{T}$. Then by Eq. (4.4) we only need to prove

$$\begin{aligned}&(x_1 \cdots x_m x'_1 \cdots x'_{m'} x''_1 \cdots x''_{m''})^{(T\diamond T')\diamond T''}\\ &= (x_1 \cdots x_m x'_1 \cdots x'_{m'} x''_1 \cdots x''_{m''})^{T\diamond(T'\diamond T'')}\end{aligned}$$

which follows from the associativity of $\diamond$ for trees obtained in Theorem 4.1.7.

(b) The Rota-Baxter property of $\lfloor\,\rfloor$ on $\text{Ш}^{\text{NC}}(X)$ follows from the Rota-Baxter property of $\lfloor\,\rfloor$ on $\mathbf{k}\mathfrak{T}$ in Theorem 4.1.7.

It can also be seen immediate from Eq. (4.28).

(c) Let $(R, *, P)$ be a unitary RBA of weight λ whose product is denoted by $*$. Let $f : X \to R$ be a set map. We will construct a $\mathbf{k}$-linear map $\overline{f} : \text{Ш}^{\text{NC}}(X) \to R$ by defining $\overline{f}(\mathbf{x})$ for $\mathbf{x} \in \tilde{\mathfrak{X}}_\infty$. We achieve this by defining $\overline{f}(\mathbf{x})$ for an RBW $\mathbf{x} \in \tilde{\mathfrak{X}}_n$, $n \geqslant 0$, by induction on n. For $\mathbf{x} \in \tilde{\mathfrak{X}}_0 := M(X)$, $\mathbf{x} = 1$ or $\mathbf{x} = \mathbf{x}_1 \cdots \mathbf{x}_r$ with $\mathbf{x}_1, \cdots, \mathbf{x}_r \in X$, define $\overline{f}(1) = 1 \in R$ or

$$\overline{f}(\mathbf{x}) = f(\mathbf{x}_1) * \cdots * f(\mathbf{x}_r). \tag{4.34}$$

Suppose $\overline{f}(\mathbf{x})$ has been defined for $\mathbf{x} \in \tilde{\mathfrak{X}}_n$ and consider $\mathbf{x}$ in $\tilde{\mathfrak{X}}_{n+1}$ which is, by its definition in Eq. (4.17),

$$\begin{aligned}\mathfrak{X}_{n+1} = &\Big(\bigsqcup_{r\geqslant 1} (\mathfrak{X}_0\lfloor\tilde{\mathfrak{X}}_n\rfloor)^r\Big) \bigsqcup \Big(\bigsqcup_{r\geqslant 0} (\mathfrak{X}_0\lfloor\tilde{\mathfrak{X}}_n\rfloor)^r \mathfrak{X}_0\Big)\\ &\bigsqcup \Big(\bigsqcup_{r\geqslant 1} (\lfloor\tilde{\mathfrak{X}}_n\rfloor\mathfrak{X}_0)^r\Big) \bigsqcup \Big(\bigsqcup_{r\geqslant 0} (\lfloor\tilde{\mathfrak{X}}_n\rfloor\mathfrak{X}_0)^r \lfloor\tilde{\mathfrak{X}}_n\rfloor\Big).\end{aligned}$$

Let $\mathbf{x}$ be in the first union component $\bigsqcup_{r\geqslant 1}(\mathfrak{X}_0 \lfloor \tilde{\mathfrak{X}}_n \rfloor)^r$ above. Then

$$\mathbf{x} = \diamond_{i=1}^{r}(\mathbf{x}_{2i-1} \lfloor \mathbf{x}_{2i} \rfloor)$$

with $\mathbf{x}_{2i-1} \in \mathfrak{X}_0$ and $\mathbf{x}_{2i} \in \tilde{\mathfrak{X}}_n$, $1 \leqslant i \leqslant r$. By the construction of the multiplication $\diamond$ and the Rota-Baxter operator P_X, we have

$$\mathbf{x} = \diamond_{i=1}^{r}(\mathbf{x}_{2i-1} \diamond \lfloor \mathbf{x}_{2i} \rfloor) = \diamond_{i=1}^{r}(\mathbf{x}_{2i-1} \diamond P_X(\mathbf{x}_{2i})).$$

Define

$$\overline{f}(\mathbf{x}) = *_{i=1}^{r}(\overline{f}(\mathbf{x}_{2i-1}) * P(\overline{f}(\mathbf{x}_{2i}))), \tag{4.35}$$

where the right hand side is well-defined by the induction hypothesis. Similarly define $\overline{f}(\mathbf{x})$ if $\mathbf{x}$ is in the other union components. For any $\mathbf{x} \in \tilde{\mathfrak{X}}_\infty$, we have $P_X(\mathbf{x}) = \lfloor \mathbf{x} \rfloor \in Ш^{NC}(X)$, and by the definition of $\overline{f}$ in Eq. (4.35), we have

$$\overline{f}(\lfloor \mathbf{x} \rfloor) = P(\overline{f}(\mathbf{x})). \tag{4.36}$$

So $\overline{f}$ commutes with the Rota-Baxter operators. Combining this equation with Eq. (4.35) we see that if $\mathbf{x} = \mathbf{x}_1 \cdots \mathbf{x}_b$ is the standard decomposition of $\mathbf{x}$, then

$$\overline{f}(\mathbf{x}) = \overline{f}(\mathbf{x}_1) * \cdots * \overline{f}(\mathbf{x}_b). \tag{4.37}$$

Note also that this is the only possible way to define $\overline{f}(\mathbf{x})$ in order for $\overline{f}$ to be a Rota-Baxter homomorphism extending f.

We remain to prove that the map $\overline{f}$ defined in Eq. (4.35) is indeed an algebra homomorphism. For this we only need to check

$$\overline{f}(\mathbf{x} \diamond \mathbf{x}') = \overline{f}(\mathbf{x}) * \overline{f}(\mathbf{x}') \tag{4.38}$$

for all RBWs $\mathbf{x}, \mathbf{x}' \in \tilde{\mathfrak{X}}_\infty$. For this we use induction on the sum of depths $n := d(\mathbf{x}) + d(\mathbf{x}')$ of $\mathbf{x}$ and $\mathbf{x}'$. Then $n \geqslant 0$.

When $n = 0$, we have $\mathbf{x}, \mathbf{x}' \in \tilde{\mathfrak{X}}_0 = M(X)$. Then Eq. (4.38) follows from the definition of $\overline{f}(\mathbf{x})$ in Eq. (4.34). Assume the multiplicativity holds for $\mathbf{x}, \mathbf{x}' \in \tilde{\mathfrak{X}}_\infty$ with $n \leqslant k$ for $k \geqslant 0$ and take $\mathbf{x}, \mathbf{x}' \in \tilde{\mathfrak{X}}_\infty$ with $n = k+1$. Let $\mathbf{x} = \mathbf{x}_1 \cdots \mathbf{x}_b$ and $\mathbf{x}' = \mathbf{x}'_1 \cdots \mathbf{x}'_{b'}$ be the standard decompositions. Then $\mathbf{x}_b$ and $\mathbf{x}'_1$ cannot be both in X. So by Eq. (4.28),

$$\begin{aligned}
&\overline{f}(\mathbf{x}_b \diamond \mathbf{x}'_1)\\
&= \begin{cases}
\overline{f}(\mathbf{x}_b \mathbf{x}'_1), & \text{if } \mathbf{x}_b \in X, \mathbf{x}'_1 \in \lfloor \tilde{\mathfrak{X}}_\infty \rfloor,\\
\overline{f}(\mathbf{x}_b \mathbf{x}'_1), & \text{if } \mathbf{x}_b \in \lfloor \tilde{\mathfrak{X}}_\infty \rfloor, \mathbf{x}'_1 \in X,\\
\overline{f}(\lfloor \lfloor \overline{\mathbf{x}}_b \rfloor \diamond \overline{\mathbf{x}}'_1 \rfloor + \lfloor \overline{\mathbf{x}}_b \diamond \lfloor \overline{\mathbf{x}}'_1 \rfloor \rfloor + \lambda \lfloor \overline{\mathbf{x}}_b \diamond \overline{\mathbf{x}}'_1 \rfloor), & \text{if } \mathbf{x}_b = \lfloor \overline{\mathbf{x}}_b \rfloor, \mathbf{x}'_1 = \lfloor \overline{\mathbf{x}}'_1 \rfloor \in \lfloor \tilde{\mathfrak{X}}_\infty \rfloor.
\end{cases}
\end{aligned}$$

In the first three cases, the right hand side is $\overline{f}(\mathbf{x}_b) * \overline{f}(\mathbf{x}'_1)$ by the definition of $\overline{f}$. In the fourth case, we have, by Eq. (4.36), the induction hypothesis and the Rota-Baxter relation of P,

$$\begin{aligned}
&\overline{f}(\lfloor\lfloor\overline{\mathbf{x}}_b\rfloor \diamond \overline{\mathbf{x}}'_1\rfloor + \lfloor\overline{\mathbf{x}}_b \diamond \lfloor\overline{\mathbf{x}}'_1\rfloor\rfloor + \lambda\lfloor\overline{\mathbf{x}}_b \diamond \overline{\mathbf{x}}'_1\rfloor) \\
&= \overline{f}(\lfloor\lfloor\overline{\mathbf{x}}_b\rfloor \diamond \overline{\mathbf{x}}'_1\rfloor) + \overline{f}(\lfloor\overline{\mathbf{x}}_b \diamond \lfloor\overline{\mathbf{x}}'_1\rfloor\rfloor) + \overline{f}(\lambda\lfloor\overline{\mathbf{x}}_b \diamond \overline{\mathbf{x}}'_1\rfloor) \\
&= P(\overline{f}(\lfloor\overline{\mathbf{x}}_b\rfloor \diamond \overline{\mathbf{x}}'_1)) + P(\overline{f}(\overline{\mathbf{x}}_b \diamond \lfloor\overline{\mathbf{x}}'_1\rfloor)) + \lambda P(\overline{f}(\overline{\mathbf{x}}_b \diamond \overline{\mathbf{x}}'_1)) \\
&= P(\overline{f}(\lfloor\overline{\mathbf{x}}_b\rfloor) * \overline{f}(\overline{\mathbf{x}}'_1)) + P(\overline{f}(\overline{\mathbf{x}}_b) * \overline{f}(\lfloor\overline{\mathbf{x}}'_1\rfloor)) + \lambda P(\overline{f}(\overline{\mathbf{x}}_b) * \overline{f}(\overline{\mathbf{x}}'_1)) \\
&= P(P(\overline{f}(\overline{\mathbf{x}}_b)) * \overline{f}(\overline{\mathbf{x}}'_1)) + P(\overline{f}(\overline{\mathbf{x}}_b) * P(\overline{f}(\overline{\mathbf{x}}'_1))) + \lambda P(\overline{f}(\overline{\mathbf{x}}_b) * \overline{f}(\overline{\mathbf{x}}'_1)) \\
&= P(\overline{f}(\overline{\mathbf{x}}_b)) * P(\overline{f}(\overline{\mathbf{x}}'_1)) \\
&= \overline{f}(\mathbf{x}_b) * \overline{f}(\mathbf{x}'_1).
\end{aligned}$$

Therefore $\overline{f}(\mathbf{x}_b \diamond \mathbf{x}'_1) = \overline{f}(\mathbf{x}_b) * \overline{f}(\mathbf{x}'_1)$. Then

$$\begin{aligned}
\overline{f}(\mathbf{x} \diamond \mathbf{x}') &= \overline{f}(\mathbf{x}_1 \cdots \mathbf{x}_{b-1}\,(\mathbf{x}_b \diamond \mathbf{x}'_1)\,\mathbf{x}'_2 \cdots \mathbf{x}'_{b'}) \\
&= \overline{f}(\mathbf{x}_1) * \cdots * \overline{f}(\mathbf{x}_{b-1}) * \overline{f}(\mathbf{x}_b \diamond \mathbf{x}'_1) * \overline{f}(\mathbf{x}'_2) * \cdots * \overline{f}(\mathbf{x}'_{b'}) \\
&= \overline{f}(\mathbf{x}_1) * \cdots * \overline{f}(\mathbf{x}_{b-1}) * \overline{f}(\mathbf{x}_b) * \overline{f}(\mathbf{x}'_1) * \overline{f}(\mathbf{x}'_2) * \cdots * \overline{f}(\mathbf{x}'_{b'}) \\
&= \overline{f}(\mathbf{x}) * \overline{f}(\mathbf{x}').
\end{aligned}$$

This is what we need. □

4.3 Unitarization of Rota-Baxter algebras

The unitarization process of associative algebras is simple and well known. For any nonunitary algebra R (even if R does have an identity), define $\tilde{R} := \mathbf{k1} \oplus R$ with componentwise addition and with product defined by

$$(a, x)(b, y) = (ab, ay + bx + xy).$$

Then $\tilde{R}$ is a unitary algebra with identity $(\mathbf{1}, 0)$ and a natural embedding

$$u_R : R \to \tilde{R}, \quad x \mapsto (0, x).$$

Further $\tilde{R}$ is the unitarization of R, characterized by the property that for any unitary algebra S and nonunitary algebra homomorphism $f : R \to S$, there is a unique unitary algebra homomorphism $\tilde{f} : \tilde{R} \to S$ such that $f = \tilde{f}u_R$. To generalize this process to Rota-Baxter algebras turns out to be much more involved since, by formally adding a unit $\mathbf{1}$ to a nonunitary algebra (R, P) , we also need to add its images under the Rota-Baxter operator P, its iterations and many more, such as $P(\mathbf{1}), P(xP(\mathbf{1}))$. Then it is not clear how these new elements should behave to form a Rota-Baxter algebra. We will start with the unitarization of free nonunitary Rota-Baxter algebras and then take care of the case of a general Rota-Baxter algebra by considering them as quotients of free Rota-Baxter algebras.

We first give the definition.

Definition 4.3.1. Fix a weight λ in the base ring $\mathbf{k}$. Let (R, P) be a nonunitary Rota-Baxter $\mathbf{k}$-algebra. A *unitarization of* R is a unitary Rota-Baxter algebra $(\tilde{R}, \tilde{P})$ with a nonunitary Rota-Baxter algebra homomorphism $u_R : R \to \tilde{R}$ such that for any unitary Rota-Baxter algebra S and a homomorphism $f : R \to S$ of nonunitary Rota-Baxter algebras, there is a unique homomorphism $\tilde{f} : \tilde{R} \to S$ of unitary Rota-Baxter algebras such that $f = \tilde{f} u_R$. $\tilde{f}$ is called the *unitarization of* f.

4.3.1 Unitarization of free Rota-Baxter algebras

Let X be a set. Let $Ш^{NC}(X)$ and $Ш^{NC,0}(X)$ be the free unitary and nonunitary Rota-Baxter algebras. Let $\tilde{j}_X : X \to Ш^{NC}(X)$ and $j_X : X \to Ш^{NC,0}(X)$ be the canonical embeddings. Regarding $Ш^{NC}(X)$ as a nonunitary Rota-Baxter algebra, then by the universal property of the free nonunitary Rota-Baxter algebra $Ш^{NC,0}(X)$, there is a unique homomorphism $u_X : Ш^{NC,0}(X) \to Ш^{NC}(X)$ of nonunitary Rota-Baxter algebras such that $\tilde{j}_X = u_X j_X$.

Theorem 4.3.2. *For any set X, the free unitary Rota-Baxter algebra $Ш^{NC}(X)$, with the nonunitary Rota-Baxter algebra homomorphism u_X: $Ш^{NC,0}(X) \to Ш^{NC}(X)$, is the unitarization of the free nonunitary Rota-Baxter algebra $Ш^{NC,0}(X)$.*

Proof. Let (S, Q) be a unitary Rota-Baxter algebra and let $f : Ш^{NC,0}(X) \to S$ be a homomorphism of nonunitary Rota-Baxter algebras. Let $f' = f j_X : X \to S$, then by the freeness of the unitary Rota-Baxter algebra $Ш^{NC}(X)$, there is a unique homomorphism $\overline{f}' : Ш^{NC}(X) \to S$ of unitary Rota-Baxter algebras (hence also of nonunitary Rota-Baxter algebras) such that $f' = \overline{f}' \tilde{j}_X$. To keep track of the maps, we show them in the following diagram

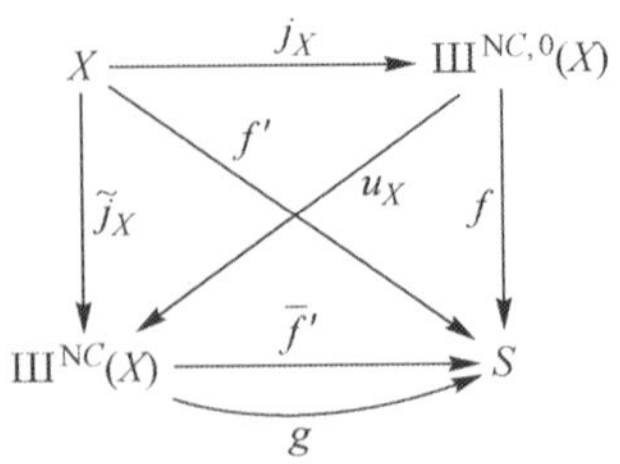

We have

$$\overline{f}\,'\circ u_X\circ j_X=\overline{f}\,'\circ\tilde{j}_X=f'=f\circ j_X.$$

By the universal property of $Ш^{NC,0}(X)$ as the nonunitary Rota-Baxter algebra on X, we have $\overline{f}\,'\circ u_X=f$. Suppose there is another unitary Rota-Baxter algebra homomorphism $g:Ш^{NC}(X)\to S$ such that $g\circ u_X=f$. Then

$$g\circ\tilde{j}_X=g\circ u_X\circ j_X=f\circ j_X=f'=\overline{f}\,'\circ\tilde{j}_X.$$

So $g=\overline{f}\,'$ by the universal property of $Ш^{NC}(X)$ as the free unitary Rota-Baxter algebra. Thus $\tilde{f}:=\overline{f}\,'$ is the unitarization of f. □

4.3.2 Unitarization of Rota-Baxter algebras

We now construct the unitarization of any given nonunitary Rota-Baxter algebra R. We use the following diagram to keep track of the maps that we will introduce below.

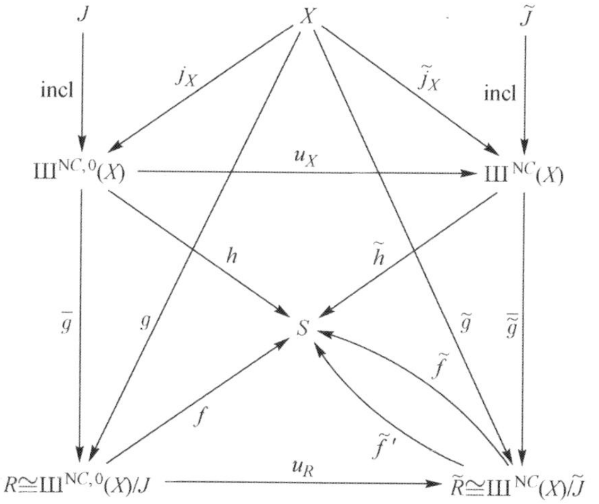

(4.39)

Let X be a generating set of R as a nonunitary Rota-Baxter algebra with $g:X\hookrightarrow R$ being the inclusion map. Let $Ш^{NC,0}(X)$ be the free nonunitary Rota-Baxter algebra on X with the canonical embedding $j_X:X\to Ш^{NC,0}(X)$. Then there is a unique nonunitary Rota-Baxter algebra homomorphism $\overline{g}:Ш^{NC,0}(X)\to R$ such that $g=\overline{g}\circ j_X$. Since X is a generating set of R, $\overline{g}$ is surjective. So $R\cong Ш^{NC,0}(X)/J$, where J as the kernel of $\overline{g}$ is a Rota-Baxter ideal of $Ш^{NC,0}(X)$. Recall from Theorem 4.3.2 that we have the unitarization $u_X:Ш^{NC,0}(X)\to Ш^{NC}(X)$. Let $\tilde{J}$ be the Rota-Baxter ideal of $Ш^{NC}(X)$ generated by $u_X(J)$, and define

$$\tilde{R}=Ш^{NC}(X)/\tilde{J}$$

with $\overline{\tilde{g}} : Ш^{NC}(X) \to \tilde{R}$ being the resulting Rota-Baxter homomorphism. Let $\tilde{g} = \overline{\tilde{g}} \circ \tilde{j}_X$. Then $\overline{\tilde{g}} : Ш^{NC}(X) \to \tilde{R}$ is the unique unitary Rota-Baxter algebra homomorphism induced from the set map $\tilde{g}$. So the notation $\overline{\tilde{g}}$ is justified.

Now since $u_X(J) \subseteq \tilde{J}$, we have $(\overline{\tilde{g}} \circ u_X)(J) = 0$. Thus $\ker(\overline{\tilde{g}} \circ u_X) \supseteq J$. Therefore, there is a unique homomorphism

$$u_R : R \cong Ш^{NC,0}(X)/J \to \tilde{R} \cong Ш^{NC}(X)/\tilde{J}$$

of nonunitary Rota-Baxter algebras such that

$$u_R \circ \overline{g} = \overline{\tilde{g}} \circ u_X.$$

Theorem 4.3.3. *With the above notation, the nonunitary Rota-Baxter algebra homomorphism*

$$u_R : R \to \tilde{R}$$

gives the unitarization of R.

By the uniqueness of Rota-Baxter unitarization, for a different choice of a generating set X of R, the unitarization we obtain will be isomorphic.

Proof. Let S be a unitary Rota-Baxter algebra and let $f : R \to S$ be a nonunitary Rota-Baxter algebra homomorphism. Let $h = f \circ \overline{g}$. By Theorem 4.3.2, there is a unique homomorphism $\tilde{h} : Ш^{NC}(X) \to S$ of unitary Rota-Baxter algebras such that $\tilde{h} \circ u_X = h$. Then

$$\ker \tilde{h} \supseteq u_X(\ker h) \supseteq u_X(\ker \overline{g}) = J.$$

Since $\ker \tilde{h}$ is a Rota-Baxter ideal of $Ш^{NC}(X)$ and $\tilde{J}$ is the Rota-Baxter ideal of $Ш^{NC}(X)$ generated by J, we must have $\ker \tilde{h} \supseteq \tilde{J}$. Therefore, there is a unique

$$\tilde{f} : \tilde{R} \to S$$

such that $\tilde{h} = \tilde{f} \circ \overline{\tilde{g}}$. Now

$$\tilde{f} \circ u_R \circ \overline{g} = \tilde{f} \circ \overline{\tilde{g}} \circ u_X = \tilde{h} \circ u_X = h = f \circ \overline{g}.$$

Since $\overline{g}$ is surjective, we have $\tilde{f} \circ u_R = f$. So the existence of $\tilde{f}$ in Definition 4.3.1 is proved.

To prove the uniqueness of $\tilde{f}$, suppose there is also a homomorphism $\tilde{f}'$:

$\tilde{R} \to S$ of unitary Rota-Baxter algebras such that $\tilde{f}' \circ u_R = f$. Then we have

$$\begin{aligned}\tilde{f}' \circ \overline{\tilde{g}} \circ u_X &= \tilde{f}' \circ u_R \circ \overline{g} \\ &= f \circ \overline{g} \\ &= \tilde{f} \circ u_R \circ \overline{g} \\ &= \tilde{f} \circ \overline{\tilde{g}} \circ u_X \\ &= \tilde{h} \circ u_X \\ &= h.\end{aligned}$$

So $\tilde{f}' \circ \overline{\tilde{g}} : Ш^{NC,0}(X) \to S$, as well as $\tilde{h}$ is the unitarization of $h : Ш^{NC,0}(X) :\to S$. By the uniqueness of this unitarization, proved in Theorem 4.3.2, we have

$$\tilde{f}' \circ \overline{\tilde{g}} = \tilde{h} = \tilde{f} \circ \overline{\tilde{g}}.$$

Since $\overline{\tilde{g}}$ is surjective, we have $\tilde{f}' = \tilde{f}$, as needed. □

4.4 Free Rota-Baxter algebra on an algebra

We now consider the free unitary and nonunitary RBA on another unitary or nonunitary algebra. Other than its theoretical significance, such constructions are important for applications, say, to study universal enveloping RBAs of dendriform algebras and tridendriform algebras, and free dendriform algebras and tridendriform algebras for which more details will be given in Chapter 5. For now the reader can think of such a free RBA on another algebra as the noncommutative variation of the free commutative RBA obtained in Chapter 3. The reader can also think of such a free RBA as the Rota-Baxter algebra analog of the tensor algebra on a module. It is well known that such a tensor algebra is essential in the study of enveloping algebras of Lie algebras.

Definition 4.4.1. Let A be a $\mathbf{k}$-algebra. A *free RBA on* A is an RBA $F(A)$ with a Rota-Baxter operator P_A and an algebra homomorphism $j_A : A \to F(A)$ such that for any RBA R and any algebra homomorphism $f : A \to R$, there is a unique RBA homomorphism $\overline{f} : F(A) \to R$ such that $\overline{f} \circ j_A = f$.

We will consider $\mathbf{k}$-algebra A with the following condition.

Assumption 1. *A is the unitarization of a nonunitary $\mathbf{k}$-algebra A^0 which has a $\mathbf{k}$-basis X such that $\tilde{X} := X \cup \{1\}$ is a $\mathbf{k}$-basis of A.*

We first display a $\mathbf{k}$-basis of the free RBA in Section 4.4.1 in the form of words. The product on the free RBA is given in Section 4.4.2 and the universal property of the free RBA is proved in Section 4.4.3.

4.4.1 Words of a free RBA on an algebra

Let A be a **k**-algebra satisfying Assumption 1 and let $\tilde{X}$ be a fixed **k**-basis given in the assumption. We will construct a **k**-basis of the free RBA on A in terms of words from the alphabet set X.

4.4.1.1 *Rota-Baxter bracketed words*

Let $X' = X \cup \{\lfloor, \rfloor\}$. Let $M(X')$ be the free monoid generated by X'. We will use concatenation to denote the product in $M(X')$ and use $\cdot$ to denote the product in A.

Definition 4.4.2. A word in $M(X')$ is called a (*strict*) *Rota-Baxter* (*bracketed*) *word* (*sRBW*) if the following conditions are satisfied.

(a) The number of $\lfloor$ in the word equals the number of $\rfloor$ in the word;

(b) Counting from the left to the right, the accumulative number of $\lfloor$ at each location is always greater than or equal to that of $\rfloor$;

(c) There is no occurrence of $\rfloor\lfloor$ in the word.

(d) There is no occurrence of x_1x_2 in the word, with $x_1, x_2 \in X$;

A word in $M(X')$ is called a *nonunitary strict Rota-Baxter word* (*sRBW*) if it satisfies the following condition in place of condition (c).

(c') there is no occurrence of $\rfloor\lfloor$, $\lfloor\rfloor$ in the word and the word is not the unit 1 of $M(X')$.

For example,

$$x_1\lfloor x_2\rfloor x_3,\ \lfloor\lfloor x_1\rfloor x_2\rfloor x_3$$

are sRBWs while

$$\lfloor x_1\lfloor x_2\rfloor x_3, x_1\rfloor x_2\lfloor,\ \lfloor x_1x_2\rfloor,\ \lfloor x_1\rfloor\lfloor x_2\rfloor$$

are not sRBWs. The word $x_1\lfloor\ \rfloor$ is an sRBW, but not a nonunitary sRBW. We note that the extra condition of sRBW over an RBW is condition (d).

Let $\tilde{\mathcal{X}}'$ (resp. $\mathcal{X}'$) denote the set of sRBWs (resp. nonunitary sRBWs) on X. Let

$$\text{Ш}^{NC}(A) := \mathbf{k}\tilde{\mathcal{X}}', \quad \text{Ш}^{NC,0}(A^0) := \mathbf{k}\mathcal{X}'$$

be the free **k**-modules generated by $\tilde{\mathcal{X}}'$ and $\mathcal{X}'$.

Note: A warning on notations: We have used $\tilde{\mathfrak{X}}_\infty$ and $\tilde{\mathfrak{X}}'$ to denote the set of RBWs. We will use $\tilde{\mathcal{X}}_\infty$ and $\tilde{\mathcal{X}}'$ to denote the set of sRBWs.

For a word $\mathbf{x} := y_1 \cdots y_m \in M(X')$, $y_i \in X', 1 \leqslant i \leqslant m$, we define the *length* $\ell(\mathbf{x})$ of $\mathbf{x}$ to be m, define the *head index* $h(\mathbf{x})$ of $\mathbf{x}$ to be 0 (resp. 1) if y_1 is in X (resp. in $\{\lfloor, \rfloor\}$). Similarly define the *tail index* $t(\mathbf{x})$ of $\mathbf{x}$ to be 0 (resp. 1) if y_m is in X (resp. in $\{\lfloor, \rfloor\}$). Let $d_i(\mathbf{x})$ be the number of $\lfloor$ in $y_1, \ldots, y_i$ minus the number of $\rfloor$ in $y_1, \ldots, y_i$. Then the first two conditions for $\tilde{X}'$ are equivalent to $d_i(\mathbf{x}) \geqslant 0$ for $1 \leqslant i \leqslant m$ and $d_m(\mathbf{x}) = 0$. Define $d(\mathbf{x}) := \max\{d_i(\mathbf{x}),\ 1 \leqslant i \leqslant m\}$ to be the *depth* of $\mathbf{x}$. Define the *breadth* $b(\mathbf{x})$ of $\mathbf{x}$ to be the number of zeros of $d_i(\mathbf{x}), 1 \leqslant i \leqslant m$.

Example 4.4.3. For $x_1, x_2, x_3 \in X$, the word $\lfloor\lfloor x_1 \rfloor x_2 \rfloor x_3$ has length 7, depth 2, head 1, tail 0 and breadth 2.

4.4.1.2 *A recursive description of sRBWs*

It will be convenient to have a recursive description of $\tilde{X}'$. For this we recall from Section 4.2 the following concept of alternating products. Let Y, Z be two subsets of $M(X')$. Define

$$\Lambda(Y,Z) = \Big(\bigcup_{r\geqslant 1} (Y\lfloor Z\rfloor)^r\Big) \bigcup \Big(\bigcup_{r\geqslant 0} (Y\lfloor Z\rfloor)^r Y\Big) \bigcup \Big(\bigcup_{r\geqslant 1} (\lfloor Z\rfloor Y)^r\Big) \bigcup \Big(\bigcup_{r\geqslant 0} (\lfloor Z\rfloor Y)^r \lfloor Z\rfloor\Big), \tag{4.40}$$

$$\tilde{\Lambda}(Y,Z) = \Lambda(Y,Z) \cup \{\mathbf{1}\}. \tag{4.41}$$

We construct two sequences $\mathcal{X}_n$ and $\tilde{\mathcal{X}}_n$ of subsets of $M(X')$ by the following recursion. Let

$$\mathcal{X}_0 = X, \quad \tilde{\mathcal{X}}_0 = \tilde{X} = X \cup \{1\},$$

and, for $n \geqslant 0$, define

$$\mathcal{X}_{n+1} = \Lambda(X, \mathcal{X}_n), \quad \tilde{\mathcal{X}}_{n+1} = \tilde{\Lambda}(X, \tilde{\mathcal{X}}_n).$$

More precisely,

$$\tilde{\mathcal{X}}_{n+1} = \Big(\bigcup_{r\geqslant 1} (X\lfloor \tilde{\mathcal{X}}_n\rfloor)^r\Big) \bigcup \Big(\bigcup_{r\geqslant 0} (X\lfloor \tilde{\mathcal{X}}_n\rfloor)^r X\Big) \bigcup \Big(\bigcup_{r\geqslant 1} (\lfloor \tilde{\mathcal{X}}_n\rfloor X)^r\Big) \bigcup \Big(\bigcup_{r\geqslant 0} (\lfloor \tilde{\mathcal{X}}_n\rfloor X)^r \lfloor \tilde{\mathcal{X}}_n\rfloor\Big) \bigcup \{1\}. \tag{4.42}$$

Further, define

$$\mathcal{X}_\infty = \bigcup_{n\geqslant 0} \mathcal{X}_n = \varinjlim \mathcal{X}_n, \tag{4.43}$$

$$\tilde{\mathcal{X}}_\infty = \bigcup_{n\geqslant 0} \tilde{\mathcal{X}}_n = \varinjlim \tilde{\mathcal{X}}_n. \tag{4.44}$$

Here the second equation in Eq. (4.43) follows since $\mathcal{X}_{n+1} \supseteq \mathcal{X}_n$. Similarly for Eq. (4.44),

Lemma 4.4.4. (a) $\mathcal{X}' = \mathcal{X}_\infty$. *In fact,* $\mathcal{X}_n$ *consists of words in* $\mathcal{X}'$ *of depth less than or equal to* n.

(b) $\tilde{\mathcal{X}}' = \tilde{\mathcal{X}}_\infty$. *In fact,* $\tilde{\mathcal{X}}_n$ *consists of words in* $\tilde{\mathcal{X}}'$ *of depth less or than equal to* n.

(c) *Every* sRBW $\mathbf{x} \neq \mathbf{1}$ *has a unique decomposition*

$$\mathbf{x} = \mathbf{x}_1 \cdots \mathbf{x}_b, \tag{4.45}$$

called the standard decomposition. Here $\mathbf{x}_i$, $1 \leqslant i \leqslant b$, *is alternatively in* X *or in* $\lfloor \tilde{\mathcal{X}}_\infty \rfloor$, *and* $b = b(\mathbf{x})$ *is the breadth of* $\mathbf{x}$. *More precisely, let*

$$\mathbf{x} = y_1 \cdots y_\ell,\ y_i \in X',\ 1 \leqslant i \leqslant \ell.$$

Then $\mathbf{x}_1 = y_1 \cdots y_{j_1}$, $\mathbf{x}_2 = y_{j_1+1} \cdots y_{j_2}, \ldots, \mathbf{x}_c = y_{j_{c-1}+1} \cdots y_{j_c}$ *where* $j_1, j_2, \ldots, j_c$ *are the values of* j, $1 \leqslant j \leqslant \ell$ *such that* $d_j(\mathbf{x}) = 0$.

(d) *For each* $n \geqslant 1$, *the union of* $\tilde{\mathcal{X}}_n = \tilde{\Lambda}(X, \tilde{\mathcal{X}}_{n-1})$ *expressed in Eq.* (4.42) *is disjoint:*

$$\begin{aligned}\tilde{\mathcal{X}}_n = &\Big(\bigsqcup_{r\geqslant 1} (X\lfloor \tilde{\mathcal{X}}_{n-1}\rfloor)^r\Big) \bigsqcup \Big(\bigsqcup_{r\geqslant 0} (X\lfloor \tilde{\mathcal{X}}_{n-1}\rfloor)^r X\Big) \\ &\bigsqcup \Big(\bigsqcup_{r\geqslant 1} (\lfloor \tilde{\mathcal{X}}_{n-1}\rfloor X)^r\Big) \bigsqcup \Big(\bigsqcup_{r\geqslant 0} (\lfloor \tilde{\mathcal{X}}_{n-1}\rfloor X)^r \lfloor \tilde{\mathcal{X}}_{n-1}\rfloor\Big).\end{aligned} \tag{4.46}$$

Proof. (a) The proof is similar to the proof of Proposition 4.2.8. We only need to prove the second statement. Let $\mathfrak{M}_n^0$ be the set of sRBWs of depth n. Then the second statement means

$$\mathcal{X}_n = \bigcup_{i\leqslant n} \mathfrak{M}_i^0,\quad 0 \leqslant n \tag{4.47}$$

for which we will verify by induction on n.

When $n = 0$, we have $\mathcal{X}_0 = X$. On the other hand, sRBWs of depth 0 do not have the occurrence of $\lfloor$ or $\rfloor$. So by condition (d) in Definition 4.4.2, they are precisely X. So $\mathcal{X}_0 = \mathfrak{M}_0^0$.

For $k \geqslant 0$, assume $\mathcal{X}_n = \bigcup_{i\leqslant n} \mathfrak{M}_i^0$ for $0 \leqslant n \leqslant k$. Then $\mathcal{X}_k = \bigcup_{i\leqslant k} \mathfrak{M}_i^0$. Since $\mathcal{X}_{k+1} = \Lambda(X, \mathcal{X}_k)$ defined in Eq. (4.42) only increases the depth of elements in $\mathcal{X}_k$ at most by 1, we have

$$\mathcal{X}_{k+1} \subseteq \bigcup_{i\leqslant k+1} \mathfrak{M}_i^0.$$

On the other hand, let $\mathbf{x}$ be an sRBW of depth at most $k+1$. We use induction on the length $m := \ell(\mathbf{x})$ of $\mathbf{x}$ to prove that $\mathbf{x}$ is in $\mathcal{X}_{k+1}$. When $m = 1$, $\mathbf{x}$ is in X and so is in $\mathcal{X}_{k+1}$. Assume this is true for $1 \leqslant m \leqslant j$, and let $\mathbf{x} = y_1 \cdots y_{j+1}$.

Then $j+1 \geqslant 2$. By the conditions on sRBWs, y_1 either is in X or is $\lfloor$. If $y_1 \in X$, then $y_2 = \lfloor$ and it is easy to check the conditions in Definition 4.4.2 and see that $y_2 \cdots y_{j+1}$ is still an RBW with depth at most $k+1$. By the induction hypothesis on length, $y_2 \cdots y_{j+1}$ is in $\mathcal{X}_{k+1}$. Since $y_2 = \lfloor$, using Eq. (4.42), we see that the word $y_2 \cdots y_{j+1}$ must be in

$$\Big(\bigsqcup_{r\geqslant 1}(\lfloor \mathcal{X}_k \rfloor X)^r\Big) \bigsqcup \Big(\bigsqcup_{r\geqslant 0}(\lfloor \mathcal{X}_k \rfloor X)^r \lfloor \mathcal{X}_k \rfloor\Big).$$

Therefore, $\mathbf{x}$ is in

$$\Big(\bigsqcup_{r\geqslant 1}(X\lfloor \mathcal{X}_k \rfloor)^r\Big) \bigsqcup \Big(\bigsqcup_{r\geqslant 0}(X\lfloor \mathcal{X}_k \rfloor)^r X\Big)$$

and so is still in $\mathcal{X}_{k+1}$ by Eq. (4.42).

Now if $y_1 = \lfloor$, then by condition (a) in Definition 4.4.2, there is $1 \leqslant \ell \leqslant j+1$ such that $d_\ell(\mathbf{x}) = 0$. Let ℓ be the minimum with this condition. Then $y_\ell = \rfloor$. So $\mathbf{x} = \lfloor y_2 \cdots y_{\ell-1} \rfloor y_{\ell+1} \cdots y_{j+1}$ and $y_2 \cdots y_{\ell-1}$ is still an sRBW. Since the depth of $\mathbf{x}$ is at most $k+1$, the depth of $y_2 \cdots y_{\ell-1}$ is at most k. So by the induction hypothesis on depth, $y_2 \cdots y_{\ell-1}$ is in $\mathcal{X}_k$ and therefore $y_1 \cdots y_\ell$ is in $\mathcal{X}_{k+1}$. If $\mathbf{x} = y_1 \cdots y_\ell$, then we are done. Otherwise, by condition (c) in Definition 4.4.2, we have $y_{\ell+1} \in X$. Further $y_{\ell+1} \cdots y_{j+1}$ is still an sRBW, of length not exceeding j and depth at most $k+1$. So the induction hypotheses on depth and length imply that $y_{\ell+1} \cdots y_{j+1}$ is in $\mathcal{X}_{k+1}$. Then the concatenation $y_1 \cdots y_\ell y_{\ell+1} \cdots y_{j+1}$ is in $\mathcal{X}_{k+1}$ by Eq. (4.42).

In summary, by induction on m, we have $\cup_{i\leqslant k+1}\mathfrak{M}_i^0 \subseteq \mathcal{X}_{k+1}$ and hence $\mathcal{X}_{k+1} = \cup_{i\leqslant k+1}\mathfrak{M}_i^0$. Then by induction on n, Eq. (4.47) holds for any $n \geqslant 0$. This completes the proof of Lemma 4.4.4. (a).

(b) The proof is similar to Item (a).

(c) Let $\mathbf{x}$ be in $\mathfrak{M}^0(X)$. The existence of the decomposition in Eq. (4.45) follows from Item (a) and the definition of $\mathcal{X}_n$. To prove the uniqueness, we use induction on $m := \ell(\mathbf{x})$. When $m = 1$, $\mathbf{x}$ is in X. So $\mathbf{x} = \mathbf{x}$ is the only decomposition. Assume the uniqueness is proved for $m \geqslant j$ and let $\mathbf{x} \in \mathfrak{M}^0(X)$ with $\ell(\mathbf{x}) = m+1$. Then $j+1 \geqslant 2$ and we have $\mathbf{x} = y_1 \cdots y_{m+1}$ with $y_i \in X' = X \cup \{\lfloor, \rfloor\}$, $1 \leqslant i \leqslant m+1$. Let $\mathbf{x} = \mathbf{x}_1 \cdots \mathbf{x}_c = \mathbf{x}'_1 \cdots \mathbf{x}'_{c'}$ be decompositions of the form in Eq. (4.45). Then $\mathbf{x}_1$ is in either X or $\lfloor \tilde{\mathcal{X}}_\infty \rfloor$ and similarly for $\mathbf{x}'_1$.

If $\mathbf{x}_1$ is in X, then $\mathbf{x}_1 = y_1$. Then $\mathbf{x}'_1$ starts with $y_1 \in X$ and hence must itself be y_1. Thus $\mathbf{x}_1 = \mathbf{x}'_1$. Then

$$y_2 \cdots y_m = \mathbf{x}_2 \cdots \mathbf{x}_c = \mathbf{x}'_2 \cdots \mathbf{x}'_{c'}.$$

By the induction hypothesis, $y_2 \cdots y_m$ has unique factorization. Therefore $c = c'$ and $\mathbf{x}_i = \mathbf{x}'_i$ for $2 \leqslant i \leqslant c$.

If $\mathbf{x}_1$ is in $\lfloor \mathcal{X}_\infty \rfloor$, then $y_1 = \lfloor$. Then $\mathbf{x}'_1$ must also be in $\lfloor \mathcal{X}_\infty \rfloor$. Let $\mathbf{x}_1 = y_1 \cdots y_u$. Then $y_u = \rfloor$ and $y_2 \cdots y_{u-1}$ is in $\mathcal{X}_\infty$. By the restriction on the sRBWs in $\mathcal{X}_\infty = \mathfrak{M}^0(X)$ in Definition 4.4.2, for $2 \leqslant i \leqslant u-1$, the number of $\lfloor$ minus the number of $\rfloor$ in $y_2, \cdots, y_i$ is greater than or equal to zero and is equal to zero when $i = u-1$. Then the number of $\lfloor$ minus the number of $\rfloor$ in $y_1 = \lfloor, y_2, \ldots, y_i$ is strictly greater than zero when $1 \leqslant i \leqslant u-1$ and is equal to zero for the first time when $i = u$. Let $\mathbf{x}'_1 = y'_1 \cdots y'_{u'}$. Then by the same argument, u' is also where this difference is zero for the first time. So we must have $u = u'$ and $\mathbf{x}_1 = \mathbf{x}'_1$. If $\mathbf{x} = \mathbf{x}_1$, then we are done. If not, then we have the decompositions $\mathbf{x}_2 \cdots \mathbf{x}_c = \mathbf{x}'_2 \cdots \mathbf{x}'_{c'}$ of the word $y_{u+1} \cdots y_{j+1}$. Then by the induction hypothesis on j, we have $c = c'$ and $\mathbf{x}_i = \mathbf{x}'_i$ for $2 \leqslant i \leqslant c$. Thus the uniqueness of the decomposition is proved.

From the above argument, the factors $\mathbf{x}_1, \ldots, \mathbf{x}_c$ of $\mathbf{x} = y_1 \cdots y_\ell$ are given by $\mathbf{x}_1 = y_1 \ldots y_{j_1}, \mathbf{x}_2 = y_{j_1+1} \cdots y_{j_2}, \ldots, \mathbf{x}_c = y_{j_{c-1}+1} \cdots y_{j_c}$ where $j = j_1, j_2, \ldots, j_c$ are exactly where $d_j(\mathbf{x}) = 0$. Recall that $d_j(\mathbf{x})$ is defined to be the difference between the number of $\lfloor$ and the number of $\rfloor$ in $y_1, \ldots, y_j$. So c is the breadth of $\mathbf{x}$ and the last statement of Item (c) is proved.

(d) The four big subsets

$$\bigsqcup_{r \geqslant 1} (X\lfloor \tilde{\mathcal{X}}_{n-1} \rfloor)^r, \ \bigsqcup_{r \geqslant 0} (X\lfloor \tilde{\mathcal{X}}_{n-1} \rfloor)^r X, \ \bigsqcup_{r \geqslant 1} (\lfloor \tilde{\mathcal{X}}_{n-1} \rfloor X)^r, \ \bigsqcup_{r \geqslant 0} (\lfloor \tilde{\mathcal{X}}_{n-1} \rfloor X)^r \lfloor \tilde{\mathcal{X}}_{n-1} \rfloor$$

are disjoint because of their distinct head and tail indices. The subsets within each of the four big subsets are disjoint because they have distinct breadth. More precisely, the head, tail and breadth of an sRBW $\mathbf{x}$ are given in the following table.

$\mathbf{x}$	$(X\lfloor \tilde{\mathcal{X}}_{n-1} \rfloor)^r$	$(X\lfloor \tilde{\mathcal{X}}_{n-1} \rfloor)^r X$	$(\lfloor \tilde{\mathcal{X}}_{n-1} \rfloor X)^r$	$(\lfloor \tilde{\mathcal{X}}_{n-1} \rfloor X)^r \lfloor \tilde{\mathcal{X}}_{n-1} \rfloor$
$h(\mathbf{x})$	0	0	1	1
$t(\mathbf{x})$	1	0	0	1
$b(\mathbf{x})$	$2r$	$2r+1$	$2r$	$2r+1$

□

4.4.2 The product in a free RBA

By Lemma 4.4.4.(a), we have

$$Ш^{NC}(A) = \bigoplus_{\mathbf{x} \in \tilde{\mathcal{X}}_\infty} \mathbf{k}\mathbf{x}, \quad Ш^{NC,0}(A^0) = \bigoplus_{\mathbf{x} \in \mathcal{X}_\infty} \mathbf{k}\mathbf{x}.$$

We now define a product $\diamond$ on $Ш^{NC}(A)$ by defining $\mathbf{x} \diamond \mathbf{x}'$ for $\mathbf{x}, \mathbf{x}' \in \tilde{\mathcal{X}}_\infty$ and then extending bilinearly. Roughly speaking, the product of $\mathbf{x}$ and $\mathbf{x}'$ is defined to be the concatenation whenever $t(\mathbf{x}) \neq h(\mathbf{x}')$. When $t(\mathbf{x}) = h(\mathbf{x}')$, the product is defined by the product in A or by a shuffle relation, as in Eq. (4.48). This is a similar free Rota-Baxter algebra on a set. The distinction is that the case when the product is in A here corresponds to the case when the product is by simple concatenation for free Rota-Baxter algebras on a set.

We first define $1 \in \tilde{\mathcal{X}}_\infty$ to be the identity of the product. To define the product of $\mathbf{x}$ and $\mathbf{x}'$ when both are in $\tilde{\mathcal{X}}_\infty \backslash \{1\}$, we use induction on the sum $n := d(\mathbf{x}) + d(\mathbf{x}')$ of the depths of $\mathbf{x}$ and $\mathbf{x}'$. Then $n \geqslant 0$. If $n = 0$, then $\mathbf{x}, \mathbf{x}'$ are in X and so are in A and we define $\mathbf{x} \diamond \mathbf{x}' = \mathbf{x} \cdot \mathbf{x}' = \sum_{x \in X} c_x^{\mathbf{x},\mathbf{x}'} x$. Here $\cdot$ is the product in A and $c_x^{\mathbf{x},\mathbf{x}'}$ are the structure constants for the product $\cdot$.

Suppose $\mathbf{x} \diamond \mathbf{x}'$ have been defined for all $\mathbf{x}, \mathbf{x}' \in \tilde{\mathcal{X}}_\infty$ with $n \leqslant k$ for a $k \geqslant 0$ and consider $\mathbf{x}, \mathbf{x}' \in \tilde{\mathcal{X}}_\infty$ with $n = k+1$. Then $n \geqslant 1$.

First assume the breadth $b(\mathbf{x}) = b(\mathbf{x}') = 1$. Then $\mathbf{x}$ and $\mathbf{x}'$ are in X or $\lfloor \tilde{\mathcal{X}}_\infty \rfloor$. Since $n \geqslant 1$, at least either $\mathbf{x}$ or $\mathbf{x}'$ is not in X. We accordingly define

$$\mathbf{x} \diamond \mathbf{x}' = \begin{cases} \mathbf{x}\mathbf{x}', & \text{if } \mathbf{x} \in X, \mathbf{x} \in \lfloor \tilde{\mathcal{X}}_\infty \rfloor, \\ \mathbf{x}\mathbf{x}', & \text{if } \mathbf{x} \in \lfloor \tilde{\mathcal{X}}_\infty \rfloor, \mathbf{x}' \in X, \\ \lfloor \lfloor \overline{\mathbf{x}} \rfloor \diamond \overline{\mathbf{x}}' \rfloor + \lfloor \overline{\mathbf{x}} \diamond \lfloor \overline{\mathbf{x}}' \rfloor \rfloor + \lambda \lfloor \overline{\mathbf{x}} \diamond \overline{\mathbf{x}}' \rfloor, & \text{if } \mathbf{x} = \lfloor \overline{\mathbf{x}} \rfloor, \mathbf{x}' = \lfloor \overline{\mathbf{x}}' \rfloor \in \lfloor \tilde{\mathcal{X}}_\infty \rfloor. \end{cases} \quad (4.48)$$

Here the product in the first and second cases is by concatenation, and in the third case is by the induction hypothesis since for each of the three products on the right hand side, the sum of the depths of the two factors is either $d(\mathbf{x}) + d(\mathbf{x}') - 1$ or $d(\mathbf{x}) + d(\mathbf{x}') - 2$ and hence is less than or equal to k.

Now assume $b(\mathbf{x}) > 1$ or $b(\mathbf{x}') > 1$. Let $\mathbf{x} = \mathbf{x}_1 \cdots \mathbf{x}_b$ and $\mathbf{x}' = \mathbf{x}'_1 \cdots \mathbf{x}'_{b'}$ be the standard decompositions from Lemma 4.4.4. We then define

$$\mathbf{x} \diamond \mathbf{x}' = \mathbf{x}_1 \cdots \mathbf{x}_{b-1} (\mathbf{x}_b \diamond \mathbf{x}'_1)\, \mathbf{x}'_2 \cdots \mathbf{x}'_{b'}, \quad (4.49)$$

where $\mathbf{x}_b \diamond \mathbf{x}'_1$ is defined by Eq. (4.48) and the rest are given by concatenation. The concatenation is well-defined since by Eq. (4.48), we have $h(\mathbf{x}_b) = h(\mathbf{x}_b \diamond \mathbf{x}'_1)$ and $t(\mathbf{x}'_1) = t(\mathbf{x}_b \diamond \mathbf{x}'_1)$. Therefore, $t(\mathbf{x}_{b-1}) \neq h(\mathbf{x}_b \diamond \mathbf{x}'_1)$ and $h(\mathbf{x}'_2) \neq t(\mathbf{x}_b \diamond \mathbf{x}'_1)$.

We record the following simple properties of $\diamond$ for later applications.

Lemma 4.4.5. *Let $\mathbf{x}, \mathbf{x}' \in \tilde{\mathcal{X}}_\infty$. We have the following statements.*
(a) *$h(\mathbf{x}) = h(\mathbf{x} \diamond \mathbf{x}')$ and $t(\mathbf{x}') = t(\mathbf{x} \diamond \mathbf{x}')$.*
(b) *If $t(\mathbf{x}) \neq h(\mathbf{x}')$, then $\mathbf{x} \diamond \mathbf{x}' = \mathbf{x}\mathbf{x}'$ (concatenation).*
(c) *If $t(\mathbf{x}) \neq h(\mathbf{x}')$, then for any $\mathbf{x}'' \in \tilde{\mathcal{X}}_\infty$, we have*

$$(\mathbf{x}\mathbf{x}') \diamond \mathbf{x}'' = \mathbf{x}(\mathbf{x}' \diamond \mathbf{x}''), \quad \mathbf{x}'' \diamond (\mathbf{x}\mathbf{x}') = (\mathbf{x}'' \diamond \mathbf{x})\mathbf{x}'.$$

Proof. (a) and (b) follow immediately from the definition of $\diamond$.

(c) Let $\mathbf{x} = \mathbf{x}_1 \cdots \mathbf{x}_b$ and $\mathbf{x}' = \mathbf{x}'_1 \cdots \mathbf{x}'_{b'}$ be the standard decompositions. Then $t(\mathbf{x}) \neq h(\mathbf{x}')$ means that $\mathbf{x}_1 \cdots \mathbf{x}_b \mathbf{x}'_1 \cdots \mathbf{x}'_{b'}$ is the standard decomposition of $\mathbf{x}\mathbf{x}'$. Let $\mathbf{x}'' = \mathbf{x}''_1 \cdots \mathbf{x}''_{b''}$ be the standard decomposition of $\mathbf{x}''$. Then by definition,

$$(\mathbf{x}\mathbf{x}') \diamond \mathbf{x}'' = \mathbf{x}_1 \cdots \mathbf{x}'_{b'-1}(\mathbf{x}'_{b'} \diamond \mathbf{x}''_1)\mathbf{x}''_2 \cdots \mathbf{x}''_{b''}.$$

Also

$$\mathbf{x}' \diamond \mathbf{x}'' = \mathbf{x}'_1 \cdots \mathbf{x}'_{b'-1}(\mathbf{x}'_{b'} \diamond \mathbf{x}''_1)\mathbf{x}''_2 \cdots \mathbf{x}''_{b''}$$

and

$$\begin{aligned}\mathbf{x}(\mathbf{x}' \diamond \mathbf{x}'') &= \mathbf{x}(\mathbf{x}'_1 \cdots \mathbf{x}'_{b'-1}(\mathbf{x}'_{b'} \diamond \mathbf{x}''_1)\mathbf{x}''_2 \cdots \mathbf{x}''_{b''}) \\ &= \mathbf{x}\mathbf{x}'_1 \cdots \mathbf{x}'_{b'-1}(\mathbf{x}'_{b'} \diamond \mathbf{x}''_1)\mathbf{x}''_2 \cdots \mathbf{x}''_{b''}.\end{aligned}$$

This proves the first equation. The proof of the second equation is the same.□

Extending $\diamond$ bilinearly, we obtain a binary operation

$$Ш^{NC}(A) \otimes Ш^{NC}(A) \to Ш^{NC}(A),$$

that is restricted to a binary operation

$$Ш^{NC,0}(A^0) \otimes Ш^{NC,0}(A^0) \to Ш^{NC,0}(A^0).$$

For $\mathbf{x} \in \tilde{\mathcal{X}}_\infty$, define

$$P_A(\mathbf{x}) = \lfloor \mathbf{x} \rfloor. \tag{4.50}$$

Obviously $\lfloor \mathbf{x} \rfloor$ is again in $\tilde{\mathcal{X}}_\infty$ and is in $\mathcal{X}_\infty$ if $\mathbf{x}$ is. Thus P_A extends to a linear operator P_A on $Ш^{NC}(A)$ and $Ш^{NC,0}(A^0)$. Let

$$j_{\tilde{X}} : \tilde{X} = X \cup \{1\} \to \tilde{\mathcal{X}}_\infty \to Ш^{NC}(A)$$

be the natural injection which extends to an algebra injection

$$j_A : A \to Ш^{NC}(A). \tag{4.51}$$

By restriction, we have a nonunitary algebra injection

$$j_{A^0} : A^0 \to Ш^{NC,0}(A^0). \tag{4.52}$$

The following is our main result of the free unitary and nonunitary Rota-Baxter algebra on a **k**-algebra or a nonunitary **k**-algebra. It will be proved in the next subsection.

Theorem 4.4.6. *Let A be a* **k***-algebra satisfying Assumption* 1.
(a) *The pair* $(Ш^{NC}(A), \diamond)$ *is a* **k***-algebra.*
(b) *The triple* $(Ш^{NC}(A), \diamond, P_A)$ *is an* RBA *of weight* λ.

(c) *The quadruple* $(Ш^{NC}(A), \diamond, P_A, j_A)$ *is the free* RBA *of weight* λ *on the algebra* A.
(d) *The pair* $(Ш^{NC,0}(A^0), \diamond)$ *is a nonunitary algebra.*
(e) *The triple* $(Ш^{NC,0}(A^0), \diamond, P_{A^0})$ *is a nonunitary* RBA *of weight* λ.
(f) *The quadruple* $(Ш^{NC,0}(A^0), \diamond, P_{A^0}, j_{A^0})$ *is the free nonunitary* RBA *of weight* λ *on the nonunitary algebra* A^0. *More precisely, for any nonunitary Rota-Baxter algebra* R *and a nonunitary algebra homomorphism* $f : A^0 \to R$, *there is a unique nonunitary Rota-Baxter algebra homomorphism* $\overline{f} : Ш^{NC,0}(A^0) \to R$ *such that* $f = \overline{f} \circ j_{A^0}$.

4.4.3 The proof of Theorem 4.4.6

Overall the proof is similar in part to the proof of Theorem 4.1.7 and in part to the proof of Theorem 4.2.16.

Since the unitary case is discussed in detail in Theorem 4.2.16 for free RBA on a set, we will give details for Items (d) — (f) in the nonunitary case for free BRA on a nonunitary algebra. The proofs of Items (a) — (c) are similar.

(d) We just need to verify the associativity. For this we only need to verify

$$(\mathbf{x}' \diamond \mathbf{x}'') \diamond \mathbf{x}''' = \mathbf{x}' \diamond (\mathbf{x}'' \diamond \mathbf{x}''') \tag{4.53}$$

for $\mathbf{x}', \mathbf{x}'', \mathbf{x}''' \in \mathcal{X}_\infty$. We will do this by induction on the sum of the depths $n := d(\mathbf{x}') + d(\mathbf{x}'') + d(\mathbf{x}''')$. If $n = 0$, then all of $\mathbf{x}', \mathbf{x}'', \mathbf{x}'''$ have depth zero and so are in X. In this case the product $\diamond$ is defined to be the product $\cdot$ in A and so is associative.

Assume the associativity holds for $n \leqslant k$ where $k \geqslant 0$ and consider $\mathbf{x}', \mathbf{x}'', \mathbf{x}'''$ in $\mathcal{X}_\infty$ with $n = d(\mathbf{x}') + d(\mathbf{x}'') + d(\mathbf{x}''') = k + 1$.

If $t(\mathbf{x}') \neq h(\mathbf{x}'')$, then by Lemma 4.4.5, we have

$$(\mathbf{x}' \diamond \mathbf{x}'') \diamond \mathbf{x}''' = (\mathbf{x}'\mathbf{x}'') \diamond \mathbf{x}''' = \mathbf{x}'(\mathbf{x}'' \diamond \mathbf{x}''') = \mathbf{x}' \diamond (\mathbf{x}'' \diamond \mathbf{x}''').$$

A similar conclusion holds if $t(\mathbf{x}'') \neq h(\mathbf{x}''')$.

Thus we only need to verify the associativity when $t(\mathbf{x}') = h(\mathbf{x}'')$ and $t(\mathbf{x}'') = h(\mathbf{x}''')$. We next reduce the breadths of the words.

Lemma 4.4.7. *If the associativity*

$$(\mathbf{x}' \diamond \mathbf{x}'') \diamond \mathbf{x}''' = \mathbf{x}' \diamond (\mathbf{x}'' \diamond \mathbf{x}''')$$

holds for all sRBWs $\mathbf{x}', \mathbf{x}''$ *and* $\mathbf{x}'''$ *of breadth one, then it holds for all sRBWs.*

Proof. We use induction on the sum of breadths $m := b(\mathbf{x}') + b(\mathbf{x}'') + b(\mathbf{x}''')$. Then $m \geqslant 3$. The case when $m = 3$ is the assumption of the lemma. Assume the associativity holds for $3 \leqslant m \leqslant j$ and take $\mathbf{x}', \mathbf{x}'', \mathbf{x}''' \in \mathcal{X}_\infty$ with $m = j+1$. Then $j+1 \geqslant 4$. So at least $\mathbf{x}'$ or $\mathbf{x}''$ or $\mathbf{x}'''$ has breadth greater than or equal to 2.

First assume $b(\mathbf{x}') \geqslant 2$. Then $\mathbf{x}' = \mathbf{x}'_1\mathbf{x}'_2$ with $\mathbf{x}'_1$, $\mathbf{x}'_2 \in \mathcal{X}_\infty$ and $t(\mathbf{x}'_1) \neq h(\mathbf{x}'_2)$. Thus

$$\begin{aligned}(\mathbf{x}' \diamond \mathbf{x}'') \diamond \mathbf{x}''' &= ((\mathbf{x}'_1\mathbf{x}'_2) \diamond \mathbf{x}'') \diamond \mathbf{x}''' \\ &= (\mathbf{x}'_1(\mathbf{x}'_2 \diamond \mathbf{x}'')) \diamond \mathbf{x}''' \quad \text{(by Lemma 4.4.5. (c))} \\ &= \mathbf{x}'_1((\mathbf{x}'_2 \diamond \mathbf{x}'') \diamond \mathbf{x}''') \quad \text{(by Lemma 4.4.5. (a) and (c))}.\end{aligned}$$

Similarly,

$$\begin{aligned}\mathbf{x}' \diamond (\mathbf{x}'' \diamond \mathbf{x}''') &= (\mathbf{x}'_1\mathbf{x}'_2) \diamond (\mathbf{x}'' \diamond \mathbf{x}''') \\ &= \mathbf{x}'_1(\mathbf{x}'_2 \diamond (\mathbf{x}'' \diamond \mathbf{x}''')).\end{aligned}$$

Thus

$$(\mathbf{x}' \diamond \mathbf{x}'') \diamond \mathbf{x}''' = \mathbf{x}' \diamond (\mathbf{x}'' \diamond \mathbf{x}''')$$

whenever

$$(\mathbf{x}'_2 \diamond \mathbf{x}'') \diamond \mathbf{x}''' = \mathbf{x}'_2 \diamond (\mathbf{x}'' \diamond \mathbf{x}''')$$

which follows from the induction hypothesis.

A similar proof works if $b(\mathbf{x}''') \geqslant 2$.

Finally if $b(\mathbf{x}'') \geqslant 2$, then $\mathbf{x}'' = \mathbf{x}''_1\mathbf{x}''_2$ with $\mathbf{x}''_1$, $\mathbf{x}''_2 \in \mathcal{X}_\infty$ and $t(\mathbf{x}''_1) \neq h(\mathbf{x}''_2)$. So using Lemma 4.4.5 repeatedly, we have

$$\begin{aligned}(\mathbf{x}' \diamond \mathbf{x}'') \diamond \mathbf{x}''' &= (\mathbf{x}' \diamond (\mathbf{x}''_1\mathbf{x}''_2)) \diamond \mathbf{x}''' \\ &= ((\mathbf{x}' \diamond \mathbf{x}''_1)\mathbf{x}''_2) \diamond \mathbf{x}''' \quad \text{(by Lemma 4.4.5. (a) and (c))} \\ &= (\mathbf{x}' \diamond \mathbf{x}''_1)(\mathbf{x}''_2 \diamond \mathbf{x}''') \quad \text{(by Lemma 4.4.5. (a) and (c))}.\end{aligned}$$

In the same way, we have

$$(\mathbf{x}' \diamond \mathbf{x}''_1)(\mathbf{x}''_2 \diamond \mathbf{x}''') = \mathbf{x}' \diamond (\mathbf{x}'' \diamond \mathbf{x}''').$$

This again proves the associativity. □

To summarize, our proof of the associativity has been reduced to the special case when $\mathbf{x}', \mathbf{x}'', \mathbf{x}''' \in \mathcal{X}_\infty$ are chosen to satisfy that

(a) $n := d(\mathbf{x}') + d(\mathbf{x}'') + d(\mathbf{x}''') = k+1 \geqslant 1$ with the assumption that the associativity holds when $n \leqslant k$.
(b) the elements are of breadth one and
(c) $t(\mathbf{x}') = h(\mathbf{x}'')$ and $t(\mathbf{x}'') = h(\mathbf{x}''')$.

By Item (b), the head index and tail index of each of the elements are the same. Therefore by Item (c), all the three elements are either in X or in $\lfloor \mathcal{X}_\infty \rfloor$. If all of $\mathbf{x}', \mathbf{x}'', \mathbf{x}'''$ are in X, then as already shown, the associativity follows from the associativity in A^0.

So it remains to consider $\mathbf{x}', \mathbf{x}'', \mathbf{x}'''$ all in $\lfloor \mathcal{X}_\infty \rfloor$. Then $\mathbf{x}' = \lfloor \overline{\mathbf{x}}' \rfloor, \mathbf{x}'' = \lfloor \overline{\mathbf{x}}'' \rfloor, \mathbf{x}''' = \lfloor \overline{\mathbf{x}}''' \rfloor$ with $\overline{\mathbf{x}}', \overline{\mathbf{x}}'', \overline{\mathbf{x}}''' \in \mathcal{X}_\infty$. Using Eq. (4.48), the bilinearity of the product $\diamond$ and the induction hypothesis, we can carry out a computation that is similar to the one in the proof of Lemma 4.1.9 and obtain

$$(\mathbf{x}' \diamond \mathbf{x}'') \diamond \mathbf{x}'' = \mathbf{x}' \diamond (\mathbf{x}'' \diamond \mathbf{x}''').$$

This completes the proof of Item (d) of Theorem 4.4.6.

(e) The proof is immediate from Eq. (4.48).

(f) Let (R, P) be a nonunitary RBA of weight λ. Let $f\colon A^0 \to R$ be a nonunitary $\mathbf{k}$-algebra morphism. We will construct a $\mathbf{k}$-linear map $\overline{f} : Ш^{NC,0}(A^0) \to R$ by defining $\overline{f}(\mathbf{x})$ for $\mathbf{x} \in \mathcal{X}_\infty$. We achieve this by defining $\overline{f}(\mathbf{x})$ for $\mathbf{x} \in \mathcal{X}_n$, $n \geqslant 0$, using induction on n. For $\mathbf{x} \in \mathcal{X}_0 := X$, define $\overline{f}(\mathbf{x}) = f(\mathbf{x})$. Suppose $\overline{f}(\mathbf{x})$ has been defined for $\mathbf{x} \in \mathcal{X}_n$ and consider $\mathbf{x}$ in $\mathcal{X}_{n+1}$ which is, by definition and Eq. (4.46),

$$\begin{aligned}\Lambda(X, \mathcal{X}_n) = &\Big(\bigsqcup_{r \geqslant 1} (X \lfloor \mathcal{X}_n \rfloor)^r \Big) \bigsqcup \Big(\bigsqcup_{r \geqslant 0} (X \lfloor \mathcal{X}_n \rfloor)^r X \Big) \\ &\bigsqcup \Big(\bigsqcup_{r \geqslant 1} (\lfloor \mathcal{X}_n \rfloor X)^r \Big) \bigsqcup \Big(\bigsqcup_{r \geqslant 0} (\lfloor \mathcal{X}_n \rfloor X)^r \lfloor \mathcal{X}_n \rfloor \Big).\end{aligned}$$

Let $\mathbf{x}$ be in the first union component $\bigsqcup_{r \geqslant 1} (X \lfloor \mathcal{X}_n \rfloor)^r$ above. Then

$$\mathbf{x} = \prod_{i=1}^{r} (\mathbf{x}_{2i-1} \lfloor \mathbf{x}_{2i} \rfloor)$$

with $\mathbf{x}_{2i-1} \in X$ and $\mathbf{x}_{2i} \in \mathcal{X}_n$, $1 \leqslant i \leqslant r$. By the construction of the multiplication $\diamond$ and the Rota-Baxter operator P_A, we have

$$\mathbf{x} = \diamond_{i=1}^{r} (\mathbf{x}_{2i-1} \diamond \lfloor \mathbf{x}_{2i} \rfloor) = \diamond_{i=1}^{r} (\mathbf{x}_{2i-1} \diamond P_A(\mathbf{x}_{2i})).$$

Define

$$\overline{f}(\mathbf{x}) = *_{i=1}^{r} \big(\overline{f}(\mathbf{x}_{2i-1}) * P\big(\overline{f}(\mathbf{x}_{2i})\big)\big), \tag{4.54}$$

where the right hand side is well defined by the induction hypothesis. Similarly define $\overline{f}(\mathbf{x})$ if $\mathbf{x}$ is in the other union components. For any $\mathbf{x} \in \mathcal{X}_\infty$, we have $P_A(\mathbf{x}) = \lfloor \mathbf{x} \rfloor \in \mathcal{X}_\infty$, and by the definition of $\overline{f}$ in Eq. (4.54), we have

$$\overline{f}(\lfloor \mathbf{x} \rfloor) = P(\overline{f}(\mathbf{x})). \tag{4.55}$$

So $\overline{f}$ commutes with the Rota-Baxter operators. Combining this equation with Eq. (4.54) we see that if $\mathbf{x} = \mathbf{x}_1 \cdots \mathbf{x}_b$ is the standard decomposition of $\mathbf{x}$, then

$$\overline{f}(\mathbf{x}) = \overline{f}(\mathbf{x}_1) * \cdots * \overline{f}(\mathbf{x}_b). \tag{4.56}$$

Note that this is the only possible way to define $\overline{f}(\mathbf{x})$ in order for $\overline{f}$ to be a Rota-Baxter homomorphism extending f.

It remains to prove that the map $\overline{f}$ defined in Eq. (4.54) is indeed an algebra homomorphism. For this we only need to check to see that

$$\overline{f}(\mathbf{x} \diamond \mathbf{x}') = \overline{f}(\mathbf{x}) * \overline{f}(\mathbf{x}') \tag{4.57}$$

for all $\mathbf{x}, \mathbf{x}' \in \mathcal{X}_\infty$. For this we use induction on the sum of depths $n := d(\mathbf{x}) + d(\mathbf{x}')$. Then $n \geqslant 0$. When $n = 0$, we have $\mathbf{x}, \mathbf{x}' \in X$. Then Eq (4.57) follows from the multiplicativity of f. Assume the multiplicativity holds for $\mathbf{x}, \mathbf{x}' \in \mathcal{X}_\infty$ with $n \geqslant k$ and take $\mathbf{x}, \mathbf{x}' \in \mathcal{X}_\infty$ with $n = k+1$. Let $\mathbf{x} = \mathbf{x}_1 \cdots \mathbf{x}_b$ and $\mathbf{x}' = \mathbf{x}'_1 \cdots \mathbf{x}'_{b'}$ be the standard decompositions. By Eq. (4.48),

$$\begin{aligned}
&\overline{f}(\mathbf{x}_b \diamond \mathbf{x}'_1)\\
&= \begin{cases} \overline{f}(\mathbf{x}_b \mathbf{x}'_1), & \text{if } \mathbf{x}_b \in X, \mathbf{x}'_1 \in \lfloor \tilde{\mathcal{X}}_\infty \rfloor,\\ \overline{f}(\mathbf{x}_b \mathbf{x}'_1), & \text{if } \mathbf{x}_b \in \lfloor \tilde{\mathcal{X}}_\infty \rfloor, \mathbf{x}'_1 \in X,\\ \overline{f}(\lfloor \lfloor \overline{\mathbf{x}}_b \rfloor \diamond \overline{\mathbf{x}}'_1 \rfloor + \lfloor \overline{\mathbf{x}}_b \diamond \lfloor \overline{\mathbf{x}}'_1 \rfloor \rfloor + \lambda \lfloor \overline{\mathbf{x}}_b \diamond \overline{\mathbf{x}}'_1 \rfloor), & \text{if } \mathbf{x}_b = \lfloor \overline{\mathbf{x}}_b \rfloor, \mathbf{x}'_1 = \lfloor \overline{\mathbf{x}}'_1 \rfloor \in \lfloor \tilde{\mathcal{X}}_\infty \rfloor. \end{cases}
\end{aligned}$$

In the first two cases, the right hand side is $\overline{f}(\mathbf{x}_b) * \overline{f}(\mathbf{x}'_1)$ by the definition of $\overline{f}$. In the third case, we have, by Eq. (4.55), the induction hypothesis and the Rota-Baxter relation of P,

$$\begin{aligned}
&\overline{f}(\lfloor \lfloor \overline{\mathbf{x}}_b \rfloor \diamond \overline{\mathbf{x}}'_1 \rfloor + \lfloor \overline{\mathbf{x}}_b \diamond \lfloor \overline{\mathbf{x}}'_1 \rfloor \rfloor + \lambda \lfloor \overline{\mathbf{x}}_b \diamond \overline{\mathbf{x}}'_1 \rfloor)\\
&= \overline{f}(\lfloor \lfloor \overline{\mathbf{x}}_b \rfloor \diamond \overline{\mathbf{x}}'_1 \rfloor) + \overline{f}(\lfloor \overline{\mathbf{x}}_b \diamond \lfloor \overline{\mathbf{x}}'_1 \rfloor \rfloor) + \overline{f}(\lambda \lfloor \overline{\mathbf{x}}_b \diamond \overline{\mathbf{x}}'_1 \rfloor)\\
&= P(\overline{f}(\lfloor \overline{\mathbf{x}}_b \rfloor \diamond \overline{\mathbf{x}}'_1)) + P(\overline{f}(\overline{\mathbf{x}}_b \diamond \lfloor \overline{\mathbf{x}}'_1 \rfloor)) + \lambda P(\overline{f}(\overline{\mathbf{x}}_b \diamond \overline{\mathbf{x}}'_1))\\
&= P(\overline{f}(\lfloor \overline{\mathbf{x}}_b \rfloor) * \overline{f}(\overline{\mathbf{x}}'_1)) + P(\overline{f}(\overline{\mathbf{x}}_b) * \overline{f}(\lfloor \overline{\mathbf{x}}'_1 \rfloor)) + \lambda P(\overline{f}(\overline{\mathbf{x}}_b) * \overline{f}(\overline{\mathbf{x}}'_1))\\
&= P(P(\overline{f}(\overline{\mathbf{x}}_b)) * \overline{f}(\overline{\mathbf{x}}'_1)) + P(\overline{f}(\overline{\mathbf{x}}_b) * P(\overline{f}(\overline{\mathbf{x}}'_1))) + \lambda P(\overline{f}(\overline{\mathbf{x}}_b) * \overline{f}(\overline{\mathbf{x}}'_1))\\
&= P(\overline{f}(\overline{\mathbf{x}}_b)) * P(\overline{f}(\overline{\mathbf{x}}'_1))\\
&= \overline{f}(\mathbf{x}_b) * \overline{f}(\mathbf{x}'_1).
\end{aligned}$$

Therefore $\overline{f}(\mathbf{x}_b \diamond \mathbf{x}'_1) = \overline{f}(\mathbf{x}_b) * \overline{f}(\mathbf{x}'_1)$. Then

$$\begin{aligned}
\overline{f}(\mathbf{x} \diamond \mathbf{x}') &= \overline{f}(\mathbf{x}_1 \cdots \mathbf{x}_{b-1}(\mathbf{x}_b \diamond \mathbf{x}'_1)\mathbf{x}'_2 \cdots \mathbf{x}'_{b'})\\
&= \overline{f}(\mathbf{x}_1) * \cdots * \overline{f}(\mathbf{x}_{b-1}) * \overline{f}(\mathbf{x}_b \diamond \mathbf{x}'_1) * \overline{f}(\mathbf{x}'_2) \cdots \overline{f}(\mathbf{x}'_{b'})\\
&= \overline{f}(\mathbf{x}_1) * \cdots * \overline{f}(\mathbf{x}_{b-1}) * \overline{f}(\mathbf{x}_b) * \overline{f}(\mathbf{x}'_1) * \overline{f}(\mathbf{x}'_2) \cdots \overline{f}(\mathbf{x}'_{b'})
\end{aligned}$$

$$= \overline{f}(\mathbf{x}) * \overline{f}(\mathbf{x}').$$

This is what we need. Now the proofs of Items (d) — (f) in Theorem 4.4.6 are completed.

4.5 Some related literature and problems

This chapter is mainly based on [56, 58, 88]. General constructions of free Rota-Baxter algebras in terms of Motzkin paths, bracketed words and leaf decorated rooted trees can be found in [88] in the framework of operated semigroups. See [70] for the noncommutative variation of Rota's construction of standard Rota-Baxter algebras and noncommutative symmetric functions.

Enumerations and generating functions of bases in free Rota-Baxter algebras have been obtained in [7, 98].

For Rota-Baxter algebra-related structures, free differential Rota-Baxter algebras were constructed in [94] and their enumerations and generating functions were studied in [99]. Free Nijenhuis algebras and free TD-algebras have been obtained in [125] and [178] respectively. Free objects on Rota-Baxter type algebras are discussed in [100].

Free Rota-Baxter algebras give Rota-Baxter operators in a large space. See [10, 46, 129] for Rota-Baxter operators on low dimensional spaces. See [95] for Rota-Baxter operators on generalized power series rings.

Part III

The Operad Aspect of Rota-Baxter Algebra

Chapter 5

Rota-Baxter Algebras and Dendriform Algebras

It is well known that the relationship between associative algebras and Lie algebras plays a fundamental role in the study of these algebraic structures and their applications. Central to this relationship are the classical theorems of Cartier-Milnor-Moore-Quillen and of Poincaré-Birkhoff-Witt. Similar connections have been recently studied for some other algebraic structures [35, 154, 139].

This chapter explores the relationship between Rota-Baxter algebras and dendriform algebras and tridendriform algebras by applying the free nonunitary Rota-Baxter algebras constructed in Chapter 4, Section 4.4 in particular.

A dendriform algebra is a module with two binary operations with certain associativity relations in Eqs. (5.1) – (5.3). Similarly a tridendriform algebra is a module with three binary operations with certain associativity relations in Eqs. (5.4) – (5.10). Free commutative dendriform (resp. tridendriform) algebras are given by shuffle (resp. quasi-shuffle) product algebras. As we have seen in Chapter 3, the latter algebras are also the main components of free commutative Rota-Baxter algebras.

The first direct link between Rota-Baxter algebras and dendriform algebras was given by Aguiar [2] who showed that a Rota-Baxter algebra of weight $\lambda = 0$ carries a dendriform algebra structure. This has been extended to further connections between linear operators and dendriform type algebras [6, 47, 127, 52]. Consequently, there are natural functors from the category of Rota-Baxter algebras of weight λ to the categories of dendriform algebras and tridendriform algebras. One of our main goals in this chapter is to study the adjoint functors. For this purpose an essential role is played by the free Rota-Baxter algebras that we constructed in Chapter 4. This is in analogy to the role played by free associative algebras in the study of the adjoint functor from the category of Lie algebras to the category of associative algebras.

We define the concepts of a dendriform algebra and a tridendriform algebra in Section 5.1, together with the first links from Rota-Baxter algebras to these algebras. Further connections between (tri-)dendriform algebra and

Rota-Baxter algebra are established in Section 5.2 through the constructions of their free objects in the commutative case in terms of shuffle algebras and quasi-shuffle algebras. The functors between the categories of Rota-Baxter algebras and the categories of dendriform and tridendriform algebras are studied in Section 5.3. There we make use of free Rota-Baxter algebras to obtain adjoint functors of the functors from Rota-Baxter algebras to dendriform algebras and tridendriform algebras. The special cases of free dendriform and tridendriform algebras are considered in Section 5.4 where we realize the free dendriform algebra and tridendriform algebra of Loday and Loday-Ronco in terms of decorated planar rooted trees as canonical subalgebras of free nonunitary Rota-Baxter algebras.

In this chapter, $\mathbf{k}$ is a commutative unitary ring which will be further assumed to be a field in Section 5.3 and Section 5.4. Let Alg be the category of unitary $\mathbf{k}$-algebras A whose unit is identified with the unit $\mathbf{1}$ of $\mathbf{k}$ by the structure homomorphism $\mathbf{k} \to A$. Let Alg^0 be the category of nonunitary $\mathbf{k}$-algebras. Similarly let RB_λ (resp. RB_λ^0) be the category of unitary (resp. nonunitary) Rota-Baxter $\mathbf{k}$-algebras of weight λ. The subscript λ will be suppressed if there is no danger of confusion. Let DD and DT be the category of dendriform $\mathbf{k}$-algebras and tridendriform $\mathbf{k}$-algebras respectively.

5.1 Concepts and basic properties

We first introduce the concepts of a dendriform algebra and a tridendriform algebra. We then set up the first connections from Rota-Baxter algebra to dendriform and tridendriform algebras.

5.1.1 Dendriform algebras and tridendriform algebras

The concept of a dendriform algebra was introduced by Loday [131] in 1995 with motivation from algebraic K-theory.

Definition 5.1.1. ([131]) A *dendriform* $\mathbf{k}$*-algebra* (previously also called a dendriform dialgebra) is a $\mathbf{k}$-module D with two binary operations $\prec$ and $\succ$ that satisfy the following relations.

$$(x \prec y) \prec z = x \prec (y \prec z + y \succ z), \tag{5.1}$$

$$(x \succ y) \prec z = x \succ (y \prec z), \quad x, y, z \in D, \tag{5.2}$$

$$(x \prec y + x \succ y) \succ z = x \succ (y \succ z). \tag{5.3}$$

Dendriform algebras have been further studied with connections to several areas in mathematics and physics, including operads [132], homological algebra [77, 78], Hopf algebra [35, 115, 138, 154], Lie and Leibniz algebra, combinatorics [8, 9, 76, 137], arithmetic [133] and quantum field theory [76].

A few years later Loday and Ronco defined tridendriform algebras in their study [138] of polytopes and Koszul duality.

Definition 5.1.2. ([138]) A *tridendriform* **k**-*algebra* (previously also called a dendriform trialgebra) is a **k**-module T equipped with three binary operations $\prec, \succ$ and $\cdot$ that satisfy the following relations.

$$(x \prec y) \prec z = x \prec (y \star z), \tag{5.4}$$
$$(x \succ y) \prec z = x \succ (y \prec z), \tag{5.5}$$
$$(x \star y) \succ z = x \succ (y \succ z), \tag{5.6}$$
$$(x \succ y) \cdot z = x \succ (y \cdot z), \quad x, y, z \in T, \tag{5.7}$$
$$(x \prec y) \cdot z = x \cdot (y \succ z), \tag{5.8}$$
$$(x \cdot y) \prec z = x \cdot (y \prec z), \tag{5.9}$$
$$(x \cdot y) \cdot z = x \cdot (y \cdot z). \tag{5.10}$$

Here we have used the notation

$$\star = \prec + \succ + \cdot\,. \tag{5.11}$$

We will see in later sections that free objects of dendriform algebras and tridendriform algebras are Hopf algebras and have interpretations as shuffles, quasi-shuffles and trees.

Proposition 5.1.3. (a) *Let* $(D, \prec, \succ)$ *be a dendriform algebra. The operation* $\star := \star_D$ *on* D *defined by*

$$x \star y := x \prec y + x \succ y, \quad x, y \in D, \tag{5.12}$$

is associative.

(b) *Let* $(T, \prec, \succ, \cdot)$ *be a tridendriform algebra. The operation* $\star := \star_T$ *on* T *defined by*

$$x \star y := x \prec y + x \succ y + x \cdot y, \quad x, y \in T, \tag{5.13}$$

is associative.

Thus a dendriform algebra and tridendriform algebra share the property that the sum of the binary operations $\star := \prec + \succ$ for a dendriform algebra or $\star := \prec + \succ + \cdot$ for a tridendriform algebra is associative. Such a property is called a "splitting the associativity". See Chapter 6 for further discussions of this property.

Proof. We just prove Item (a). The proof of Item (b) is similar.

Adding the left hand sides of Eqs. (5.1) – (5.3), we obtain

$$\begin{aligned}
&(x \prec y) \prec z + (x \succ y) \prec z + (x \prec y + x \succ y) \succ z \\
&= (x \prec y + x \succ y) \prec z + (x \prec y + x \succ y) \succ z \\
&= (x \star y) \prec z + (x \star y) \succ z \\
&= (x \star y) \star z.
\end{aligned}$$

Similarly, adding the right hand sides of these equations, we obtain $x \star (y \star z)$. Thus we have proved the associativity of $\star$. □

5.1.2 From Rota-Baxter algebras to dendriform algebras

Theorem 5.1.4. (a) [3] *A Rota-Baxter algebra* (R, P) *of weight zero defines a dendriform algebra* $(R, \prec_P, \succ_P)$, *where*

$$x \prec_P y = xP(y), \ x \succ_P y = P(x)y, \quad \forall x, y \in R. \tag{5.14}$$

(b) [47] *A Rota-Baxter algebra* (R, P) *of weight* λ *defines a tridendriform algebra* $(R, \prec_P, \succ_P, \cdot_P)$, *where*

$$x \prec_P y = xP(y), \ x \succ_P y = P(x)y, \ x \cdot_P y = \lambda xy, \quad \forall x, y \in R. \tag{5.15}$$

(c) [47] *A Rota-Baxter algebra* (R, P) *of weight* λ *defines a dendriform algebra* $(R, \prec'_P, \succ'_P)$, *where*

$$x \prec'_P y = xP(y) + \lambda xy, \ x \succ'_P y = P(x)y, \quad \forall x, y \in R. \tag{5.16}$$

Proof. We just verify Item (a). Items (b) and (c) can be verified in the same way. To verify the three relations in Eqs. (5.1) – (5.3), we have

$$\begin{aligned}
(x \prec_P y) \prec_P z &= xP(y)P(z) \\
&= xP(yP(z) + P(y)z) \\
&= x \prec_P (y \prec_P z + y \succ_P z). \\
(x \succ_P y) \prec_P z &= (P(x)y)P(z) \\
&= P(x)(yP(z)) \\
&= x \succ_P (y \prec_P z). \\
(x \prec_P y + x \succ_P y) \succ_P z &= P(xP(y) + P(x)y)z \\
&= P(x)P(y)z \\
&= x \succ_P (y \succ_P z).
\end{aligned}$$

□

We note that Item (c) specializes in Item (a) when $\lambda = 0$. The same can be said of Item (b) since when $\lambda = 0$, the product $\cdot_P$ is zero and the relations of the tridendriform algebra reduce to the relations of a dendriform algebra.

Thus we obtain functors

$$\mathcal{E} : \mathrm{RB}_\lambda^0 \to \mathrm{DT}, \ \mathcal{F} : \mathrm{RB}_\lambda^0 \to \mathrm{DD}.$$

We will study their adjoint functors. The two functors $\mathcal{E}$ and $\mathcal{F}$ are related by the following simple observation.

Lemma 5.1.5. *Let $(D, \prec, \succ, \cdot)$ be in* DT. *Then $(D, \prec', \succ')$ is in* DD. *Here $\prec' = \prec + \cdot$ and $\succ' = \succ$.*

Proof. Let $\star' = \prec' + \succ$. Then we have $\star' = \star$. We have

$$\begin{aligned}
(a \prec' b) \prec' c &= (a \cdot b + a \prec b) \prec' c \\
&= (a \cdot b + a \prec b) \cdot c + (a \cdot b + a \prec b) \prec c \\
&= (a \cdot b) \cdot c + (a \prec b) \cdot c + (a \cdot b) \prec c + (a \prec b) \prec c \\
&= a \cdot (b \cdot c) + a \cdot (b \succ c) + a \cdot (b \prec c) + a \prec (b \star c) \\
&\qquad\qquad\qquad \text{(by Eq. (5.7))} \\
&= a \prec' (b \star' c).
\end{aligned}$$

This verifies the first relation for the dendriform algebra. The other two relations are also easy to verify:

$$a \succ' (b \succ' c) = a \succ (b \succ c) = (a \star b) \succ c = (a \star' b) \succ' c.$$

$$(a \succ' b) \prec' c = (a \succ b) \cdot c + (a \succ b) \prec c = a \succ (b \cdot c) + a \succ (b \prec c) = a \succ' (b \prec' c).$$

□

Let $\mathcal{G} : \mathrm{DT} \to \mathrm{DD}$ be the functor obtained from Lemma 5.1.5. Then we have $\mathcal{F} = \mathcal{G} \circ \mathcal{E}$. So we have the following commutative diagram of functors where $\lambda \neq 0$.

$$\begin{array}{ccc}
\mathrm{RB}_\lambda & \xrightarrow{\ \mathcal{E}\ } & \mathrm{DT} \\
& \searrow^{\mathcal{F}} & \downarrow{\scriptstyle \mathcal{G}} \\
& & \mathrm{DD}
\end{array} \tag{5.17}$$

5.2 Free objects in the commutative case and shuffle products

In this section we study the relationship between free commutative dendriform algebras and tridendriform algebras on one hand, and commutative Rota-

Baxter algebras of weight zero and non-zero weight on the other hand, with the connection made by the shuffle product and quasi-shuffle product.

Definition 5.2.1. (a) A dendriform algebra $(D, \prec, \succ)$ is called *commutative* if

$$x \succ y = y \prec x, \quad x, y \in D.$$

(b) A tridendriform algebra $(T, \prec, \succ, \cdot)$ is called *commutative* if

$$x \succ y = y \prec x, \quad x \cdot y = y \cdot x, \quad x, y \in T.$$

Proposition 5.2.2. (a) *If a dendriform algebra $(D, \prec, \succ)$ is commutative, then $\prec$ satisfies the relation*

$$(x \prec y) \prec z = x \prec (y \prec z + z \prec y). \tag{5.18}$$

Conversely, given a $(D, \prec)$ where D is a **k***-module and $\prec$ is binary operation on D that satisfies Eq.* (5.18)*, then the triple $(D, \prec, \succ)$, where $x \succ y := y \prec x$, is a commutative dendriform algebra.*

(b) *If a tridendriform algebra $(T, \prec, \succ, \cdot)$ is commutative, then the binary operations $\prec$ and $\cdot$ satisfy the relations*

$$x \cdot y = y \cdot x, \tag{5.19}$$
$$(x \prec y) \prec z = x \prec (y \prec z + z \prec y + y \cdot z), \tag{5.20}$$
$$(x \cdot y) \prec z = x \cdot (y \prec z), \tag{5.21}$$
$$(x \cdot y) \cdot z = x \cdot (y \cdot z). \tag{5.22}$$

Conversely, given a triple $(T, \prec, \cdot)$ where T is a **k***-module and $\prec, \cdot$ are binary operations on T that satisfy the relations in Eqs.* (5.20)–(5.22)*, then the quadruple $(T, \prec, \succ, \cdot)$, where $x \succ y := y \prec x$, is a commutative tridendriform algebra.*

Proof. (a) The first statement is clear. To prove the converse, suppose a binary operation $\prec$ on D satisfies Eq. (5.18) and define $x \succ y := y \prec x$ for $x, y \in D$. Then Eq. (5.18) is Eq. (5.1). Also

$$\begin{aligned}(x \succ y) \prec z &= (y \prec x) \prec z \\ &= y \prec (x \prec z + z \prec x) \\ &= y \prec (z \succ x + z \prec x) \\ &= (y \prec z) \prec x \\ &= x \succ (y \prec z),\end{aligned}$$

proving Eq. (5.2). Further,

$$\begin{aligned}(x \prec y + x \succ y) \succ z &= z \prec (x \prec y + y \prec x)\\ &= (z \prec y) \prec x\\ &= x \succ (y \succ z),\end{aligned}$$

proving Eq. (5.3).

(b) The first statement is again clear. For the converse, consider a triple $(T, \prec, \cdot)$ that satisfies Eqs. (5.19) – (5.22) and define $x \succ y = y \prec x$. We just need to derive Eqs. (5.5) – (5.8). The proofs of Eqs. (5.5) and (5.6) are similar to the proofs of Eqs. (5.2) and (5.3) in Item (a) respectively. We also have

$$\begin{aligned}(x \succ y) \cdot z &= (y \prec x) \cdot z\\ &= z \cdot (y \prec x)\\ &= (z \cdot y) \prec x\\ &= x \succ (z \cdot y)\\ &= x \succ (y \cdot z),\end{aligned}$$

proving Eq. (5.7). From this, we then have

$$(x \prec y) \cdot z = (y \succ x) \cdot z = y \succ (x \cdot z) = y \succ (z \cdot x) = (y \succ z) \cdot x = x \cdot (y \succ z),$$

proving Eq. (5.8). □

Let V be a **k**-module. Denote

$$T(V) = \bigoplus_{k \geqslant 0} V^{\otimes k}, \quad T^+(V) = \bigoplus_{k \geqslant 1} V^{\otimes k}.$$

Recall from Section 2.2.2 and Section 3.1.1 that the shuffle product ш on $T(V)$ is defined by Eq. (3.5):

$$\mathfrak{a} ш \mathfrak{b} = a_1 \otimes (\mathfrak{a}' ш \mathfrak{b}) + b_1 \otimes (\mathfrak{a} ш \mathfrak{b}'),$$

where

$$\mathfrak{a} = a_1 \otimes \cdots \otimes a_m = a_1 \otimes \mathfrak{a}' \in V^{\otimes m}, \quad \mathfrak{b} = b_1 \otimes \cdots \otimes b_n = b_1 \otimes \mathfrak{b}' \in V^{\otimes n}.$$

This product is restricted to an associative product on $T^+(V)$.

On the other hand, let A be a nonunitary **k**-algebra. Denote

$$T(A) = \bigoplus_{k \geqslant 0} A^{\otimes k}, \quad T^+(A) = \bigoplus_{k \geqslant 1} A^{\otimes k}.$$

Recall from Section 3.1.2.2 that the quasi-shuffle product $*_\lambda$ of weight λ on $T(A)$ is defined by Eq. (3.9):

$$\mathfrak{a} *_\lambda \mathfrak{b} = a_1 \otimes (\mathfrak{a}' *_\lambda \mathfrak{b}) + b_1 \otimes (\mathfrak{a} *_\lambda \mathfrak{b}') + \lambda(a_1 b_1) \otimes (\mathfrak{a}' *_\lambda \mathfrak{b}').$$

This product is restricted to an associative product $*_\lambda$ on $T^+(A)$.

We also use the algebra A to define a relative version of tridendriform algebras.

Definition 5.2.3. (a) Let V be a $\mathbf{k}$-module. The *symmetric algebra* over V is

$$S(V) := \bigoplus_{n \geqslant 0} S^n(V) \tag{5.23}$$

with the concatenation product. Here $S^n(V)$ denotes the quotient of $V^{\otimes n}$ modulo the action of the symmetric group. Also let $S^+(V)$ denote the nonunitary subalgebra $\bigoplus_{n\geqslant 1} S^n(V)$ of $S(V)$.

(b) Let A be a $\mathbf{k}$-algebra. A tridendriform $\mathbf{k}$-algebra $(T, \prec, \succ, \cdot)$ is called a *tridendriform algebra over* A if there is an algebra homomorphism $A \to (T, \cdot)$.

We note that $S(V)$ is isomorphic to the quotient algebra of the tensor algebra $T(V)$ modulo the commutator relation. As a result, $S(V)$ (resp. $\bar{S}(V)$) is the free commutative unitary (resp. nonunitary) $\mathbf{k}$-algebra generated by V. When V is the free $\mathbf{k}$-module with a basis X, $S(V)$ is naturally isomorphic to $\mathbf{k}[X]$.

Theorem 5.2.4. (a) *In the shuffle product algebra* $(T^+(V), ш)$ *define operations* $\prec$ *and* $\succ$ *by*

$$\mathfrak{a} \prec_V \mathfrak{b} := a_1 \otimes (\mathfrak{a}' ш \mathfrak{b}), \ \ \mathfrak{a} \succ_V \mathfrak{b} := b_1 \otimes (\mathfrak{a} ш \mathfrak{b}'), \tag{5.24}$$

for pure tensors $\mathfrak{a} = a_1 \otimes \mathfrak{a}'$, $\mathfrak{b} = b_1 \otimes \mathfrak{b}'$. *Then* $(T^+(V), \prec_V, \succ_V)$ *is the free commutative dendriform algebra on* V.

(b) *Let* A *be a nonunitary* $\mathbf{k}$*-algebra. In the quasi-shuffle product algebra* $(T^+(A), *)$ *define operations* $\prec_A, \succ_A, \cdot_A$ *by*

$$\mathfrak{a} \prec_A \mathfrak{b} := a_1 \otimes (\mathfrak{a}' * \mathfrak{b}), \ \ \mathfrak{a} \succ_A \mathfrak{b} := b_1 \otimes (\mathfrak{a} * \mathfrak{b}'),$$
$$\mathfrak{a} \cdot_A \mathfrak{b} := (a_1 b_1) \otimes (\mathfrak{a}' * \mathfrak{b}') \tag{5.25}$$

for pure tensors $\mathfrak{a} = a_1 \otimes \mathfrak{a}'$ *and* $\mathfrak{b} = b_1 \otimes \mathfrak{b}'$. *Then* $(T^+(A), \prec_A, \succ_A, \cdot_A)$, *together with the natural nonunitary algebra homomorphism* $i_A : A \to T^+(A)$, *is the free commutative tridendriform algebra on* A. *More precisely, we have the following universal property:*

For any tridendriform algebra $(T, \prec, \succ, \cdot)$ and nonunitary **k***-algebra homomorphism $f : A \to T$, there is unique homomorphism of tridendriform algebras $\bar{f} : T^+(A) \to T$ such that $\bar{f} \circ i_A = f$.*

(c) *Let V be a* **k***-module. The tridendriform algebra $(T^+(S^+(V)), \prec_{S^+(V)}, \succ_{S^+(V)}, \cdot_{S^+(V)})$ is the free commutative tridendriform algebra on V.*

Proof. We will only prove Items (b) and (c). The proof of Item (a) is similar and is left as an exercise.

(b) We first verify that we do have a tridendriform algebra from Eq. (5.25). Instead of checking the axioms directly, as in [136], we apply the relationship between quasi-shuffle algebra and free commutative Rota-Baxter algebra.

Recall from Section 3.2 that the free commutative Rota-Baxter algebra $Ш(A)$ of weight 1 is defined by

$$Ш(A) := Ш_{\mathbf{k},1}(A) := A \otimes Ш^+_{\mathbf{k}}(A) = \bigoplus_{k \geqslant 1} A^{\otimes k}, \tag{5.26}$$

namely, the tensor product algebra of the algebra A with the mixable shuffle (i.e., quasi-shuffle) algebra $(Ш^+_{\mathbf{k}}(A), ш_1)$. The multiplication in this tensor algebra denoted by $\diamond$ is usually suppressed. Thus we have

$$(a_0 \otimes \mathfrak{a}) \diamond (b_0 \otimes \mathfrak{b}) := (a_0 b_0) \otimes (\mathfrak{a} ш_\lambda \mathfrak{b}), \quad a_0,\ b_0 \in A,\ \mathfrak{a},\ \mathfrak{b} \in Ш^+(A). \tag{5.27}$$

Also recall that the Rota-Baxter operator P_A on $Ш(A)$ is defined by $P_A(\mathfrak{a}) = 1 \otimes \mathfrak{a}$ for $\mathfrak{a} \in Ш(A)$.

Then for the operations defined in Eq. (5.25) we have

$$\mathfrak{a} \prec_A \mathfrak{b} = a_1 \otimes (\mathfrak{a}' * \mathfrak{b}) = \mathfrak{a} \diamond P_A(\mathfrak{b}), \tag{5.28}$$

$$\mathfrak{a} \succ_A \mathfrak{b} = b_1 \otimes (\mathfrak{a} * \mathfrak{b}') = P_A(\mathfrak{a}) \diamond \mathfrak{b}, \tag{5.29}$$

$$\mathfrak{a} \cdot_A \mathfrak{b} = (a_1 b_1) \otimes (\mathfrak{a}' * \mathfrak{b}') = \mathfrak{a} \diamond \mathfrak{b}. \tag{5.30}$$

We therefore conclude that $(Ш(A), \prec, \succ, \cdot)$ is a tridendriform algebra by Theorem 5.1.4. (b).

We next prove that $(T^+(A), \prec, \succ, \cdot)$ is the free commutative tridendriform algebra over A by verifying its universal property. Let $(T, \prec, \succ, \cdot)$ be a tridendriform algebra over A and let $f : A \to T$ be a nonunitary algebra homomorphism. We note that for any pure tensor $\mathfrak{a} := a_1 \otimes \cdots \otimes a_n \in A^{\otimes n} \subseteq T^+(A)$, rewriting $\mathfrak{a} = a_1 \otimes \mathfrak{a}'$ with $\mathfrak{a}' = a_2 \otimes \cdots \otimes a_n$, we have

$$\mathfrak{a} = a_1 \prec (a_2 \prec \cdots) = a_1 \prec \mathfrak{a}'.$$

Thus in order for a map $\bar{f} : T^+(A) \to T$ to be a homomorphism of tridendriform algebras, we must have

$$\bar{f}(\mathfrak{a}) = \bar{f}(a_1 \otimes \mathfrak{a}') = \bar{f}(a_1 \prec_A \mathfrak{a}') = \bar{f}(a_1) \prec \bar{f}(\mathfrak{a}') = f(a_1) \prec \bar{f}(\mathfrak{a}'). \tag{5.31}$$

This recursion uniquely determines $\bar{f}$. It remains to show that $\bar{f}$ is indeed a homomorphism of tridendriform algebras over A. So for pure tensors $\mathfrak{a}$ and $\mathfrak{b}$ in $T^+(A)$, we need to verify

$$\bar{f}(\mathfrak{a} \prec_A \mathfrak{b}) = \bar{f}(\mathfrak{a}) \prec \bar{f}(\mathfrak{b}), \quad \bar{f}(\mathfrak{a} \cdot_A \mathfrak{b}) = \bar{f}(\mathfrak{a}) \cdot \bar{f}(\mathfrak{b}). \tag{5.32}$$

Let $\mathfrak{a} \in A^{\otimes m}, \mathfrak{b} \in A^{\otimes n}, m, n \geqslant 1$. We verify Eq. (5.32) by induction on $m+n$. When $m+n=2$, we have $m=n=1$. So $\mathfrak{a}=a$ and $\mathfrak{b}=b$ are in A. Then

$$\begin{aligned}\bar{f}(\mathfrak{a} \prec \mathfrak{b}) &= \bar{f}(a \otimes b)\\ &= f(a) \prec f(b) \quad \text{(by Eq. (5.31))}\\ &= \bar{f}(\mathfrak{a}) \prec \bar{f}(\mathfrak{b}),\end{aligned}$$

and

$$\begin{aligned}\bar{f}(\mathfrak{a} \cdot_A \mathfrak{b}) &= \bar{f}(ab)\\ &= f(ab)\\ &= f(a) \cdot f(b).\end{aligned}$$

Assume that Eq. (5.32) has been verified for $m+n=k$ where $k \geqslant 2$. Then since

$$\mathfrak{a} * \mathfrak{b} = \mathfrak{a} \prec_A \mathfrak{b} + \mathfrak{b} \prec_A \mathfrak{a} + \mathfrak{a} \cdot_A \mathfrak{b},$$

and

$$u \star v := u \prec v + v \prec u + u \cdot v, \quad \forall u, v \in T,$$

we also have

$$\begin{aligned}&\bar{f}(\mathfrak{a} * \mathfrak{b})\\ &= \bar{f}(\mathfrak{a} \prec_A \mathfrak{b} + \mathfrak{b} \prec_A \mathfrak{a} + \mathfrak{a} \cdot_A \mathfrak{b})\\ &= \bar{f}(\mathfrak{a}) \prec \bar{f}(\mathfrak{b}) + \bar{f}(\mathfrak{b}) \prec \bar{f}(\mathfrak{a}) + \bar{f}(\mathfrak{a}) \cdot \bar{f}(\mathfrak{b})\\ &= \bar{f}(\mathfrak{a}) \star \bar{f}(\mathfrak{b}).\end{aligned} \tag{5.33}$$

Next consider $\mathfrak{a} \in A^{\otimes m}$ and $\mathfrak{b} \in A^{\otimes n}$ with $m+n=k+1$. With the notations $\mathfrak{a} = a_1 \otimes \mathfrak{a}'$ and $\mathfrak{b} = b_1 \otimes \mathfrak{b}'$, we have

$$\begin{aligned}\bar{f}(\mathfrak{a} \prec_A \mathfrak{b}) &= \bar{f}(a_1 \otimes (\mathfrak{a}' * \mathfrak{b})) \quad \text{(by Eq. (5.28))}\\ &= \bar{f}(a_1) \prec \bar{f}(\mathfrak{a}' * \mathfrak{b}) \quad \text{(by Eq. (5.31))}\\ &= \bar{f}(a_1) \prec (\bar{f}(\mathfrak{a}') \star \bar{f}(\mathfrak{b})) \quad \text{(by induction hypothesis)}\\ &= (\bar{f}(a_1) \prec \bar{f}(\mathfrak{a}')) \prec \bar{f}(\mathfrak{b}) \quad \text{(by Eq. (5.20))}\\ &= \bar{f}(\mathfrak{a}) \prec \bar{f}(\mathfrak{b}). \quad \text{(by Eq. (5.31))},\end{aligned}$$

and also,

$$\begin{aligned}
\bar{f}(\mathfrak{a} \cdot_A \mathfrak{b}) &= \bar{f}((a_1b_1) \otimes (\mathfrak{a}' * \mathfrak{b}')) \quad \text{(by Eq. (5.30))}\\
&= \bar{f}(a_1b_1) \prec \bar{f}(\mathfrak{a}' * \mathfrak{b}') \quad \text{(by Eq. (5.31))}\\
&= (f(a_1) \cdot f(b_1)) \prec (\bar{f}(\mathfrak{a}') \star \bar{f}(\mathfrak{b}')) \quad \text{(by Eq. (5.33))}\\
&= ((f(a_1) \cdot f(b_1)) \prec \bar{f}(\mathfrak{a}')) \prec \bar{f}(\mathfrak{b}') \quad \text{(by Eq. (5.20))}\\
&= ((f(b_1) \cdot f(a_1)) \prec \bar{f}(\mathfrak{a}')) \prec \bar{f}(\mathfrak{b}') \quad \text{(by Eq. (5.19))}\\
&= (f(b_1) \cdot (f(a_1) \prec \bar{f}(\mathfrak{a}'))) \prec \bar{f}(\mathfrak{b}') \quad \text{(by Eq. (5.21))}\\
&= ((f(a_1) \prec \bar{f}(\mathfrak{a}')) \cdot f(b_1)) \prec \bar{f}(\mathfrak{b}') \quad \text{(by Eq. (5.19))}\\
&= (f(a_1) \prec \bar{f}(\mathfrak{a}')) \cdot (f(b_1) \prec \bar{f}(\mathfrak{b}')) \quad \text{(by Eq. (5.21))}\\
&= \bar{f}(\mathfrak{a}) \cdot_A \bar{f}(\mathfrak{b}). \quad \text{(by Eq. (5.31))}
\end{aligned}$$

This completes the induction and hence the proof of Item (b).

(c). The forgetful functor from the category of commutative tridendriform algebras $(T, \prec, \succ, \cdot)$ to the category of **k**-modules T is the composition of the forgetful functor from the category of commutative tridendriform algebras $(T, \prec, \succ, \cdot)$ to the category of commutative nonunitary **k**-algebras $(T, \cdot)$ and the forgetful functor from the category of commutative nonunitary **k**-algebras to the category of **k**-modules. Further the adjoint (free) functor of the forgetful functor from the category of commutative nonunitary **k**-algebras to the category of **k**-modules is the functor that assigns a **k**-module V to the nonunitary symmetric algebra $S^+(V)$. Then Item (c) follows from Item (b). □

5.3 Universal enveloping algebras of tridendriform algebras

In analogy to the adjoint functor of the functor from the category of associative algebras to the category of Lie algebras, we naturally give the following definition.

Definition 5.3.1. Let $D \in$ DT (resp. DD) and let $\lambda \in \mathbf{k}$. A *universal enveloping Rota-Baxter algebra* of weight λ of D is a Rota-Baxter algebra $U_{\mathrm{RB}}(D) := U_{\mathrm{RB},\lambda}(D) \in \mathrm{RB}_\lambda^0$ with a morphism $\rho : D \to U_{\mathrm{RB}}(D)$ in DT (resp. DD) such that for any $R \in \mathrm{RB}_\lambda^0$ and morphism $f : D \to R$ in DT (resp. DD), there is a unique $\check{f} : U_{\mathrm{RB}}(D) \to R$ in RB_λ^0 such that $\check{f} \circ \rho = f$.

By the universal property of $U_{\mathrm{RB}}(D)$, it is unique up to isomorphisms in RB_λ^0. We will separately consider the enveloping algebras for dendriform algebras and tridendriform algebras.

5.3.1 The tridendriform algebra case

Let $D = (D, \prec, \succ, \cdot) \in \mathrm{DT}$. Then $(D, \cdot)$ is a nonunitary $\mathbf{k}$-algebra. Let $\lambda \in \mathbf{k}$ be given. Let $\mathrm{III}^{NC,0}(D) := \mathrm{III}^{NC,0}{}_{\lambda}(D)$ be the free Rota-Baxter algebra on D of weight λ constructed in Section 4.4.2. Identify D as a subalgebra of $\mathrm{III}^{NC,0}(D)$ by the natural injection j_D in Eq. (4.52). Let I_{RB} be the Rota-Baxter ideal of $\mathrm{III}^{NC,0}(D)$ generated by the set

$$\big\{x \prec y - x\lfloor y\rfloor,\ x \succ y - \lfloor x\rfloor y \ \big|\ x, y \in D\big\}. \qquad (5.34)$$

Let $\pi : \mathrm{III}^{NC,0}(D) \to \mathrm{III}^{NC,0}(D)/I_{\mathrm{RB}}$ be the quotient map.

Theorem 5.3.2. *Let $(D, \prec, \succ, \cdot)$ be a tridendriform algebra. The quotient Rota-Baxter algebra $\mathrm{III}^{NC,0}(D)/I_{\mathrm{RB}}$, together with $\rho := \pi \circ j_D$, is the universal enveloping Rota-Baxter algebra of D.*

Proof. Let $(R, P) \in \mathrm{RB}^0_\lambda$ be given. It gives an object in DT by Theorem 5.1.4 which we still denote by R. Let $f : D \to R$ be a morphism in DT. We will complete the following commutative diagram

$$\begin{array}{ccc} D & \xrightarrow{\ j_D\ } & \mathrm{III}^{NC,0}(D) \\ {\scriptstyle f}\downarrow & \overset{\bar f}{\swarrow} & \downarrow{\scriptstyle \pi} \\ R & \xleftarrow{\ \check f\ } & \mathrm{III}^{NC,0}(D)/I_{\mathrm{RB}} \end{array} \qquad (5.35)$$

By the freeness of $\mathrm{III}^{NC,0}(D)$, there is a morphism $\bar f : \mathrm{III}^{NC,0}(D) \to R$ in RB^0 such that the upper left triangle commutes. So for any $x, y \in D$, we have

$$\begin{aligned} \bar f(x \prec y - x\lfloor y\rfloor) &= \bar f(x \prec y) - \bar f(x)P(\bar f(y)) \\ &= f(x \prec y) - f(x)P(f(y)) \\ &= f(x \prec y) - f(x) \prec_P f(y) \\ &= f(x \prec y) - f(x \prec y) = 0. \end{aligned}$$

Therefore, $x \prec y - x\lfloor y\rfloor$ is in $\ker(\bar f)$. Similarly, $x \succ y - \lfloor x\rfloor y$ is in $\ker(\bar f)$. Thus I_{RB} is in $\ker(\bar f)$ and there is a morphism $\check f : \mathrm{III}^{NC,0}(D)/I_{\mathrm{RB}} \to R$ in RB^0 such that $\bar f = \check f \circ \pi$. Then

$$\check f \circ \rho = \check f \circ \pi \circ j_D = \bar f \circ j_D = f.$$

This proves the existence of $\check f$.

Suppose $\check f' : \mathrm{III}^{NC,0}(D)/I_{\mathrm{RB}} \to A$ is a morphism in RB^0 such that $\check f' \circ \rho = f$. Then

$$(\check f' \circ \pi) \circ j_D = f = (\check f \circ \pi) \circ j_D.$$

By the universal property of the free Rota-Baxter algebra $Ш^{NC,0}(D)$ on D, we have $\check{f}' \circ \pi = \check{f} \circ \pi$ in RB. Since π is surjective, we have $\check{f}' = \check{f}$. This proves the uniqueness of $\check{f}$. □

5.3.2 The dendriform algebra case

Now let $D = (D, \prec, \succ) \in$ DD. Let $T^+(D) = \bigoplus_{n \geqslant 1} D^{\otimes n}$ be the nonunitary tensor product algebra on D. Then $T^+(D)$ is the free nonunitary algebra generated by the **k**-module D. A **k**-basis of $T^+(D)$ is given by

$$X = \{x_1 \otimes \cdots \otimes x_n \big| x_i \in \Omega, \quad 1 \leqslant i \leqslant n, 1 \leqslant n\}$$

with the product given by the tensor concatenation:

$$(x_1 \otimes \cdots \otimes x_n)(y_1 \otimes \cdots \otimes y_m) = x_1 \otimes \cdots \otimes x_n \otimes y_1 \otimes \cdots \otimes y_m.$$

By Theorem 4.4.6,

$$Ш^{NC,0}{}_{\lambda}(T^+(D))$$

is the free nonunitary Rota-Baxter algebra on $T^+(D)$ of weight λ constructed in Section 4.4.2.

Exercise 5.3.3. Identify D as a **k**-submodule of $Ш^{NC,0}(T^+(D))$ by the natural injection $j_D : D \to T^+(D) \to Ш^{NC,0}(T^+(D))$. Then $Ш^{NC,0}(T^+(D))$ is also the free nonunitary Rota-Baxter algebra on the **k**-module D, defining the universal property that, for any nonunitary Rota-Baxter **k**-algebra R and **k**-module homomorphism $f : D \to R$, there is unique homomorphism of nonunitary Rota-Baxter algebras $\bar{f} : Ш^{NC,0}(T^+(D)) \to R$ such that $\bar{f} \circ j_D = f$. (Hint: Follow the proof of Proposition 3.2.2.)

Let J_{RB} be the Rota-Baxter ideal of $Ш^{NC,0}(T^+(D))$ generated by the set

$$\{x \prec y - x\lfloor y\rfloor - \lambda x \otimes y,\ x \succ y - \lfloor x\rfloor y \mid x, y \in D\} \tag{5.36}$$

Let $\pi : Ш^{NC,0}(T^+(D)) \to Ш^{NC,0}(T^+(D))/J_{\mathrm{RB}}$ be the quotient map.

Theorem 5.3.4. *Let $(D, \prec, \succ)$ be a dendriform **k**-algebra. The quotient Rota-Baxter algebra $Ш^{NC,0}(T^+(D))/J_{\mathrm{RB}}$, together with $\pi \circ j_{T^+(D)}$, is the universal enveloping Rota-Baxter algebra of D of weight λ.*

Proof. Let (R, P) be a Rota-Baxter algebra of weight λ and let $f : D \to R$ be a morphism in DD. More precisely, we have $f : D \to \mathcal{G}R$ where $\mathcal{G}R = (R, \prec'_P, \succ'_P)$ is the dendriform algebra in Theorem 5.1.4.

We will complete the following commutative diagram

$$
\begin{array}{ccc}
D & \xrightarrow{\ j_{T^+(D)}\ } & Ш^{NC,0}(T^+(D)) \\
{\scriptstyle f}\downarrow & \overset{\bar{f}}{\swarrow} & \downarrow{\scriptstyle \pi} \\
R & \xleftarrow{\ \check{f}\ } & Ш^{NC,0}(T^+(D))/J_{\mathrm{RB}}
\end{array}
\tag{5.37}
$$

By the universal property of $Ш^{NC,0}(T^+(D))$ from Exercise 5.3.3, there is a unique morphism $\bar{f} : Ш^{NC,0}(T^+(D)) \to R$ in RB^0 such that the upper left triangle commutes. So for any $x, y \in D$, we have

$$
\begin{aligned}
\bar{f}(x \prec y - x\lfloor y\rfloor - \lambda x \otimes y) &= \bar{f}(x \prec y) - \bar{f}(x)P(\bar{f}(y)) - \lambda \bar{f}(x \otimes y) \\
&= f(x \prec y) - f(x)P(f(y)) - \lambda f(x)f(y) \\
&= f(x \prec y) - f(x) \prec'_P f(y) \\
&= f(x \prec y) - f(x \prec y) = 0.
\end{aligned}
$$

Therefore, $x \prec y - x\lfloor y\rfloor - \lambda x \otimes y$ is in $\ker(\bar{f})$. Similarly, $x \succ y - \lfloor x\rfloor y$ is in $\ker(\bar{f})$. Thus J_{RB} is in $\ker(\bar{f})$ and there is a morphism $\check{f} : Ш^{NC,0}(T^+(D))/J_{\mathrm{RB}} \to R$ in RB^0 such that $\bar{f} = \check{f} \circ \pi$. Then

$$
\check{f} \circ \rho = \check{f} \circ \pi \circ j_{T^+(D)} = \bar{f} \circ j_\omega = f.
$$

This proves the existence of $\check{f}$.

The rest of the proof is the same as for Theorem 5.3.2. □

5.4 Free dendriform, tridendriform algebras and free Rota-Baxter algebras

The results in this section can be regarded as more precise forms of results in Section 5.3 in special cases. Our emphasis here is to interpret free dendriform algebras and free tridendriform algebras as natural subalgebras of free Rota-Baxter algebras. This interpretation also gives a planar tree correspondence on free Rota-Baxter algebras.

5.4.1 The dendriform algebra case

5.4.1.1 *Free dendriform algebras*

Let $\mathbf{k}$ be a field. We briefly recall the construction of free dendriform algebra $\mathrm{DD}(V)$ on a $\mathbf{k}$-vector space V as colored planar binary trees. For details, see [131, 154].

Let X be a basis of V. For $n \geqslant 0$, let Y_n be the set of planar binary trees with $n+1$ leaves and one root such that the valence of each internal vertex is exactly two. Let $Y_{n,X}$ be the set of planar binary trees with $n+1$ leaves and with vertices decorated by elements of X. The unique tree with one leave is denoted by $|$. So we have $Y_0 = Y_{0,X} = \{|\}$. Let $\mathbf{k}Y_{n,X}$ be the $\mathbf{k}$-vector space generated by $Y_{n,X}$. Here are the first few of them without decoration.

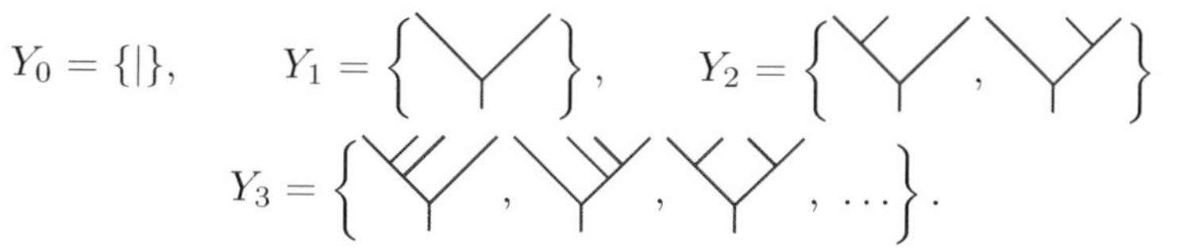

For $T \in Y_{m,X}, U \in Y_{n,X}$ and $x \in X$, the grafting of T and U on x is $T \vee_x U \in Y_{m+n+1,X}$. Let $\mathrm{DD}(V)$ be the graded vector space $\bigoplus_{n\geqslant 1} \mathbf{k}Y_{n,X}$. Define binary operations $\prec$ and $\succ$ on $\mathrm{DD}(V)$ recursively by

(a) $| \succ T = T \prec | = T$ and $| \prec T = T \succ | = 0$ for $T \in Y_{n,X}, n \geqslant 1$.

(b) For $T = T^\ell \vee_x T^r$ and $U = U^\ell \vee_y U^r$, define

$$T \prec U = T^\ell \vee_x (T^r \prec U + T^r \succ U), \quad T \succ U = (T \prec U^\ell + T \succ U^\ell) \vee_y U^r.$$

Since $| \prec |$ and $| \succ |$ are not defined, the binary operations $\prec$ and $\succ$ are only defined on $\mathrm{DD}(V)$ though the operation $\star := \prec + \succ$ can be extended to $H_{\mathrm{LR}} := \mathbf{k}Y_0 \oplus \mathrm{DD}(V)$ by defining $| \star T = T \star | = T$.

Theorem 5.4.1. **(Loday)**[131] $(\mathrm{DD}(V), \prec, \succ)$ *is the free dendriform algebra on* V.

Theorem 5.4.2. *Let* V *be a* $\mathbf{k}$*-vector space. The free dendriform algebra on* V *is a sub dendriform algebra of the free Rota-Baxter algebra* $Ш^{NC,0}(V)$ *of weight zero.*

The proof will be given in the next subsection.

5.4.1.2 *Proof of Theorem* 5.4.2

For the given vector space V, make V into a $\mathbf{k}$-algebra without identity by given V the zero product. Let $Ш^{NC,0}(V)$ be the free nonunitary Rota-Baxter algebra of weight zero on V constructed in Theorem 4.4.6. Since $Ш^{NC,0}(V)$ is a dendriform algebra, the natural map $j_V : V \to \mathrm{III}^{\mathrm{NC},0}(V)$ extends uniquely to a dendriform algebra morphism $D(j) : \mathrm{DD}(V) \to \mathrm{III}^{\mathrm{NC},0}(V)$. We will prove that this map is injective and identifies $\mathrm{DD}(V)$ as a subalgebra of $\mathrm{III}^{\mathrm{NC},0}(V)$ in the category of dendriform algebras. We first define a map

$$\phi : \mathrm{DD}(V) \to \mathrm{III}^{\mathrm{NC},0}(V)$$

and then show in Theorem 5.4.4 below that it agrees with $D(j)$. We construct ϕ by defining $\phi(T)$ for $T \in Y_{n,X}, n \geqslant 1$, inductively on n. Any $T \in Y_{n,X}, n \geqslant 1$ can be uniquely written as $T = T^\ell \vee_x T^r$ with $x \in X$ and $T^\ell, T^r \in \cup_{0 \leqslant i < n} Y_{i,X}$. We then define

$$\phi(T) = \begin{cases} \lfloor \phi(T^\ell) \rfloor x \lfloor \phi(T^r) \rfloor, & T^\ell \neq 1, T^r \neq 1, \\ x \lfloor \phi(T^r) \rfloor, & T^\ell = 1, T^r \neq 1, \\ \lfloor \phi(T^\ell) \rfloor x, & T^\ell \neq 1, T^r = 1, \\ x, & T^\ell = 1, T^r = 1. \end{cases} \tag{5.38}$$

For example,

$\phi\Big($  $\Big) = x$, $\phi\Big($ 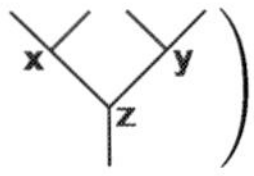 $\Big) = \lfloor x \rfloor z \lfloor y \rfloor$.

We recall [131] that $\mathrm{DD}(V)$ with the operation $\star := \prec + \succ$ is an associative algebra.

We now describe a submodule of $\mathrm{III}^{\mathrm{NC},0}(V)$ to be identified with the image of ϕ in Theorem 5.4.4.

Definition 5.4.3. An sRBW $y \in \mathcal{X}_\infty$ is called a *dendriform word* (*DW*) if it satisfies the following *additional* properties.

(a) y is not in $\lfloor \mathcal{X}_\infty \rfloor$.

(b) There is no subword $\lfloor \lfloor \mathbf{x} \rfloor \rfloor$ with $\mathbf{x} \in \mathcal{X}_\infty$ in the word.

(c) There is no subword of the form $\mathbf{x}_1 \lfloor \mathbf{x}_2 \rfloor \mathbf{x}_3$ with $\mathbf{x}_1, \mathbf{x}_3 \in X$ and $\mathbf{x}_2 \in \mathcal{X}_\infty$.

We let $\mathrm{DW}(V)$ be the subspace of $\mathrm{III}^{\mathrm{NC},0}(V)$ generated by the dendriform words.

For example

$$x_0 \lfloor x_1 \lfloor x_2 \rfloor \rfloor, \ \lfloor x_0 \rfloor x_1 \lfloor x_2 \rfloor$$

are dendriform words; while

$$\lfloor \lfloor x_1 \rfloor \rfloor, \ \lfloor \lfloor x_1 \rfloor x_2 \lfloor x_3 \rfloor \rfloor, x_1 \lfloor x_2 \rfloor x_3$$

are sRBWs but not dendriform words.

Equivalently, $\mathrm{DW}(V)$ can be characterized in terms of the decomposition in Eq. (4.46). For subsets Y, Z of $\mathcal{X}_\infty$, define

$$D(Y,Z) = (Y\lfloor Z\rfloor)\bigcup(\lfloor Z\rfloor Y)\bigcup\lfloor Z\rfloor Y\lfloor Z\rfloor.$$

Then define $D_0(V) = X$ and, for $n \geqslant 0$, inductively define

$$\begin{aligned} D_{n+1}(V) &= D(X, D_n(V)) \\ &= (X\lfloor D_n(V)\rfloor)\bigcup(\lfloor D_n(V)\rfloor X)\bigcup\lfloor D_n(V)\rfloor X\lfloor D_n(V)\rfloor. \end{aligned} \quad (5.39)$$

Then $D_\infty := \cup_{n\geqslant 0} D_n(V)$ is the set of dendriform words and $\mathrm{DW}(V) = \oplus_{\mathbf{x}\in D_\infty}\mathbf{kx}$.

Theorem 5.4.2 follows from the following theorem.

Theorem 5.4.4. (a) $\phi : \mathrm{DD}(V) \to \mathrm{III}^{\mathrm{NC},0}(V)$ *is a homomorphism of dendriform algebras.*
(b) $\phi = D(j)$, *the morphism of dendriform algebras induced by* $j : V \to \mathrm{III}^{\mathrm{NC},0}(V)$.
(c) $\phi(\mathrm{DD}(V)) = \mathrm{DW}(V)$.
(d) ϕ *is injective.*

Proof. (a) we first note that the operations $\prec$ and $\succ$ can be equivalently defined as follows without using $\mid \prec T, \mid \succ T, T \prec \mid$ and $T \succ \mid$. For $T \in Y_{m,X}, U \in Y_{n,X}$ with $m \geqslant 1, n \geqslant 1$. Then $T = T^\ell \vee_x T^r, U = U^\ell \vee_y U^r$ with $x, y \in X$ and $T^\ell, T^r, U^\ell, U^r \in \cup_{i\geqslant 0} Y_{i,X}$. Define

$$T \prec U := \begin{cases} T^\ell \vee_x (T^r \prec U + T^r \succ U), & \text{if } T^r \neq \mid, \\ T^\ell \vee_x U, & \text{if } T^r = \mid. \end{cases} \quad (5.40)$$

$$T \succ U := \begin{cases} (T \prec U^\ell + T \succ U^\ell) \vee_y U^r, & \text{if } U^\ell \neq \mid, \\ T \vee_y U^r, & \text{if } U^\ell = \mid. \end{cases} \quad (5.41)$$

Thus we have

$$\begin{aligned} \phi(T \prec U) &= \begin{cases} \phi(T^\ell \vee_x (T^r \prec U + T^r \succ U)), & \text{if } T^r \neq \mid, \\ \phi(T^\ell \vee_x U), & \text{if } T^r = \mid. \end{cases} \\ &= \begin{cases} \lfloor\phi(T^\ell)\rfloor x\lfloor\phi(T^r \prec U + T^r \succ U)\rfloor, & \text{if } T^r \neq \mid, T^\ell \neq \mid, \\ x\lfloor\phi(T^r \prec U + T^r \succ U)\rfloor, & \text{if } T^r \neq \mid, T^\ell = \mid, \\ \lfloor\phi(T^\ell)\rfloor x\lfloor\phi(U)\rfloor, & \text{if } T^r = \mid, T^\ell \neq \mid, \\ x\lfloor\phi(U)\rfloor, & \text{if } T^r = \mid, T^\ell = \mid. \end{cases} \\ &\qquad \text{(by definition of } \phi) \\ &= \begin{cases} \lfloor\phi(T^\ell)\rfloor x\lfloor\phi(T^r) \prec_R \phi(U) + \phi(T^r) \succ_R \phi(U)\rfloor, & \text{if } T^r \neq \mid, T^\ell \neq \mid, \\ x\lfloor(\phi(T^r) \prec_R \phi(U) + \phi(T^r) \succ_R \phi(U))\rfloor, & \text{if } T^r \neq \mid, T^\ell = \mid, \\ \lfloor\phi(T^\ell)\rfloor x\lfloor\phi(U)\rfloor, & \text{if } T^r = \mid, T^\ell \neq \mid, \\ x\lfloor\phi(U)\rfloor, & \text{if } T^r = \mid, T^\ell = \mid. \end{cases} \end{aligned}$$

(by induction hypothesis)

On the other hand, we have

$$\phi(T) \prec_R \phi(U) = \phi(T^\ell \vee_x T^r)\lfloor \phi(U) \rfloor$$
$$= \begin{cases} \lfloor \phi(T^\ell) \rfloor x \lfloor \phi(T^r) \rfloor \lfloor \phi(U) \rfloor, & \text{if } T^r \neq |, T^\ell \neq |, \\ x \lfloor \phi(T^r) \rfloor \lfloor \phi(U) \rfloor, & \text{if } T^r \neq |, T^\ell = |, \\ \lfloor \phi(T^\ell) \rfloor x \lfloor \phi(U) \rfloor, & \text{if } T^r = |, T^\ell \neq |, \\ x \lfloor \phi(U) \rfloor, & \text{if } T^r = |, T^\ell = |. \end{cases}$$
(by definition of ϕ)
$$= \begin{cases} \lfloor \phi(T^\ell) \rfloor x \lfloor \phi(T^r) \lfloor \phi(U) \rfloor + \lfloor \phi(T^r) \rfloor \phi(U) \rfloor, & \text{if } T^r \neq |, T^\ell \neq |, \\ x \lfloor \phi(T^r) \lfloor \phi(U) \rfloor + \lfloor \phi(T^r) \rfloor \phi(U) \rfloor, & \text{if } T^r \neq |, T^\ell = |, \\ \lfloor \phi(T^\ell) \rfloor x \lfloor \phi(U) \rfloor, & \text{if } T^r = |, T^\ell \neq |, \\ x \lfloor \phi(U) \rfloor, & \text{if } T^r = |, T^\ell = |. \end{cases}$$
(by Rota – Baxter relation of $P(T) = \lfloor T \rfloor$).

This proves $\phi(T \prec U) = \phi(T) \prec_R \phi(U)$. We similarly prove $\phi(T \succ U) = \phi(T) \succ_R \phi(U)$. Thus ϕ is a homomorphism in DD.

(b) follows from the uniqueness of the dendriform algebra morphism $\mathrm{DD}(V) \to Ш^{NC,0}(V)$ extending the map $j_V : V \to Ш^{NC,0}(V)$.

(c) We only need to prove $\mathrm{DW}(V) \subseteq \phi(\mathrm{DD}(V))$ and $\phi(\mathrm{DD}(V)) \subseteq \mathrm{DW}(V)$. To prove the former, we prove $D_n \subseteq \phi(\mathrm{DD}(V))$ by induction on n.

When $n = 0$, $D_n = X$ so the inclusion is clear. Suppose the inclusion holds for n. Then by the definition of $D_{n+1}(V)$ in Eq. (5.39), an element of $D_{n+1}(V)$ is of the following three forms:

i) It is $\mathbf{x}\lfloor \mathbf{x}' \rfloor$ with $\mathbf{x} \in X$, $\mathbf{x}' \in D_n(V)$. Then it is $\mathbf{x} \prec_R \mathbf{x}'$ which is in $\phi(\mathrm{DD}(V))$ by the induction hypothesis and the fact that $\phi(\mathrm{DD}(V))$ is a sub dendriform algebra.

ii) It is $\lfloor \mathbf{x} \rfloor \mathbf{x}'$ with $\mathbf{x} \in D_n(V)$ and $\mathbf{x}' \in X$. Then the same proof works.

iii) It is $\lfloor \mathbf{x} \rfloor \mathbf{x}' \lfloor \mathbf{x}'' \rfloor$ with $\mathbf{x}, \mathbf{x}'' \in D_n(V)$ and $\mathbf{x}' \in X$. Then it is

$$(\mathbf{x} \succ_R \mathbf{x}') \prec_R \mathbf{x}'' = \mathbf{x}' \succ_R (\mathbf{x}' \prec_R \mathbf{x}'').$$

By induction, $\mathbf{x}$ and $\mathbf{x}''$ are in the sub dendriform algebra $\phi(\mathrm{DD}(V))$. So the element itself is in $\phi(\mathrm{DD}(V))$.

The second inclusion follows easily by induction on degrees of trees in $\mathrm{DD}(V)$.

(d) By the definition of ϕ and Item (c), ϕ gives a one-one correspondence between $\cup_{n \geqslant 0} Y_{n,X}$ as a basis of $\mathrm{DD}(V)$ and $\mathrm{DW}(V)$ as a basis of $\phi(\mathrm{DD}(V))$. Therefore ϕ is injective. □

5.4.2 The tridendriform algebra case

5.4.2.1 *Free tridendriform algebras*

We describe the construction of free tridendriform algebra $\mathrm{DT}(V)$ on a vector space V as decorated planar trees. For details, when V is of rank one over $\mathbf{k}$, see [137].

Let X be a basis of V. For $n \geqslant 0$, let T_n be the set of planar trees with $n+1$ leaves and one root such that the valence of each internal vertex is at least two. Let $T_{n,X}$ be the set of planar trees with $n+1$ leaves and with vertices *valently decorated* by elements of X, in the sense that if a vertex has valence k, then the vertex is decorated by a vector in X^{k-1}. For example, the vertex of  is decorated by $x \in X$ while the vertex of  is decorated by $(x,y) \in X^2$. The unique tree with one leaf is denoted by $|$. So we have $T_0 = T_{0,X} = \{|\}$. Let $\mathbf{k}[T_{n,X}]$ be the $\mathbf{k}$-vector space generated by $T_{n,X}$.

Here are the first few of them without decoration.

$T_0 = \{|\}$, $T_1 = \{$  $\}$, $T_2 = \{$  $\}$

$T_3 = \{$  $, \ldots\}$.

For $T^{(i)} \in T_{n_i,X}$, $0 \leqslant i \leqslant k$, and $x_i \in X$, $1 \leqslant i \leqslant k$, the grafting of $T^{(i)}$ over $(x_1, \cdots, x_k)$ is

$$T^{(0)} \vee_{x_1} T^{(1)} \vee_{x_2} \cdots \vee_{x_k} T^{(k)}.$$

Any tree can be uniquely expressed as such a grafting of lower degree trees. For example,

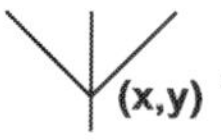

$$= | \vee_x | \vee_y |.$$

Let $\mathrm{DT}(V)$ be the graded vector space $\bigoplus_{n\geqslant 1} \mathbf{k}T_{n,X}$. Define binary operations $\prec$, $\succ$ and $\cdot$ on $\mathrm{DT}(V)$ recursively by

(a) $| \succ T = T \prec | = T$, $| \prec T = T \succ | = 0$ and $| \cdot T = T \cdot | = 0$ for $T \in T_{n,X}, n \geqslant 1$;

(b) For $T = T^{(0)} \vee_{x_1} \cdots \vee_{x_m} T^{(m)}$ and $U = U^{(0)} \vee_{y_1} \cdots \vee_{y_n} U^{(n)}$, define

$$\begin{aligned} T \prec U &= T^{(0)} \vee_{x_1} \cdots \vee_{x_m} (T^{(m)} \star U), \\ T \succ U &= (T \star U^{(0)}) \vee_{y_1} \cdots \vee_{y_n} U^{(n)}, \\ T \cdot U &= T^{(0)} \vee_{x_1} \cdots \vee_{x_m} (T^{(m)} \star U^{(0)}) \vee_{y_1} \cdots \vee_{y_n} U^{(n)}. \end{aligned}$$

Here $\star := \prec + \succ + \cdot$ Since $| \prec |$, $| \succ |$ and $| \cdot |$ are not defined, the binary operations $\prec$, $\succ$ and $\cdot$ are only defined on $\mathrm{DT}(V)$ though the operation $\star$ can be extended to $H_{\mathrm{DT}} := \mathbf{k}T_0 \oplus \mathrm{DT}(V)$ by defining $| \star T = T \star | = T$.

Theorem 5.4.5. $(\mathrm{DT}(V), \prec, \succ, \cdot)$ *is the free tridendriform algebra on* V.

Proof. The proof is given by Loday and Ronco in [137] when V is of dimension one. The proof for the general case is similar. □

Our goal is to prove

Theorem 5.4.6. *Let* V *be a* **k**-*vector space. The free tridendriform algebra on* V *is a canonical sub-tridendriform algebra of the free Rota-Baxter algebra* $Ш^{NC,0}(V)$ *of weight one.*

We restrict the weight of the Rota-Baxter algebra to one to ease the notation. The proof will be given in the next subsection.

5.4.2.2 *Proof of Theorem* 5.4.6

Let V be the given **k**-vector space and let $T^+(V) = \bigoplus_{n\geqslant 1} V^{\otimes n}$ denote the nonunitary tensor algebra on V. By Theorem 4.4.6, $Ш^{NC,0}{}_1(T^+(V))$ is the free nonunitary Rota-Baxter algebra on $T^+(V)$ of weight 1 constructed in Section 4.4.2.

Since $Ш^{NC,0}(T^+(V))$ is a tridendriform algebra, the natural map

$$j_V : V \to Ш^{NC,0}(T^+(V))$$

extends uniquely to a tridendriform algebra homomorphism

$$T(j) : \mathrm{DT}(V) \to Ш^{NC,0}(T^+(V)).$$

We will prove that this map is injective and identifies $\mathrm{DT}(V)$ as a subalgebra of $Ш^{NC,0}(\Omega)$ in the category of tridendriform algebras. We first define a map

$$\psi : \mathrm{DT}(V) \to Ш^{NC,0}(T^+(V))$$

and then show in Theorem 5.4.8 below that it agrees with $T(j)$. We construct ψ by defining $\psi(T)$ for $T \in T_{n,X}, n \geqslant 1$, inductively on n. Any $T \in T_{n,X}, n \geqslant 1$, can be uniquely written as $T = T^{(0)} \vee_{x_1} \cdots \vee_{x_k} T^{(k)}$ with $x_i \in X$ and $T^{(i)} \in \cup_{0\leqslant i<n} T_{i,X}$. We then define

$$\psi(T) = \overline{\lfloor \psi(T^{(0)}) \rfloor} x_1 \overline{\lfloor \psi(T^{(1)}) \rfloor} \cdots \overline{\lfloor \psi(T^{(k-1)}) \rfloor} x_k \overline{\lfloor \psi(T^{(k)}) \rfloor}, \tag{5.42}$$

where $\overline{\lfloor \psi(T^{(i)}) \rfloor} = \lfloor \psi(T^{(i)}) \rfloor$ if $\psi(T^{(i)}) \neq |$. If $\psi(T^{(i)}) = |$, then the factor $\lfloor \psi(T^{(i)}) \rfloor$ is dropped when $i = 0$ or k, and is replaced by $\otimes$ when $0 < i < k$. For example,

$$\overline{\lfloor \psi(|) \rfloor} x_1 \overline{\lfloor \psi(T^{(1)}) \rfloor} x_2 \cdots x_k \overline{\lfloor \psi(T^{(k)}) \rfloor} = x_1 \overline{\lfloor \psi(T^{(1)}) \rfloor} x_2 \cdots x_k \overline{\lfloor \psi(T^{(k)}) \rfloor}$$

and

$$\overline{\lfloor \psi(T^{(0)}) \rfloor} x_1 \overline{\lfloor \psi(|) \rfloor} x_2 \overline{\lfloor \psi(T^{(2)}) \rfloor} \cdots x_k \overline{\lfloor \psi(T^{(k)}) \rfloor}$$

$$= \overline{\lfloor \psi(T^{(0)}) \rfloor}(x_1 \otimes x_2)\overline{\lfloor \psi(T^{(2)}) \rfloor} \cdots x_k \overline{\lfloor \psi(T^{(k)}) \rfloor}.$$

In particular,

$\psi\Big($ 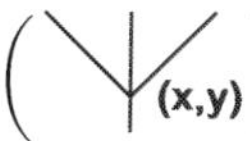 $\Big) = \psi(| \vee_x | \vee_y |) = \overline{\lfloor \psi(|) \rfloor} \vee_x \overline{\lfloor \psi(|) \rfloor} \vee_y \overline{\lfloor \psi(|) \rfloor} = x \otimes y.$

We now describe a submodule of $Ш^{NC,0}(T^+(V))$ to be identified with the image of ψ in Theorem 5.4.8.

Definition 5.4.7. An sRBW $y \in \mathcal{X}_\infty$ is called a *tridendriform word* (*TW*) if it satisfies the following *additional* properties.

(a) y is not in $\lfloor \mathcal{X}_\infty \rfloor$.

(b) There is no subword $\lfloor \lfloor \mathbf{x} \rfloor \rfloor$ with $\mathbf{x} \in \mathcal{X}_\infty$ in the word.

We let $\mathrm{TW}(V)$ be the subspace of $Ш^{NC,0}(T^+(V))$ generated by tridendriform words.

For example,

$$x_0 \lfloor x_1 \lfloor x_2 \rfloor \rfloor, \lfloor x_0 \rfloor x_1 \lfloor x_2 \rfloor, \lfloor x_0 \rfloor x_1 \lfloor x_2 \rfloor x_3 \lfloor x_4 \rfloor, x_0 \otimes x_1$$

are tridendriform words; while

$$\lfloor \lfloor x_1 \rfloor \rfloor, \lfloor x_1 \lfloor x_2 \rfloor x_3 \rfloor$$

are sRBWs but not tridendriform words.

Equivalently, tridendriform words can be characterized in terms of the decomposition in Eq. (4.46). For subsets Y, Z of $\mathcal{X}_\infty$, define

$$\begin{aligned} S(Y,Z) = \Big(\bigcup_{r \geqslant 1} (Y \lfloor Z \rfloor)^r\Big) \bigcup \Big(\bigcup_{r \geqslant 0} (Y \lfloor Z \rfloor)^r Y\Big) \\ \bigcup \Big(\bigcup_{r \geqslant 1} \lfloor Z \rfloor (Y \lfloor Z \rfloor)^r\Big) \bigcup \Big(\bigcup_{r \geqslant 0} \lfloor Z \rfloor (Y \lfloor Z \rfloor)^r Y\Big). \end{aligned} \tag{5.43}$$

Then define $S_0(V) = S(X)$, the free semigroup generated by X, and identify it as a basis of the free nonunitary algebra $T^+(V)$ on V. For $n \geqslant 0$, inductively define

$$S_{n+1}(V) = S(M(X), S_n(V)). \tag{5.44}$$

Then $S_\infty := \cup_{n \geqslant 0} S_n(V)$ is the set of tridendriform words and $\mathrm{TW}(V) = \oplus_{\mathbf{x} \in S_\infty} \mathbf{k}\mathbf{x}$.

Theorem 5.4.6 follows from the following theorem.

Theorem 5.4.8. (a) *The map* $\psi : \mathrm{DT}(V) \to Ш^{NC,0}(T^+(V))$ *is a homomorphism of tridendriform algebras.*

(b) *The map* ψ *equals* $T(j)$, *the homomorphism of tridendriform algebras induced by* $j : V \to \text{Ш}^{\text{NC},0}(T^+(V))$.

(c) $\psi(\text{DT}) = \text{DT}(V)$.

(d) ψ *is injective.*

Proof. (a) We first note that the operations $\prec$, $\succ$ and $\cdot$ can be equivalently defined as follows without using $| \prec T$, etc. For $T \in T_{i,X}, U \in T_{j,X}$ with $i \geqslant 1, j \geqslant 1$. Then $T = T^{(0)} \vee_{x_1} \cdots \vee_{x_m} T^{(m)}$ and $U = U^{(0)} \vee_{y_1} \cdots \vee_{y_n} U^{(n)}$, define

$$T \prec U = \begin{cases} T^{(0)} \vee_{x_1} \cdots \vee_{x_m} (T^{(m)} \star U), & \text{if } T^{(m)} \neq |, \\ T^{(0)} \vee_{x_1} \cdots \vee_{x_m} U, & \text{if } T^{(m)} = |. \end{cases}$$

$$T \succ U = \begin{cases} (T \star U^{(0)}) \vee_{y_1} \cdots \vee_{y_n} U^{(n)}, & \text{if } U^{(0)} \neq |, \\ T \vee_{y_1} \cdots \vee_{y_n} U^{(n)}, & \text{if } U^{(0)} = |. \end{cases}$$

$$T \cdot U = \begin{cases} T^{(0)} \vee_{x_1} \cdots \vee_{x_m} (T^{(m)} \star U^{(0)}) \vee_{y_1} \cdots \vee_{y_n} U^{(n)}, & \text{if } T^{(m)} \neq |, U^{(0)} \neq |, \\ T^{(0)} \vee_{x_1} \cdots \vee_{x_m} U^{(0)} \vee_{y_1} \cdots \vee_{y_n} U^{(n)}, & \text{if } T^{(m)} = |, U^{(0)} \neq |, \\ T^{(0)} \vee_{x_1} \cdots \vee_{x_m} T^{(m)} \vee_{y_1} \cdots \vee_{y_n} U^{(n)}, & \text{if } T^{(m)} \neq |, U^{(0)} = |. \end{cases}$$

Now we use induction on $i + j$ to prove

$$\psi(T \prec U) = \psi(T) \prec_P \psi(U), \quad \psi(T \succ U) = \psi(T) \succ_P \psi(U), \tag{5.45}$$
$$\psi(T \cdot U) = \psi(T) \cdot_P \psi(U). \tag{5.46}$$

Here $P := P_{T^+(V)}$ is the Rota-Baxter operator on $\text{Ш}^{\text{NC},0}(T^+(V))$. Since $i+j \geqslant 2$, we first assume $i + j = 2$. Then $T = | \vee_x |$, $U = | \vee_y |$. So by Eq. (5.42),

$$\psi(T \prec U) = \psi((| \vee_x |) \prec U) = \psi(| \vee_x U) = x\lfloor \psi(U) \rfloor = x\lfloor y \rfloor = x \prec_P y.$$

We similarly have $\psi(T \succ U) = x \succ_P y$ and

$$\psi(T \cdot U) = \psi((| \vee_x |) \cdot (| \vee_y |)) = \psi(| \vee_x | \vee_y |) = x \otimes y = x \cdot_P y.$$

Assume Eq. (5.45) holds for $T \in T_{i,X}$, $U \in T_{j,X}$ with $i + j \leqslant k$ with $k \geqslant 2$. Then we also have

$$\begin{aligned} \psi(T \star U) &= \psi(T \prec U + T \succ U + T \cdot U) \\ &= \psi(T) \prec_P \psi(U) + \psi(T) \succ_P \psi(U) + \psi(T) \cdot_P \psi(U) \\ &= \psi(T) \star_P \psi(U). \end{aligned} \tag{5.47}$$

Here $\star_P = \prec_P + \succ_P + \cdot_P$. Consider T, U with $i + j = k + 1$. We consider two cases of $T = T^{(0)} \vee_{x_1} \cdots \vee_{x_i} T^{(i)}$. Since $U \neq |$, we have $\overline{\lfloor T^{(i)} \star U \rfloor} = \lfloor T^{(i)} \star U \rfloor$ if $T^{(i)} \neq |$, and $\overline{\lfloor U \rfloor} = \lfloor U \rfloor$ if $T^{(i)} = |$.

Case 1. If $T^{(i)} \neq |$, then

$$\begin{aligned}
&\psi(T \prec U)\\
&= \psi(T^{(0)} \vee_{x_1} \cdots \vee_{x_i} (T^{(i)} \star U)) \quad \text{(definition of } \prec)\\
&= \overline{\lfloor \psi(T^{(0)}) \rfloor} x_1 \cdots x_i \lfloor \psi(T^{(i)} \star U) \rfloor \quad \text{(definition of } \psi)\\
&= \overline{\lfloor \psi(T^{(0)}) \rfloor} x_1 \cdots x_i \lfloor \psi(T^{(i)}) \star_P \psi(U) \rfloor \quad \text{(induction hypothesis (5.47))}\\
&= \overline{\lfloor \psi(T^{(0)}) \rfloor} x_1 \cdots x_i \lfloor \psi(T^{(i)}) \rfloor \lfloor \psi(U) \rfloor \quad \text{(relation (1.1))}\\
&= \psi(T^{(0)} \vee_{x_1} \cdots \vee_{x_i} T^{(i)}) \prec_P \psi(U) \quad \text{(defintion of } \psi)\\
&= \psi(T) \prec_P \psi(U).
\end{aligned}$$

Case 2. If $T^{(m)} = |$, then

$$\begin{aligned}
\psi(T \prec U) &= \psi(T^{(0)} \vee_{x_1} \cdots \vee_{x_i} U) \quad \text{(definition of } \prec)\\
&= \overline{\lfloor \psi(T^{(0)}) \rfloor} x_1 \cdots x_i \lfloor \psi(U) \rfloor \quad \text{(definition of } \psi)\\
&= \psi(T^{(0)} \vee_{x_1} \cdots \vee_{x_i} T^{(m)}) \lfloor \psi(U) \rfloor \quad \text{(defintion of } \psi)\\
&= \psi(T) \prec_P \psi(U).
\end{aligned}$$

This proves $\psi(T \prec U) = \psi(T) \prec_P \psi(U)$. We similarly prove $\psi(T \succ U) = \psi(T) \succ_P \psi(U)$ and $\psi(T \cdot U) = \psi(T) \cdot_P \psi(U)$. Thus ψ is a homomorphism in DT.

(b) follows from the uniqueness of the homomorphism $\mathrm{DT}(V) \to \text{III}^{NC,0}(T^+(V))$ of tridendriform algebras extending the map $i : V \to \text{III}^{NC,0}(T^+(V))$.

(c) We only need to prove $\mathrm{TW}(V) \subseteq \psi(\mathrm{DT}(V))$ and $\psi(\mathrm{DT}(V)) \subseteq \mathrm{TW}(V)$. To prove the former, we prove $S_n(V) \subseteq \psi(\mathrm{DT}(V))$ by induction on n.

When $n = 0$, $S_n(V) = X$, so the inclusion is clear. Suppose the inclusion holds for $1 \leqslant n \leqslant k$. Then by the definition of $S_{k+1}(V)$ in Eq. (5.44), an element of $S_{k+1}(V)$ has length greater than or equal to 2. We apply induction on its length. If the length is 2, then it is one of the following two cases.

i) It is $\mathbf{x}\lfloor \mathbf{x}' \rfloor$ with $\mathbf{x} \in X$, $\mathbf{x}' \in S_k(V)$. Then it is $\mathbf{x} \prec_P \mathbf{x}'$ which is in $\psi(\mathrm{DT}(V))$ by the induction hypothesis and the consequence from part (a) that $\psi(\mathrm{DT}(V))$ is a sub dendriform algebra.

ii) It is $\lfloor \mathbf{x} \rfloor \mathbf{x}'$ with $\mathbf{x} \in S_k(V)$ and $\mathbf{x}' \in X$. Then the same proof works.

Suppose all elements of S_{k+1} with length $\leqslant q$ and $\geqslant 2$ are in $\psi(\mathrm{DT}(V))$. Consider an element $\mathbf{x}$ of S_{k+1} with length $q+1$. Then $q+1 \geqslant 3$. If $q+1 = 3$, we again have two cases.

i) $\mathbf{x} = \lfloor \overline{\mathbf{x}}_1 \rfloor \mathbf{x}_2 \lfloor \overline{\mathbf{x}}_3 \rfloor$ with $\overline{\mathbf{x}}_1, \overline{\mathbf{x}}_2 \in S_n(V)$ and $\mathbf{x}_1 \in X$. Then it is $(\overline{\mathbf{x}}_1 \succ_P \mathbf{x}_2) \prec_P \overline{\mathbf{x}}_3$. By induction hypothesis on n, $\overline{\mathbf{x}}_1$ and $\overline{\mathbf{x}}_3$ are in the sub dendriform algebra $\psi(\mathrm{DT}(V))$. So the element itself is in $\psi(\mathrm{DT}(V))$.

ii) $\mathbf{x} = \mathbf{x}_1 \lfloor \overline{\mathbf{x}}_2 \rfloor \mathbf{x}_3$ with $\mathbf{x}_1, \mathbf{x}_3 \in X$ and $\overline{\mathbf{x}}_2 \in S_n(V)$. Then $\mathbf{x} = \mathbf{x}_1 \cdot_P (\overline{\mathbf{x}}_2 \succ \mathbf{x}_3)$ which is in $\psi(\mathrm{DT}(V))$.

If $q+1 \geqslant 4$, then $\mathbf{x}$ can be expressed as the concatenation of $\mathbf{x}_1$ and $\mathbf{x}_2$ of lengths at least two and hence are in $\mathrm{TW}(V)$. By induction hypotheses, $\mathbf{x}_1$ and $\mathbf{x}_2$ are in $\psi(\mathrm{DT}(V))$. Therefore $\mathbf{x} = \mathbf{x}_1 \cdot_P \mathbf{x}_2$ is in $\psi(\mathrm{DT}(V))$.

This completes the proof of the first inclusion. The proof of the second inclusion follows from a similar induction on the degree of trees in $\mathrm{DT}(V)$.

(d) By the definition of ψ and Item (c), ψ gives a one-one correspondence between $\cup_{n\geqslant 0} T_{n,X}$ as a basis of $\mathrm{DT}(V)$ and $\mathrm{TW}(V)$ as a basis of $\psi(\mathrm{DT}(V))$. Therefore ψ is injective. □

5.5 Some related literature and problems

The main references of this chapter are [56, 58].

Recall that the Poincaré-Birkhoff-Witt Theorem for the universal enveloping algebra $U(\mathfrak{g})$ of a Lie algebra $\mathfrak{g}$ displays a basis of $U(\mathfrak{g})$ in terms of a basis of $\mathfrak{g}$. An analogue of the Poincaré-Birkhoff-Witt Theorem for the universal enveloping Rota-Baxter algebra $U_{\mathrm{RB}}(D)$ of a dendriform algebra D would be the statement that constructs a canonical basis of $U_{\mathrm{RB}}(D)$ from a basis of D, thus embedding D as a dendriform subalgebra of $U_{\mathrm{RB}}(D)$. Even though finding such an analogue remains an open problem, the embedding problem has been solved by Chen and Mo [37].

It has been shown [8, 76, 114] that the noncommutative version of the Connes-Kreimer Hopf algebra NCK of rooted trees is isomorphic to the Hopf algebra of Loday-Ronco. Composing the injection ψ in Theorem 5.4.8 with such an isomorphism, as the one explicitly defined in [8], we obtain an injection from NCK into free noncommutative Rota-Baxter algebras.

Some recent studies of equations and identities on dendriform algebras can be found in [65, 66, 69].

For Lie algebra analogues of dendriform algebra, such as pre-Lie algebras, PostLie algebras and L-dendriform algebra, see [22, 129, 130, 169].

The relationship between a Rota-Baxter operator and a dendriform or tridendriform algebra can be extended to a relative version of the Rota-Baxter operator, called a relative Rota-Baxter operator or an $\mathcal{O}$-operator [21], a name adapted from its Lie algebra origin. It is also called a generalized Rota-Baxter operator in the weight zero case [167]. $\mathcal{O}$-operators are shown to not only derive dendriform and tridendriform algebras, but derive all of them. Further there are natural bijections between certain equivalent classes of $\mathcal{O}$-operators and those of dendriform and tridendriform algebras. Tridendriform algebras can also be derived from other operators, such as TD-operators [128] and Rota-Baxter operators on totally compatible dialgebras [176].

Chapter 6

Rota-Baxter Operators on Operads and Manin Products

After the introduction of dendriform algebras and tridendriform algebras that carries two and three binary operations respectively (see Chapter 5), quite a few more similar algebra structures have been introduced, such as the quadri-algebra of Aguiar and Loday [6], the ennea-algebra, the NS-algebra, the dendriform-Nijenhuis algebra and the octo-algebra of Leroux [126, 127, 128]. These algebras have a common property of "splitting the associativity", that is, expressing the multiplication of an associative algebra as the sum of a string of binary operations. The operations in the string satisfy a set of relations and the associativity of the multiplication follows from the sum of these relations. The first instance of such algebras, the dendriform algebra, has a string of two operators. The later constructions were largely inspired by the connection [3, 47, 126] with Rota-Baxter operators.

Two themes can be found in these recent constructions. One is the construction of a new type of algebras that has the combined features of types of two or more algebras that were previously known. The other is the use of a linear operator with certain features, such as a Rota-Baxter operator, on a known type of algebras to obtain another type of algebras with richer structures. Even though the ideas of the themes are simple, to carry them out for a particular construction can be quite complicated.

In this chapter we study these constructions in the framework of operads and their products, given by generators and relations. This enables us to clarify, simplify and further generalize the constructions and properties of these recent algebra structures. This discussion also serves as an easy entry, through a special case, to the rich general theory of operads, as presented in the systematic treatments [140, 147].

In Section 6.1, we give the general formulation of nonsymmetric operads defined by partial compositions and then specialize in the case of being binary quadratic. Such operads are defined by generators and relations and give rise to all the above algebra structures [135]. To ease the notation, we call them binary quadratic nonsymmetric operads and the corresponding algebras binary quadratic nonsymmetric algebras. The generator-relation construction of binary quadratic nonsymmetric operads allows a concrete description which is quite simple and can be found in the existing literature [134]. We use this description to formally define the types of such operads.

We then define in Section 6.2 products of binary quadratic nonsymmetric operads that are similar to (but different from) the operad products of Manin-Ginzburg-Kapranov [80, 133, 142, 143]. We show, formalizing the first theme, that some recently obtained binary quadratic nonsymmetric algebras [6, 126, 127, 128] are products of simpler algebras. Properties of products of binary quadratic nonsymmetric algebras are also studied in this section.

A full understanding of the second theme mentioned above depends on sufficient knowledge of the linear operators that give rise to binary quadratic nonsymmetric algebras. Using the framework introduced here, we can make the second theme precise applying the operators to any types of binary quadratic nonsymmetric algebras. This is presented in Section 6.3.

6.1 Nonsymmetric operads

Let A be a vector space over $\mathbf{k}$. The dendriform algebra and its analogues, such as the tridendriform algebra, quadri-algebra and ennea-algebra, are constructed from a finite set of binary operations together with a set of "associativity" relations. These constructions fit into the general framework of operads that are binary, quadratic and nonsymmetric. We will first briefly recall these concepts and refer to the general references [134, 138] for details.

6.1.1 Binary quadratic nonsymmetric operads

There are several equivalent definitions of nonsymmetric operads. We will use the definition by partial compositions. See [140, 147] for full treatment.

This definition of nonsymmetric operads is modeled after multi-linear maps of vector spaces. This is similar to the definition of $\mathbf{k}$-algebras modeling after the linear maps.

Let V be a $\mathbf{k}$-vector space. Then $\mathrm{Hom}(V, V)$ is a $\mathbf{k}$-vector space with a composition operator that satisfies the associativity rule. Generalizing this, we call a $\mathbf{k}$-vector space A a $\mathbf{k}$-algebra if A has a binary operation that satisfies the associativity. So $\mathrm{Hom}(V, V)$ is a $\mathbf{k}$-algebra, and V is called a A-module (and V

a representation of A) if there is a $\mathbf{k}$-algebra homomorphism $A \to \mathrm{Hom}(V, V)$. This analogy generalizes to operads.

Extending linear maps to multi-linear maps, consider the spaces

$$\mathcal{E}_n := \mathcal{E}_{V,n} := \mathrm{Hom}(V^{\otimes n}, V), \quad n \geqslant 1.$$

For $\mu \in \mathcal{E}_m$, $\nu \in \mathcal{E}_n$ and $1 \leqslant i \leqslant m$, define the composition

$$(\mu \circ_i \nu) := (\mu \circ_{m,n,i} \nu) : V^{\otimes(m+n-1)} \to V$$

by

$$\begin{aligned}&(\mu \circ_i \nu)(x_1, \ldots, x_{m+n-1})\\ &= \mu(x_1, \ldots, x_{i-1}, \nu(x_i, \ldots, x_{i+m-1}), x_{i+m}, \ldots, x_{m+n-1}).\end{aligned}$$

So we have the composition maps

$$\circ_i : \mathcal{E}_m \otimes \mathcal{E}_n \to \mathcal{E}_{m+n-1}, \quad 1 \leqslant i \leqslant m.$$

Further for $\lambda \in \mathcal{E}_\ell, \mu \in \mathcal{E}_m$ and $\nu \in \mathcal{E}_n$, we have

(a) (*Sequential composition*)

$$(\lambda \circ_i \mu) \circ_{i-1+j} \nu = \lambda \circ_i (\mu \circ_j \nu), \quad 1 \leqslant i \leqslant \ell, 1 \leqslant j \leqslant m, \tag{6.1}$$

(b) (*Parallel composition*)

$$(\lambda \circ_i \mu) \circ_{k-1+m} \nu = (\lambda \circ_k \nu) \circ_i \mu, \quad 1 \leqslant i < k \leqslant \ell. \tag{6.2}$$

(c) (*Identity*) There is an element $\mathrm{id} \in \mathcal{E}_1$ such that $\mathrm{id} \circ \mu = \mu$ and $\mu \circ \mathrm{id} = \mu$ for $\mu \in \mathcal{P}_n, n \geqslant 0$.

Exercise 6.1.1. Verify that these conditions hold.

Extracting this structure for abstract sequences of vector spaces, we give the following definition.

Definition 6.1.2. Let $\mathbf{k}$ be a field.

(a) A *graded vector space* is a sequence $\mathcal{P} := \{\mathcal{P}_n\}_{n\geqslant 0}$ of $\mathbf{k}$-vector spaces $\mathcal{P}_n, n \geqslant 0$.

(b) A *nonsymmetric* (*ns*) *operad* is a graded vector space $\mathcal{P} = \{\mathcal{P}_n\}_{n\geqslant 0}$ equipped with *partial compositions*:

$$\circ_i := \circ_{m,n,i} : \mathcal{P}_m \otimes \mathcal{P}_n \longrightarrow \mathcal{P}_{m+n-1}, \quad 1 \leqslant i \leqslant m, \tag{6.3}$$

so that, for $\lambda \in \mathcal{P}_\ell, \mu \in \mathcal{P}_m$ and $\nu \in \mathcal{P}_n$, relations (a)–(c) above hold.

An ns operad $\mathcal{P} = \{\mathcal{P}_n\}$ is called *binary* if $\mathcal{P}_1 = \mathbf{k}.\ \mathrm{id}$ and $\mathcal{P}_n, n \geqslant 3$ are induced from $\mathcal{P}_2$ by composition. Then particularly,

$$\mathcal{P}_3 = (\mathcal{P}_2 \circ_1 \mathcal{P}_2) \oplus (\mathcal{P}_2 \circ_2 \mathcal{P}_2), \tag{6.4}$$

which can be identified with $\mathcal{P}_2^{\otimes 2} \oplus \mathcal{P}_2^{\otimes 2}$. A binary ns operad $\mathcal{P}$ is called *quadratic* if all relations among the binary operations in $\mathcal{P}_2$ are derived from $\mathcal{P}_3$.

Thus a binary, quadratic, ns operad is determined by a pair (V, R) where $V = \mathcal{P}_2$, called the *space of generators*, and R is a subspace of $V^{\otimes 2} \oplus V^{\otimes 2}$, called the *space of relations*. So we can denote $\mathcal{P} = \mathcal{P}(V)/(R)$ or $\mathcal{P}(V, R)$. Note that a typical element of $V^{\otimes 2} \oplus V^{\otimes 2}$ is of the form

$$\left(\sum_{i=1}^{k} \odot_i^{(1)} \otimes \odot_i^{(2)}, \sum_{j=1}^{m} \odot_j^{(3)} \otimes \odot_j^{(4)}\right), \quad \odot_i^{(1)}, \odot_i^{(2)}, \odot_j^{(3)}, \odot_j^{(4)} \in V,$$
$$1 \leqslant i \leqslant k, 1 \leqslant j \leqslant m, k, m \geqslant 1.$$

For a given binary quadratic ns operad $\mathcal{P} = \mathcal{P}(V)/(R) = \mathcal{P}(V, R)$, a **k**-vector space A is called a $\mathcal{P}$*-algebra* if A has binary operations (indexed by) V and if, for

$$\left(\sum_{i=1}^{k} \odot_i^{(1)} \otimes \odot_i^{(2)}, \sum_{j=1}^{m} \odot_j^{(3)} \otimes \odot_j^{(4)}\right) \in R \subseteq V^{\otimes 2} \oplus V^{\otimes 2}$$

with $\odot_i^{(1)}, \odot_i^{(2)}, \odot_j^{(3)}, \odot_j^{(4)} \in V$, $1 \leqslant i \leqslant k$, $1 \leqslant j \leqslant m$, we have

$$\sum_{i=1}^{k} (x \odot_i^{(1)} y) \odot_i^{(2)} z = \sum_{j=1}^{m} x \odot_j^{(3)} (y \odot_j^{(4)} z), \quad \forall\, x, y, z \in A. \tag{6.5}$$

Since a binary quadratic ns operad $\mathcal{P}(V, R)$ is determined by $(\vec{v}, \vec{r})$ where $\vec{v}$ is a basis of V and $\vec{r}$ is a basis of R, we also use $\mathcal{P}(\vec{v}, \vec{r})$ to denote this operad.

Let $\mathcal{P}(V, R)$ and $\mathcal{P}(V', R')$ be binary quadratic nonsymmetric operads with associative operations $\star$ and $\star'$ respectively. A *morphism* $f : \mathcal{P}(V, R) \to \mathcal{P}(V', R')$ is a linear map $V \to V'$ sending $\star$ to $\star'$ and inducing a linear map $R \to R'$. An invertible morphism is called an *isomorphism*, and called an *automorphism* if $\mathcal{P}(V, R) = \mathcal{P}(V', R')$.

6.1.2 Examples

We now describe known examples of dendriform-related algebras in the context of operads.

1. (Associative algebra) An associative **k**-algebra is a **k**-vector space A with an associative product $\cdot$. Its corresponding operad is $\mathcal{P}(\vec{v}_A, \vec{r}_A)$ with $\vec{v}_A = \{\cdot\}$ and $\vec{r}_A = \{(\cdot \otimes \cdot, \cdot \otimes \cdot)\}$.

2. (Dendriform algebra) Recall that the *dendriform algebra* is defined to be a **k**-vector space D with binary operations $\prec$ and $\succ$ such that

$$(x \prec y) \prec z = x \prec (y \prec z + y \succ z), \quad (x \succ y) \prec z = x \succ (y \prec z),$$

$$(x \prec y + x \succ y) \succ z = x \succ (y \succ z).$$

This means that D is a $\mathcal{P}$-algebra for the *dendriform operad* $\mathcal{P}_D := \mathcal{P}(\vec{v}_D, \vec{r}_D)$ with $\vec{v}_D = \{\prec, \succ\}$ and

$$\vec{r}_D = \{(\prec \otimes \prec, \prec \otimes (\prec + \succ)), \quad (\succ \otimes \prec, \succ \otimes \prec), ((\prec + \succ) \otimes \succ, \succ \otimes \succ)\}. \tag{6.6}$$

3. (Tridendriform algebra) A *tridendriform algebra* is a $\mathbf{k}$-vector space T equipped with binary operations $\prec, \succ$ and $\cdot$ that satisfy the relations

$$\begin{gathered} (x \prec y) \prec z = x \prec (y \star z), \quad (x \succ y) \prec z = x \succ (y \prec z), \\ (x \star y) \succ z = x \succ (y \succ z), \quad (x \succ y) \cdot z = x \succ (y \cdot z), \\ (x \prec y) \cdot z = x \cdot (y \succ z), \quad (x \cdot y) \prec z = x \cdot (y \prec z), (x \cdot y) \cdot z = x \cdot (y \cdot z) \end{gathered}$$

for $x, y, z \in D$. Here $\star = \prec + \succ + \cdot$. Thus a tridendriform algebra is a $\mathcal{P}$-algebra for the *tridendriform operad* $\mathcal{P}_T := \mathcal{P}(\vec{v}_T, \vec{r}_T)$ with $\vec{v}_T = \{\prec, \succ, \cdot\}$ and

$$\begin{aligned} \vec{r}_T = \{&(\prec \otimes \prec, \prec \otimes \star), (\succ \otimes \prec, \succ \otimes \prec), (\star \otimes \succ, \succ \otimes \succ), (\succ \otimes \cdot, \succ \otimes \cdot), \\ &(\prec \otimes \cdot, \cdot \otimes \succ), (\cdot \otimes \prec, \cdot \otimes \prec), (\cdot \otimes \cdot, \cdot \otimes \cdot)\}. \end{aligned} \tag{6.7}$$

Note that the tridendriform algebra contains an associative operation $\cdot$ which is part of the splitting of the associative operation $\star$.

4. (NS-algebra) An *NS-algebra* of Leroux [127] is a $\mathbf{k}$-vector space N with three binary operations $\prec, \succ, \bullet$ that satisfy the relations

$$(x \prec y) \prec z = x \prec (y \star z), \quad (x \succ y) \prec z = x \succ (y \prec z), \quad (x \star y) \succ z = x \succ (y \succ z),$$

$$(x \star y) \bullet z + (x \bullet y) \prec z = x \succ (y \bullet z) + x \bullet (y \star z),$$

for $x, y, z \in D$. Here $\star = \prec + \succ + \bullet$ gives an associative operation. This is a $\mathcal{P}_N$-algebra for the operad $\mathcal{P}_N := \mathcal{P}(\vec{v}_N, \vec{r}_N)$ with $\vec{v}_N = \{\prec, \succ \bullet\}$ and

$$\begin{aligned} \vec{r}_N = \{&(\prec \otimes \prec, \prec \otimes \star), (\succ \otimes \prec, \succ \otimes \prec), (\star \otimes \succ, \succ \otimes \succ), \\ &(\star \otimes \bullet + \bullet \otimes \prec, \succ \otimes \bullet + \bullet \otimes \star)\}. \end{aligned} \tag{6.8}$$

5. (L-dipterous algebras) An *L-dipterous algebra* [139] is a $\mathcal{P}_L$-algebra for the operad $\mathcal{P}_L := (\vec{v}_L, \vec{r}_L)$, where $\vec{v}_L = \{\star, \succ\}$ and

$$\vec{r}_L = \{(\star \otimes \star, \star \otimes \star), (\star \otimes \succ, \succ \otimes \succ), (\succ \otimes \star, \succ \otimes \star)\}. \tag{6.9}$$

An L-anti-dipterous algebra is a $\mathcal{P}_P$-algebra for operad $\mathcal{P}_P := (\vec{v}_P, \vec{r}_P)$, where $\vec{v}_P = \{\star, \prec\}$ and

$$\vec{r}_P = \{(\star \otimes \star, \star \otimes \star), (\prec \otimes \prec, \prec \otimes \star), (\star \otimes \prec, \star \otimes \prec)\}. \tag{6.10}$$

6.2 Products of binary quadratic NS operads

6.2.1 Definitions

Manin [142, 143] defined the black dot product $\bullet$ for quadratic algebras. It was later generalized for operads [80]. We define a similar product, called the black square product, for binary quadratic ns operads.

Let $\mathcal{P}_1 = (V_1, R_1)$ and $\mathcal{P}_2 = (V_2, R_2)$ be two binary quadratic ns operads. Define

$$\mathcal{P}(V_1, R_1)\blacksquare\,\mathcal{P}(V_2, R_2) = \mathcal{P}(V_1 \otimes V_2, S_{(23)}(R_1 \otimes R_2)).$$

Here $S_{(23)}$ means the exchange of factors 2 and 3 in the tensor product. We will rephrase this product more precisely for later applications.

For $\mathcal{P}(V_1, R_1)$ and $\mathcal{P}(V_2, R_2)$ as above, and for $\odot^{(i)} \in V_i,\ i = 1, 2$, we use a column vector $\begin{bmatrix} \odot_1 \\ \odot_2 \end{bmatrix}$ to denote the tensor product $\odot_1 \otimes \odot_2 \in V_1 \otimes V_2$. The purpose of this is to distinguish it from the tensor product in $R_i \subseteq V_i^{\otimes 2} \oplus V_i^{\otimes 2}$. Further, for $f_i = (\odot_i^{(1)} \otimes \odot_i^{(2)}, \odot_i^{(3)} \otimes \odot_i^{(4)}) \in V_i^{\otimes 2} \oplus V_i^{\otimes 2},\ i = 1, 2$, define

$$\begin{bmatrix} f_1 \\ f_2 \end{bmatrix} = \left(\begin{bmatrix} \odot_1^{(1)} \\ \odot_2^{(1)} \end{bmatrix} \otimes \begin{bmatrix} \odot_1^{(2)} \\ \odot_2^{(2)} \end{bmatrix}, \begin{bmatrix} \odot_1^{(3)} \\ \odot_2^{(3)} \end{bmatrix} \otimes \begin{bmatrix} \odot_1^{(4)} \\ \odot_2^{(4)} \end{bmatrix} \right) \in (V_1 \otimes V_2)^{\otimes 2} \oplus (V_1 \otimes V_2)^{\otimes 2}.$$

This notation extends by bilinearity to all $f_i \in V_i^{\otimes 2} \oplus V_i^{\otimes 2},\ i = 1, 2$. More precisely, elements of $V_i^{\otimes 2} \oplus V_i^{\otimes 2}$ are finite sums of the form

$$f_1 = \sum_j (\odot_{1,j}^{(1)} \otimes \odot_{1,j}^{(2)}, \odot_{1,j}^{(3)} \otimes \odot_{1,j}^{(4)}), \quad \odot_{1,j}^{(r)} \in V_1, \quad r = 1, \ldots, 4$$

and

$$f_2 = \sum_k (\odot_{2,k}^{(1)} \otimes \odot_{2,k}^{(2)}, \odot_{2,k}^{(3)} \otimes \odot_{2,k}^{(4)}), \quad \odot_{2,k}^{(r)} \in V_2, \quad r = 1, \ldots, 4.$$

We then define

$$\begin{bmatrix} f_1 \\ f_2 \end{bmatrix} = \sum_{j,k} \left(\begin{bmatrix} \odot_{1,j}^{(1)} \\ \odot_{2,k}^{(1)} \end{bmatrix} \otimes \begin{bmatrix} \odot_{1,j}^{(2)} \\ \odot_{2,k}^{(2)} \end{bmatrix}, \begin{bmatrix} \odot_{1,j}^{(3)} \\ \odot_{2,k}^{(3)} \end{bmatrix} \otimes \begin{bmatrix} \odot_{1,j}^{(4)} \\ \odot_{2,k}^{(4)} \end{bmatrix} \right).$$

We define a subspace of $(V_1 \otimes V_2)^{\otimes 2} \oplus (V_1 \otimes V_2)^{\otimes 2}$ by

$$R_1 \blacksquare\, R_2 = \left\{ \begin{bmatrix} f_1 \\ f_2 \end{bmatrix} \middle| f_i \in R_i, i = 1, 2 \right\}.$$

So $R_1 \blacksquare\, R_2$ can be regarded as the set of relations for the operator set $V_1 \otimes V_2$.

Definition 6.2.1. The *black square product* of $\mathcal{P}_1 = \mathcal{P}(V_1, R_1)$ and $\mathcal{P}_2 = \mathcal{P}(V_2, R_2)$ is the operad

$$\mathcal{P}(V_1, R_1) \blacksquare \mathcal{P}(V_2, R_2) := \mathcal{P}(V_1 \otimes V_2, R_1 \blacksquare R_2).$$

Let $\vec{v}_i = \{v_{i,j}\}$ be a basis of V_i, $i = 1, 2$. Then

$$\vec{v}_1 \blacksquare \vec{v}_2 := \left\{ \begin{bmatrix} v_{1,j} \\ v_{2,k} \end{bmatrix} \right\}_{j,k}$$

is a basis of $V_1 \otimes V_2$. Similarly, let $\vec{r}_i = \{r_{i,j}\}$ be a basis of R_i, $i = 1, 2$. Then

$$\vec{r}_1 \blacksquare \vec{r}_2 := \left\{ \begin{bmatrix} r_{1,j} \\ r_{2,k} \end{bmatrix} \right\}_{j,k}$$

is a basis of $R_1 \blacksquare R_2$ and takes the shape of a box (matrix).

6.2.2 Basic properties

6.2.2.1 *Transposes and isomorphisms*

Let $\mathcal{P}(V_1 \otimes V_2, R_1 \blacksquare R_2)$ be the black square product of binary quadratic ns operads $\mathcal{P}_1 = \mathcal{P}(V_1, R_1)$ and $\mathcal{P}_2 = \mathcal{P}(V_2, R_2)$. We define the **transpose** of $\mathcal{P}(V_1 \otimes V_2, R_1 \blacksquare R_2)$ by $\mathcal{P}(V_2 \otimes V_1, R_2 \blacksquare R_1)$. It is obtained by changing the two factors in $\begin{bmatrix} \odot_1 \\ \odot_2 \end{bmatrix}$ throughout $V_1 \otimes V_2$ and $R_1 \blacksquare R_2$.

Lemma 6.2.2. *Let $\mathcal{P}(V_i, R_i)$, $i = 1, 2$, be two binary quadratic ns operads.*
(a) *$\mathcal{P}(V_1 \otimes V_2, R_1 \blacksquare R_2)$ is isomorphic to its transpose $\mathcal{P}(V_2 \otimes V_1, R_2 \blacksquare R_1)$.*
(b) *If $\mathcal{P}(V_1, R_1)$ is isomorphic to $\mathcal{P}(V_1', R_1')$ and $\mathcal{P}(V_2, R_2)$ is isomorphic to $\mathcal{P}(V_2', R_2')$, then $\mathcal{P}(V_1 \otimes V_2, R_1 \blacksquare R_2)$ is isomorphic to $\mathcal{P}(V_1' \otimes V_2', R_1' \blacksquare R_2')$.*

Proof. (a). The linear map $V_1 \otimes V_2 \to V_2 \otimes V_1$ sending $\begin{bmatrix} v_1 \\ v_2 \end{bmatrix}$ to $\begin{bmatrix} v_2 \\ v_1 \end{bmatrix}$ is bijective that sends $\begin{bmatrix} \star_1 \\ \star_2 \end{bmatrix}$ to $\begin{bmatrix} \star_2 \\ \star_1 \end{bmatrix}$ and induces a bijective linear map $R_1 \otimes R_2 \to R_2 \otimes R_1$. This proves the claim.

The proof of (b) is similar. □

6.2.2.2 *Tensor products*

Proposition 6.2.3. *Let D_i be a $\mathcal{P}(V_i, R_i)$-algebra for $i = 1, 2$. For $a_1 \otimes a_2, b_1 \otimes b_2 \in D_1 \otimes D_2$ and $\odot_1 \in V_1, \odot_2 \in V_2$, define*

$$(a_1 \otimes a_2) \begin{bmatrix} \odot_1 \\ \odot_2 \end{bmatrix} (b_1 \otimes b_2) = (a_1 \odot_1 b_1) \otimes (a_2 \odot_2 b_2).$$

This makes $D_1 \otimes D_2$ into an algebra over the operad $\mathcal{P}(V_1 \otimes V_2, R_1 \blacksquare R_2)$.

For example, when both $\mathcal{P}(V_1, R_1)$ and $\mathcal{P}(V_2, R_2)$ are the operads for the dendriform algebras, then $D_1 \otimes D_2$ is a quadri-algebra. See [6, 1.5]. When both of them are of the operad for the tridendriform algebras, then $D_1 \otimes D_2$ is an ennea-algebra. When the two operads are for tridendriform algebras and of NS-algebras, respectively, then $D_1 \otimes D_2$ is a dendriform-Nijenhuis algebra [127].

Proof. We only need to verify that the relations in $R_1 \blacksquare R_2$ are satisfied by $D_1 \otimes D_2$.

Recall that $R_1 \blacksquare R_2$ consists of elements of the form $\begin{bmatrix} f_1 \\ f_2 \end{bmatrix}$ with $f_i \in R_i,\ i = 1, 2$. Further, if

$$f_i = \sum_{j_i} \left(\odot^{(1)}_{i,j_i} \otimes \odot^{(2)}_{i,j_i}, \odot^{(3)}_{i,j_i} \otimes \odot^{(4)}_{i,j_i} \right) \in R_i, \quad i = 1, 2,$$

then

$$\begin{bmatrix} f_1 \\ f_2 \end{bmatrix} = \sum_{j_1, j_2} \left(\begin{bmatrix} \odot^{(1)}_{1,j_1} \\ \odot^{(1)}_{2,j_2} \end{bmatrix} \otimes \begin{bmatrix} \odot^{(2)}_{1,j_1} \\ \odot^{(2)}_{2,j_2} \end{bmatrix}, \begin{bmatrix} \odot^{(3)}_{1,j_1} \\ \odot^{(3)}_{2,j_2} \end{bmatrix} \otimes \begin{bmatrix} \odot^{(4)}_{1,j_1} \\ \odot^{(4)}_{2,j_2} \end{bmatrix} \right).$$

For $x_i, y_i, z_i \in D_i,\ i = 1, 2$, we have

$$\begin{aligned}
&\sum_{j_1, j_2} \left((x_1 \otimes x_2) \begin{bmatrix} \odot^{(1)}_{1,j_1} \\ \odot^{(1)}_{2,j_2} \end{bmatrix} (y_1 \otimes y_2) \right) \begin{bmatrix} \odot^{(2)}_{1,j_1} \\ \odot^{(2)}_{2,j_2} \end{bmatrix} (z_1 \otimes z_2) \\
&= \sum_{j_1, j_2} \left((x_1 \odot^{(1)}_{1,j_1} y_1) \odot^{(2)}_{1,j_1} z_1 \right) \otimes \left((x_2 \odot^{(1)}_{2,j_2} y_2) \odot^{(2)}_{2,j_2} z_2 \right) \\
&= \left(\sum_{j_1} (x_1 \odot^{(1)}_{1,j_1} y_1) \odot^{(2)}_{1,j_1} z_1 \right) \otimes \left(\sum_{j_2} (x_2 \odot^{(1)}_{2,j_2} y_2) \odot^{(2)}_{2,j_2} z_2 \right) \\
&= \left(\sum_{j_1} x_1 \odot^{(3)}_{1,j_1} (y_1 \odot^{(4)}_{1,j_1} z_1) \right) \otimes \left(\sum_{j_2} x_2 \odot^{(3)}_{2,j_2} (y_2 \odot^{(4)}_{2,j_2} z_2) \right) \\
&= \sum_{j_1, j_2} x \begin{bmatrix} \odot^{(3)}_{1,j_1} \\ \odot^{(3)}_{2,j_2} \end{bmatrix} \left(y \begin{bmatrix} \odot^{(4)}_{1,j_1} \\ \odot^{(4)}_{2,j_2} \end{bmatrix} z \right).
\end{aligned}$$

This is what we want. □

6.2.3 Examples

We now show that some analogues of dendriform algebras are products of more basic algebras. Therefore, their generators and relations can be easily described.

6.2.3.1 *Quadri-algebra*

The *quadri-algebra* of Aguiar and Loday [6] is defined by four binary operations $\{\nearrow, \nwarrow, \searrow, \swarrow\}$ and 9 relations. Using the auxiliary operations

$$\wedge = \nearrow + \nwarrow, \vee = \searrow + \swarrow, \prec = \nwarrow + \swarrow, \succ = \nearrow + \searrow, \star = \wedge + \vee = \prec + \succ,$$

the 9 relations are given in the following 3×3 matrix.

$$\begin{array}{lll} (x \nwarrow y) \nwarrow z = x \nwarrow (y \star z) & (x \nearrow y) \nwarrow z = x \nearrow (y \prec z) & (x \wedge y) \nearrow z = x \nearrow (y \succ z) \\ (x \swarrow y) \nwarrow z = x \swarrow (y \wedge z) & (x \searrow y) \nwarrow z = x \searrow (y \nwarrow z) & (x \vee y) \nearrow z = x \searrow (y \nearrow z) \\ (x \prec y) \swarrow z = x \swarrow (y \vee z) & (x \succ y) \swarrow z = x \searrow (y \swarrow z) & (x \star y) \searrow z = x \searrow (y \searrow z) \end{array} \tag{6.11}$$

To put quadri-algebra in the context of the black square product, recall that the dendriform algebra of Loday is the algebra for the operad $\mathcal{P}_D := \mathcal{P}(\vec{v}_D, \vec{r}_D)$ with $\vec{v}_D = \{\prec, \succ\}$ and

$$\vec{r}_D = \{(\prec \otimes \prec, \prec \otimes \star), (\succ \otimes \prec, \succ \otimes \prec), (\star \otimes \succ, \succ \otimes \succ)\},$$

where $\star = \succ + \prec$.

Proposition 6.2.4. *The operad $\mathcal{P}_Q$ of the quadri-algebra is isomorphic to the operad*

$$\mathcal{P}(V_D, R_D) \blacksquare \mathcal{P}(V_D, R_D) = \mathcal{P}(V_D \otimes V_D, R_D \blacksquare R_D).$$

Since the two factors of the product are the same, the transpose of a quadri-algebra is the same algebra. Since the opposite of a dendriform algebra is still a dendriform algebra, by Lemma 6.2.2, if we exchange one or both pairs of $\prec$ and $\succ$, we still obtain a quadri-algebra. The opposite quadri-algebra in [6] is obtained when both pairs are exchanged. As a consequence, applying every element from the dihedral group D_4 of order 8 to a quadri-algebra again gives a quadri-algebra.

Proof. We first define a bijection between the binary operations of the quadri-algebra and the binary operations

$$\vec{v}_D \otimes \vec{v}_D = \left\{ \begin{bmatrix} \prec \\ \prec \end{bmatrix}, \begin{bmatrix} \prec \\ \succ \end{bmatrix}, \begin{bmatrix} \succ \\ \prec \end{bmatrix}, \begin{bmatrix} \succ \\ \succ \end{bmatrix} \right\}$$

of the black square product which is given in the following table.

$$\begin{array}{lll} \nwarrow \leftrightarrow \begin{bmatrix} \prec \\ \prec \end{bmatrix}, & \nearrow \leftrightarrow \begin{bmatrix} \prec \\ \succ \end{bmatrix}, & \wedge \leftrightarrow \begin{bmatrix} \prec \\ \star \end{bmatrix} \\ \swarrow \leftrightarrow \begin{bmatrix} \succ \\ \prec \end{bmatrix}, & \searrow \leftrightarrow \begin{bmatrix} \succ \\ \succ \end{bmatrix}, & \vee \leftrightarrow \begin{bmatrix} \succ \\ \star \end{bmatrix} \\ \prec \leftrightarrow \begin{bmatrix} \star \\ \prec \end{bmatrix}, & \succ \leftrightarrow \begin{bmatrix} \star \\ \succ \end{bmatrix}, & \star \leftrightarrow \begin{bmatrix} \star \\ \star \end{bmatrix} \end{array} \tag{6.12}$$

Here the entries on the third row and column are defined to be the sums along the corresponding projections.

To help to visualize this bijection, the reader can imagine the xy plane as sitting in the 3-dimensional coordinate system in the usual way. Thus the plane is laying flat with the x-axis pointing outwards to the reader and the y-axis pointing to the right. Also $\prec$ (resp. $\succ$) points to the negative (resp. positive) direction of the axis. Then, for example, the northwest arrow $\nwarrow$ should be visualized not pointing up and left, but rather inward and left to the third quadrant — the negative direction in both the x and y coordinates. This agrees with the meaning of $\left[\begin{smallmatrix}\prec\\ \prec\end{smallmatrix}\right]$.

Under the bijection in Eq. (6.12), the relation matrix (6.11) is sent to

$$
\begin{aligned}
&(x\left[\begin{smallmatrix}\prec\\ \prec\end{smallmatrix}\right]y)\left[\begin{smallmatrix}\prec\\ \prec\end{smallmatrix}\right]z = x\left[\begin{smallmatrix}\prec\\ \prec\end{smallmatrix}\right](y\left[\begin{smallmatrix}\star\\ \star\end{smallmatrix}\right]z), \quad (x\left[\begin{smallmatrix}\prec\\ \succ\end{smallmatrix}\right]y)\left[\begin{smallmatrix}\prec\\ \prec\end{smallmatrix}\right]z = x\left[\begin{smallmatrix}\prec\\ \succ\end{smallmatrix}\right](y\left[\begin{smallmatrix}\star\\ \prec\end{smallmatrix}\right]z),\\
&(x\left[\begin{smallmatrix}\prec\\ \star\end{smallmatrix}\right]y)\left[\begin{smallmatrix}\prec\\ \succ\end{smallmatrix}\right]z = x\left[\begin{smallmatrix}\prec\\ \succ\end{smallmatrix}\right](y\left[\begin{smallmatrix}\star\\ \succ\end{smallmatrix}\right]z), \quad (x\left[\begin{smallmatrix}\succ\\ \prec\end{smallmatrix}\right]y)\left[\begin{smallmatrix}\prec\\ \prec\end{smallmatrix}\right]z = x\left[\begin{smallmatrix}\succ\\ \prec\end{smallmatrix}\right](y\left[\begin{smallmatrix}\prec\\ \star\end{smallmatrix}\right]z),\\
&(x\left[\begin{smallmatrix}\succ\\ \succ\end{smallmatrix}\right]y)\left[\begin{smallmatrix}\prec\\ \prec\end{smallmatrix}\right]z = x\left[\begin{smallmatrix}\succ\\ \succ\end{smallmatrix}\right](y\left[\begin{smallmatrix}\prec\\ \prec\end{smallmatrix}\right]z), \quad (x\left[\begin{smallmatrix}\succ\\ \star\end{smallmatrix}\right]y)\left[\begin{smallmatrix}\prec\\ \succ\end{smallmatrix}\right]z = x\left[\begin{smallmatrix}\succ\\ \succ\end{smallmatrix}\right](y\left[\begin{smallmatrix}\prec\\ \succ\end{smallmatrix}\right]z),\\
&(x\left[\begin{smallmatrix}\star\\ \prec\end{smallmatrix}\right]y)\left[\begin{smallmatrix}\succ\\ \prec\end{smallmatrix}\right]z = x\left[\begin{smallmatrix}\succ\\ \prec\end{smallmatrix}\right](y\left[\begin{smallmatrix}\succ\\ \star\end{smallmatrix}\right]z), \quad (x\left[\begin{smallmatrix}\star\\ \succ\end{smallmatrix}\right]y)\left[\begin{smallmatrix}\succ\\ \prec\end{smallmatrix}\right]z = x\left[\begin{smallmatrix}\succ\\ \succ\end{smallmatrix}\right](y\left[\begin{smallmatrix}\succ\\ \prec\end{smallmatrix}\right]z),\\
&(x\left[\begin{smallmatrix}\star\\ \star\end{smallmatrix}\right]y)\left[\begin{smallmatrix}\succ\\ \succ\end{smallmatrix}\right]z = x\left[\begin{smallmatrix}\succ\\ \succ\end{smallmatrix}\right](y\left[\begin{smallmatrix}\succ\\ \succ\end{smallmatrix}\right]z)
\end{aligned}
$$

which is simply the matrix of $\vec{r}_D \blacksquare \vec{r}_D$. □

6.2.3.2 *Ennea-algebra*

The ennea-algebra of Leroux [126] has 9 binary operations

$$\nwarrow, \uparrow, \nearrow, \prec, \cdot, \succ, \swarrow, \downarrow, \searrow$$

and 49 relations.

Proposition 6.2.5. *The operad $\mathcal{P}_E$ of the ennea-algebra is isomorphic to the black square product*

$$(V_T, R_T) \blacksquare (V_T, R_T),$$

where $\mathcal{P}(V_T, R_T)$ is the operad of the tridendriform algebra.

Again it is apparent that exchanging $\begin{bmatrix}\odot_1\\ \odot_2\end{bmatrix} \leftrightarrow \begin{bmatrix}\odot_2\\ \odot_1\end{bmatrix}$ gives the transpose of the relation matrix. So the transpose of an ennea-algebra is the same algebra. It is also clear that the opposite algebra is the self-product of the opposite tridendriform algebra, and elements in D_4 give algebras isomorphic to the ennea-algebra.

Proof. Recall that the operad of the tridendriform algebra is $\mathcal{P}_T := \mathcal{P}(\vec{v}_T, \vec{r}_T)$ with $\vec{v}_T = \{\prec, \succ \cdot\}$ and

$$\vec{r}_T = \{(\prec \otimes \prec, \prec \otimes \star), (\succ \otimes \prec, \succ \otimes \prec), (\star \otimes \succ, \succ \otimes \succ), (\succ \otimes \cdot, \succ \otimes \cdot),$$

$$(\prec\otimes\cdot, \cdot\otimes\succ), (\cdot\otimes\prec, \cdot\otimes\prec), (\cdot\otimes\cdot, \cdot\otimes\cdot)\}.$$

Similar to the quadri-algebra, we first give a bijection between the operations of the operad of the ennea-algebra and those in $V_T\otimes V_T$ in the following table. The entries on the fourth row and column are defined to be the sums along the corresponding projections.

$$\begin{array}{cccc}
\nwarrow\leftrightarrow\left[\begin{smallmatrix}\prec\\ \prec\end{smallmatrix}\right], & \uparrow\leftrightarrow\left[\begin{smallmatrix}\prec\\ \cdot\end{smallmatrix}\right], & \nearrow\leftrightarrow\left[\begin{smallmatrix}\prec\\ \succ\end{smallmatrix}\right], & \wedge\leftrightarrow\left[\begin{smallmatrix}\prec\\ \star\end{smallmatrix}\right]\\
\prec\leftrightarrow\left[\begin{smallmatrix}\cdot\\ \prec\end{smallmatrix}\right], & \cdot\leftrightarrow\left[\begin{smallmatrix}\cdot\\ \cdot\end{smallmatrix}\right], & \succ\leftrightarrow\left[\begin{smallmatrix}\cdot\\ \succ\end{smallmatrix}\right], & \star\leftrightarrow\left[\begin{smallmatrix}\cdot\\ \star\end{smallmatrix}\right]\\
\swarrow\leftrightarrow\left[\begin{smallmatrix}\succ\\ \prec\end{smallmatrix}\right], & \downarrow\leftrightarrow\left[\begin{smallmatrix}\succ\\ \cdot\end{smallmatrix}\right], & \searrow\leftrightarrow\left[\begin{smallmatrix}\succ\\ \succ\end{smallmatrix}\right], & \vee\leftrightarrow\left[\begin{smallmatrix}\succ\\ \star\end{smallmatrix}\right]\\
\triangleleft\leftrightarrow\left[\begin{smallmatrix}\star\\ \prec\end{smallmatrix}\right], & \bar{\cdot}\leftrightarrow\left[\begin{smallmatrix}\star\\ \cdot\end{smallmatrix}\right], & \triangleright\leftrightarrow\left[\begin{smallmatrix}\star\\ \succ\end{smallmatrix}\right], & \bar{\star}\leftrightarrow\left[\begin{smallmatrix}\star\\ \star\end{smallmatrix}\right]
\end{array}$$

Applying this bijection to the 49 relations for the ennea-algebra in [126], we see that the relations become the entries of the 7×7 matrix $\vec{r}_T\blacksquare\vec{r}_T$. □

6.2.3.3 *Dendriform-Nijenhuis algebra*

The dendriform-Nijenhuis algebra [127] is equipped with 9 binary operations

$$\nearrow,\ \searrow,\ \swarrow,\ \nwarrow,\ \uparrow,\ \downarrow,\ \tilde{\prec},\ \tilde{\succ},\ \tilde{\bullet} \tag{6.13}$$

satisfying 28 relations.

Proposition 6.2.6. *The operad of the dendriform-Nijenhuis algebra is isomorphic to the operad $\mathcal{P}(V_T,R_T)\blacksquare\mathcal{P}(V_N,R_N)$ where $\mathcal{P}(V_T,R_T)$ is the operad of the tridendriform algebra and $\mathcal{P}(V_N,R_N)$ is the operad of the NS-algebra in Eq.* (6.8).

Proof. We give a correspondence between the binary operations of the dendriform-Nijenhuis algebra and those of the black square product. To distinguish operators in the tridendriform algebra and NS-algebra, we denote $V_N=\{<,>,\bullet\}$ and $\diamondsuit=<+>+\bullet$. A bijection between the binary operations in Eq. (6.13) and $\vec{v}_T\otimes\vec{v}_N$ is given by

$$\begin{array}{cccc}
\nwarrow\leftrightarrow\left[\begin{smallmatrix}\prec\\ <\end{smallmatrix}\right], & \uparrow\leftrightarrow\left[\begin{smallmatrix}\prec\\ \bullet\end{smallmatrix}\right], & \nearrow\leftrightarrow\left[\begin{smallmatrix}\prec\\ >\end{smallmatrix}\right], & \wedge\leftrightarrow\left[\begin{smallmatrix}\prec\\ \diamondsuit\end{smallmatrix}\right]\\
\tilde{\prec}\leftrightarrow\left[\begin{smallmatrix}\cdot\\ <\end{smallmatrix}\right], & \tilde{\bullet}\leftrightarrow\left[\begin{smallmatrix}\bullet\\ \cdot\end{smallmatrix}\right], & \tilde{\succ}\leftrightarrow\left[\begin{smallmatrix}\cdot\\ >\end{smallmatrix}\right], & \tilde{\star}\leftrightarrow\left[\begin{smallmatrix}\cdot\\ \diamondsuit\end{smallmatrix}\right]\\
\swarrow\leftrightarrow\left[\begin{smallmatrix}\succ\\ <\end{smallmatrix}\right], & \downarrow\leftrightarrow\left[\begin{smallmatrix}\succ\\ \bullet\end{smallmatrix}\right], & \searrow\leftrightarrow\left[\begin{smallmatrix}\succ\\ >\end{smallmatrix}\right], & \vee\leftrightarrow\left[\begin{smallmatrix}\succ\\ \diamondsuit\end{smallmatrix}\right]\\
\triangleleft\leftrightarrow\left[\begin{smallmatrix}\star\\ <\end{smallmatrix}\right], & \bar{\bullet}\leftrightarrow\left[\begin{smallmatrix}\star\\ \bullet\end{smallmatrix}\right], & \triangleright\leftrightarrow\left[\begin{smallmatrix}\star\\ >\end{smallmatrix}\right], & \bar{\star}\leftrightarrow\left[\begin{smallmatrix}\star\\ \diamondsuit\end{smallmatrix}\right]
\end{array}$$

The operations on the fourth row and column are defined to be the sum along the corresponding projections. Then the relations in the operad of the dendriform-Nijenhuis algebra [127] is identified with the 7×4 matrix $\vec{r}_T\blacksquare\vec{r}_N$.

□

6.2.3.4 *Octo-algebra*

The operad of the octo-algebra of Leroux [128] is isomorphic to the black square product

$$\mathcal{P}_Q \blacksquare \mathcal{P}_D := \mathcal{P}(V_Q \otimes V_D, R_Q \blacksquare R_D),$$

where $\mathcal{P}(V_Q, R_Q)$ is the operad of the quadri-algebra and $\mathcal{P}(V_D, R_D)$ is the operad of the dendriform algebra. As we will see later in Section 6.2.4, it is also the third power of operad of the dendriform algebra.

6.2.4 Powers of a binary quadratic ns operad

Inductively, we define the black square product of any finite number of binary quadratic NS operads: given $\mathcal{P}(V_i, R_i), 1 \leqslant i \leqslant n$, define

$$\mathop{\blacksquare}_{i=1}^{n} \mathcal{P}(V_i, R_i) := \left(\mathop{\blacksquare}_{i=1}^{n-1} \mathcal{P}(V_i, R_i) \right) \blacksquare \mathcal{P}(V_n, R_n).$$

Particularly, we define the powers of a specific type of binary quadratic nonsymmetric operad. For simplicity, we only describe the powers of a tridendriform algebra.

Definition 6.2.7. An N-th power of the tridendriform algebra is a **k**-vector space D equipped with 3^N binary operations

$$\begin{bmatrix} \odot_1 \\ \vdots \\ \odot_N \end{bmatrix}, \quad \odot_i \in \{\succ_i, \prec_i, \circ_i\}, \quad 1 \leqslant i \leqslant N$$

such that, for any choice of $(\odot_i^{(1)} \otimes \odot_i^{(2)}, \odot_i^{(3)} \otimes \odot_i^{(4)})$ in

$$R_i := \left\{ \begin{array}{l} (\prec_i \otimes \prec_i, \prec_i \otimes \star_i), (\succ_i \otimes \prec_i, \succ_i \otimes \prec_i), (\star_i \otimes \succ_i, \succ_i \otimes \succ_i), (\succ_i \otimes \circ_i, \\ \succ_i \otimes \circ_i), (\prec_i \otimes \circ_i, \circ_i \otimes \succ_i), (\circ_i \otimes \prec_i, \circ_i \otimes \prec_i), (\circ_i \otimes \circ_i, \circ_i \otimes \circ_i) \end{array} \right\}$$

as $1 \leqslant i \leqslant N$, we have

$$\left(x \begin{bmatrix} \odot_1^{(1)} \\ \vdots \\ \odot_N^{(1)} \end{bmatrix} y \right) \begin{bmatrix} \odot_1^{(2)} \\ \vdots \\ \odot_N^{(2)} \end{bmatrix} z = x \begin{bmatrix} \odot_1^{(3)} \\ \vdots \\ \odot_N^{(3)} \end{bmatrix} \left(y \begin{bmatrix} \odot_1^{(4)} \\ \vdots \\ \odot_N^{(4)} \end{bmatrix} z \right).$$

Here $\star_i = \succ_i + \prec_i + \circ_i$. When $\circ_i = 0$ for $1 \leqslant i \leqslant N$, we call the algebra the N-th power of the dendriform algebra.

Of course, the dendriform algebra and quadri-algebra (resp. tridendriform algebra and ennea-algebra) are just the first and second powers of dendriform algebra (resp. tridendriform algebra). The octo-algebra introduced by Leroux [128] is the third power of the dendriform algebra.

We then have

Proposition 6.2.8. *The operad for the N-th power of the tridendriform algebra is the N-th black square power of the operad $\mathcal{P}_T$ of the tridendriform algebra: $\mathcal{P}_T^{\blacksquare n}$.*

For example, the octo-algebra was introduced by Leroux [128]. It is defined using 8 operations

$$\nearrow_i, \nwarrow_i, \swarrow_i, \searrow_i, \quad i = 1, 2$$

and 27 relations.

Proposition 6.2.9. *The operad for the octo-algebra is the third power of the operad $(\vec{v}_D, \vec{r}_D)$ for the dendriform algebra.*

Proof. The following table gives a correspondence between these 8 operations and the operations in $\vec{v}_D \otimes \vec{v}_D \otimes \vec{v}_D = \vec{v}_Q \otimes \vec{v}_D$ with $\vec{v}_Q$ being the generators of the quadri-algebra. As in the case of the quadri-algebra, each operation on the third row and third column in the first two blocks is the sum of the corresponding projection, and each operation in the third block is the sum of the corresponding operations on the first and second blocks.

$$\begin{array}{lll}
\nwarrow_1 \leftrightarrow \begin{bmatrix}\prec\\ \prec\\ \prec\end{bmatrix}, & \nearrow_1 \leftrightarrow \begin{bmatrix}\prec\\ \succ\\ \prec\end{bmatrix}, & \wedge_1 \leftrightarrow \begin{bmatrix}\prec\\ \star\\ \prec\end{bmatrix},\\
\swarrow_1 \leftrightarrow \begin{bmatrix}\succ\\ \prec\\ \prec\end{bmatrix}, & \searrow_1 \leftrightarrow \begin{bmatrix}\succ\\ \succ\\ \prec\end{bmatrix}, & \vee_1 \leftrightarrow \begin{bmatrix}\succ\\ \star\\ \prec\end{bmatrix},\\
\prec_1 \leftrightarrow \begin{bmatrix}\star\\ \prec\\ \prec\end{bmatrix}, & \succ_1 \leftrightarrow \begin{bmatrix}\star\\ \succ\\ \prec\end{bmatrix}, & \star_1 \leftrightarrow \begin{bmatrix}\star\\ \star\\ \prec\end{bmatrix},\\
\hline
\nwarrow_2 \leftrightarrow \begin{bmatrix}\prec\\ \prec\\ \succ\end{bmatrix}, & \nearrow_2 \leftrightarrow \begin{bmatrix}\prec\\ \succ\\ \succ\end{bmatrix}, & \wedge_2 \leftrightarrow \begin{bmatrix}\prec\\ \star\\ \succ\end{bmatrix},\\
\swarrow_2 \leftrightarrow \begin{bmatrix}\succ\\ \prec\\ \succ\end{bmatrix}, & \searrow_2 \leftrightarrow \begin{bmatrix}\succ\\ \succ\\ \succ\end{bmatrix}, & \vee_2 \leftrightarrow \begin{bmatrix}\succ\\ \star\\ \succ\end{bmatrix},\\
\prec_2 \leftrightarrow \begin{bmatrix}\star\\ \prec\\ \succ\end{bmatrix}, & \succ_2 \leftrightarrow \begin{bmatrix}\star\\ \succ\\ \succ\end{bmatrix}, & \star_2 \leftrightarrow \begin{bmatrix}\star\\ \star\\ \succ\end{bmatrix},\\
\hline
\nwarrow_{12} \leftrightarrow \begin{bmatrix}\prec\\ \prec\\ \star\end{bmatrix}, & \nearrow_{12} \leftrightarrow \begin{bmatrix}\prec\\ \succ\\ \star\end{bmatrix}, & \bigwedge_{12} \leftrightarrow \begin{bmatrix}\prec\\ \star\\ \star\end{bmatrix},\\
\swarrow_{12} \leftrightarrow \begin{bmatrix}\succ\\ \prec\\ \star\end{bmatrix}, & \searrow_{12} \leftrightarrow \begin{bmatrix}\succ\\ \succ\\ \star\end{bmatrix}, & \bigvee_{12} \leftrightarrow \begin{bmatrix}\succ\\ \star\\ \star\end{bmatrix},\\
\ll \leftrightarrow \begin{bmatrix}\star\\ \prec\\ \star\end{bmatrix}, & \gg \leftrightarrow \begin{bmatrix}\star\\ \succ\\ \star\end{bmatrix}, & \bar{\star} \leftrightarrow \begin{bmatrix}\star\\ \star\\ \star\end{bmatrix},
\end{array}$$

Then the 27 axioms of an octo-algebra in [128] correspond to the entries in

$$\vec{r}_D \blacksquare \vec{r}_D \blacksquare \vec{r}_D = \vec{r}_Q \blacksquare \vec{r}_D.$$

Here $\vec{r}_Q$ is the relation vector of the operad for the quadri-algebra. □

Recall that the opposite of the dendriform algebra is still a dendriform algebra. Considering also permutations of the three coordinates in $\begin{bmatrix} \odot_1 \\ \odot_2 \\ \odot_3 \end{bmatrix}$, we find that each element in the group of rigid motions of the cube gives an algebra isomorphic to the octo-algebra.

6.3 Binary quadratic ns operads from linear operators

In this section we study the relationship between operations of certain linear operators and black square products.

6.3.1 Operators on binary quadratic ns operads

A common method used to obtain a new operad structure from a known operad structure is by means of a linear operator on the known operad. Such linear operators include (left, right and two-sided) Rota-Baxter operators and Nijenhuis operators. Examples of such constructions can be found in [6, 138, 126, 127, 128]. The constructions were usually verified by checking the relations for each case of operators and operads. We will verify the construction for (left, right and two-sided) Rota-Baxter and Nijenhuis operators on all binary quadratic NS operads. We first give the definitions.

Definition 6.3.1. Let D be a $\mathcal{P}$-algebra for a binary quadratic ns operad (V, R). A linear operator P on D is called a *Rota-Baxter operator of weight* λ if, for each $\odot \in V$, we have

$$P(x) \odot P(y) = P(P(x) \odot y + x \odot P(y) + \lambda x \odot y), \quad x,\ y \in D \quad (6.14)$$

A linear operator N on D is called a *Nijenhuis operator* if, for each $\odot \in V$, we have

$$N(x) \odot N(y) = N(N(x) \odot y + x \odot N(y) - N(x \odot y)), \quad x,\ y \in D. \quad (6.15)$$

Theorem 6.3.2. *Let D be a $\mathcal{P}$-algebra for a binary quadratic NS operad $\mathcal{P}(V, R)$.*

(a) *Let P be a Rota-Baxter operator of weight λ on the $\mathcal{P}$-algebra D. For each $\odot \in V$, define binary operations on D by*

$$x \begin{bmatrix} \odot \\ \prec \end{bmatrix} y = x \odot P(y), \quad x \begin{bmatrix} \odot \\ \succ \end{bmatrix} y = P(x) \odot y, \quad x \begin{bmatrix} \odot \\ \cdot \end{bmatrix} y = \lambda x \odot y.$$

Then these operations make D into an algebra for the operad $\mathcal{P} \blacksquare \mathcal{P}_T := \mathcal{P}(V \otimes V_T, R \blacksquare R_T)$ where $\mathcal{P}_T := \mathcal{P}(V_T, R_T)$ is the operad of the tridendriform algebra.

(b) *Let N be a Nijenhuis operator on the binary quadratic nonsymmetric algebra D. For each $\odot \in V$, define binary operations on D by*

$$x \begin{bmatrix} \odot \\ \prec \end{bmatrix} y = x \odot N(y), \quad x \begin{bmatrix} \odot \\ \succ \end{bmatrix} y = N(x) \odot y, \quad x \begin{bmatrix} \odot \\ \bullet \end{bmatrix} y = -N(x \odot y).$$

Then these operations make D into an algebra for the operad $\mathcal{P} \blacksquare \mathcal{P}_N := \mathcal{P}(V \otimes V_N, R \blacksquare R_N)$ where $\mathcal{P}_N := (V_N, R_N)$ is the operad for the NS algebra in Eq. (6.8).

Proof. (a). Recall that

$$\vec{r}_T = \{(\prec \otimes \prec, \prec \otimes \star), (\succ \otimes \prec, \succ \otimes \prec), (\star \otimes \succ, \succ \otimes \succ), (\succ \otimes \cdot, \succ \otimes \cdot), \\ (\prec \otimes \cdot, \cdot \otimes \succ), (\cdot \otimes \prec, \cdot \otimes \prec), (\cdot \otimes \cdot, \cdot \otimes \cdot)\}.$$

To prove our claim we only need to prove that, for each

$$f = \sum_{j=1}^{r} (\odot_j^{(1)} \otimes \odot_j^{(2)}, \odot_j^{(3)} \otimes \odot_j^{(4)}) \in R$$

and each of the 7 pairs in $\vec{r}_T$, the "tensor product" of the two is a relation on $V \otimes V_T$. We consider the 7 cases separately.

Case 1. The pair is $(\prec \otimes \prec, \prec \otimes \star)$. We need to verify that

$$f \blacksquare (\prec \otimes \prec, \prec \otimes \star) = \sum_{j=1}^{r} \left(\begin{bmatrix} \odot_j^{(1)} \\ \prec \end{bmatrix} \otimes \begin{bmatrix} \odot_j^{(2)} \\ \prec \end{bmatrix}, \begin{bmatrix} \odot_j^{(3)} \\ \prec \end{bmatrix} \otimes \begin{bmatrix} \odot_j^{(4)} \\ \star \end{bmatrix} \right)$$

is a relation on D. We have

$$\begin{aligned} &\sum_{j=1}^{r} \left(x \begin{bmatrix} \odot_j^{(1)} \\ \prec \end{bmatrix} y \right) \begin{bmatrix} \odot_j^{(2)} \\ \prec \end{bmatrix} z \\ &= \sum_{j=1}^{r} (x \odot_j^{(1)} P(y)) \odot_j^{(2)} P(z) \\ &= \sum_{j=1}^{r} x \odot_j^{(3)} (P(y) \odot_j^{(4)} P(z)) \end{aligned}$$

$$
\begin{aligned}
&= \sum_{j=1}^{r} x \odot_j^{(3)} P(P(y) \odot_j^{(4)} z + y \odot_j^{(4)} P(z) + \lambda y \odot_j^{(4)} z) \\
&= \sum_{j=1}^{r} x \begin{bmatrix} \odot_j^{(3)} \\ \prec \end{bmatrix} (P(y) \odot_j^{(4)} z + y \odot_j^{(4)} P(z) + \lambda y \odot_j^{(4)} z) \\
&= \sum_{j=1}^{r} x \begin{bmatrix} \odot_j^{(3)} \\ \prec \end{bmatrix} \left(y \left(\begin{bmatrix} \odot_j^{(4)} \\ \prec \end{bmatrix} + \begin{bmatrix} \odot_j^{(4)} \\ \succ \end{bmatrix} + \begin{bmatrix} \odot_j^{(4)} \\ \cdot \end{bmatrix} \right) z \right) \\
&= \sum_{j=1}^{r} x \begin{bmatrix} \odot_j^{(3)} \\ \prec \end{bmatrix} \left(y \begin{bmatrix} \odot_j^{(4)} \\ \star \end{bmatrix} z \right),
\end{aligned}
$$

as desired.

Case 2. The pair is $(\succ \otimes \prec, \succ \otimes \prec)$. Then we have

$$
\begin{aligned}
&\sum_{j=1}^{r} \left(x \begin{bmatrix} \odot_j^{(1)} \\ \succ \end{bmatrix} y \right) \begin{bmatrix} \odot_j^{(2)} \\ \prec \end{bmatrix} z \\
&= \sum_{j=1}^{r} (P(x) \odot_j^{(1)} y) \odot_j^{(2)} P(z) \\
&= \sum_{j=1}^{r} P(x) \odot_j^{(3)} (y \odot_j^{(4)} P(z)) \\
&= \sum_{j=1}^{r} x \begin{bmatrix} \odot_j^{(3)} \\ \succ \end{bmatrix} \left(y \begin{bmatrix} \odot_j^{(4)} \\ \prec \end{bmatrix} z \right).
\end{aligned}
$$

Case 3. The pair is $(\star \otimes \succ, \succ \otimes \succ)$. We have

$$
\begin{aligned}
&\sum_{j=1}^{r} \left(x \begin{bmatrix} \odot_j^{(1)} \\ \star \end{bmatrix} y \right) \begin{bmatrix} \odot_j^{(2)} \\ \succ \end{bmatrix} z \\
&= \sum_{j=1}^{r} P(x \odot_j^{(1)} P(y) + P(x) \odot_j^{(1)} y + \lambda x \odot_j^{(1)} y) \odot_j^{(2)} z \\
&= \sum_{j=1}^{r} (P(x) \odot_j^{(1)} P(y)) \odot_j^{(2)} z \\
&= \sum_{j=1}^{r} P(x) \odot_j^{(3)} (P(y) \odot_j^{(4)} z) \\
&= \sum_{j=1}^{r} x \begin{bmatrix} \odot_j^{(3)} \\ \succ \end{bmatrix} \left(y \begin{bmatrix} \odot_j^{(4)} \\ \succ \end{bmatrix} z \right).
\end{aligned}
$$

Case 4. The pair is $(\succ\otimes\cdot,\succ\otimes\cdot)$. We have

$$\begin{aligned}
&\sum_{j=1}^{r}\left(x\begin{bmatrix}\odot_j^{(1)}\\ \succ\end{bmatrix}y\right)\begin{bmatrix}\odot_j^{(2)}\\ \cdot\end{bmatrix}z\\
&=\sum_{j=1}^{r}(P(x)\odot_j^{(1)}y)\odot_j^{(2)}\lambda z\\
&=\sum_{j=1}^{r}P(x)\odot_j^{(3)}(y\odot_j^{(4)}\lambda z)\\
&=\sum_{j=1}^{r}x\begin{bmatrix}\odot_j^{(3)}\\ \succ\end{bmatrix}\left(y\begin{bmatrix}\odot_j^{(4)}\\ \cdot\end{bmatrix}z\right).
\end{aligned}$$

The cases for the pairs $(\prec\otimes\cdot,\cdot\otimes\succ)$, $(\cdot\otimes\prec,\cdot\otimes\prec)$ and $(\cdot\otimes\cdot,\cdot\otimes\cdot)$ are proved in the same way as Case 4.

(b). Now we consider a Nijenhuis operator N. The relation set of the NS-algebra is

$$\begin{aligned}
\vec{r}_N=\{&(\prec\otimes\prec,\prec\otimes\star),(\succ\otimes\prec,\succ\otimes\prec),(\star\otimes\succ,\succ\otimes\succ),\\
&(\star\otimes\bullet+\bullet\otimes\prec,\succ\otimes\bullet+\bullet\otimes\star)\}.
\end{aligned}$$

So we only need to prove that, for each

$$f=\sum_{j=1}^{r}(\odot_j^{(1)}\otimes\odot_j^{(2)},\odot_j^{(3)}\otimes\odot_j^{(4)})\in R$$

and each of the 4 pairs in $\vec{r}_N$, the "tensor product" of the two is a relation for D. The verification of the first three pairs is the same as the first three cases in the Rota-Baxter operator case. For the fourth case, taking f as above, we have

$$\begin{aligned}
&\sum_{j=1}^{r}\left(\left(x\begin{bmatrix}\odot_j^{(1)}\\ \star\end{bmatrix}y\right)\begin{bmatrix}\odot_j^{(2)}\\ \bullet\end{bmatrix}z+\left(x\begin{bmatrix}\odot_j^{(1)}\\ \bullet\end{bmatrix}y\right)\begin{bmatrix}\odot_j^{(2)}\\ \prec\end{bmatrix}z\right)\\
&=\sum_{j=1}^{r}\Big(-N((N(x)\odot_j^{(1)}y+\odot_j^{(1)}N(y)-N(x\odot_j^{(1)}y))\odot_j^{(2)}z)-N(x\odot_j^{(1)}y)\odot_j^{(2)}N(z)\Big)\\
&=\sum_{j=1}^{r}\Big(-N((N(x)\odot_j^{(1)}y+x\odot_j^{(1)}N(y)-N(x\odot_j^{(1)}y))\odot_j^{(2)}z)\\
&\quad-N(N(x\odot_j^{(1)}y)\odot_j^{(2)}z+(x\odot_j^{(1)}y)\odot_j^{(2)}N(z)-N((x\odot_j^{(1)}y)\odot_j^{(2)}z))\Big)\\
&=\sum_{j=1}^{r}\Big(-N((N(x)\odot_j^{(1)}y)\odot_j^{(2)}z+(x\odot_j^{(1)}N(y))\odot_j^{(2)}z
\end{aligned}$$

$$
\begin{aligned}
&+(x \odot_j^{(1)} y) \odot_j^{(2)} N(z) - N((x \odot_j^{(1)} y) \odot_j^{(2)} z))\Big) \\
= &-N\left(\sum_{j=1}^{r} (N(x) \odot_j^{(1)} y) \odot_j^{(2)} z + \sum_{j=1}^{r} (x \odot_j^{(1)} N(y)) \odot_j^{(2)} z\right. \\
&\left.+\sum_{j=1}^{r} (x \odot_j^{(1)} y) \odot_j^{(2)} N(z) - N\left(\sum_{j=1}^{r} (x \odot_j^{(1)} y) \odot_j^{(2)} z\right)\right) \\
= &-N\left(\sum_{j=1}^{r} N(x) \odot_j^{(3)} (y \odot_j^{(4)} z) + \sum_{j=1}^{r} x \odot_j^{(3)} (N(y) \odot_j^{(4)} z)\right. \\
&\left.+\sum_{j=1}^{r} x \odot_j^{(3)} (y \odot_j^{(4)} N(z)) - N\left(\sum_{j=1}^{r} x \odot_j^{(3)} (y \odot_j^{(4)} z)\right)\right).
\end{aligned}
$$

On the other hand,

$$
\begin{aligned}
&\sum_{j=1}^{r}\left(x \begin{bmatrix} \odot_j^{(3)} \\ \succ \end{bmatrix} \left(y \begin{bmatrix} \odot_j^{(4)} \\ \bullet \end{bmatrix} z\right) + x \begin{bmatrix} \odot_j^{(3)} \\ \bullet \end{bmatrix} \left(y \begin{bmatrix} \odot_j^{(4)} \\ \star \end{bmatrix} z\right)\right) \\
= &\sum_{j=1}^{r}\Big(-N(x) \odot_j^{(3)} N(y \odot_j^{(4)} z) - N(x \odot_j^{(3)} (N(y) \odot_j^{(4)} z + y \odot_j^{(4)} N(z) - N(y \odot_j^{(4)} z)))\Big) \\
= &\sum_{j=1}^{r}\Big(- N(N(x) \odot_j^{(3)} (y \odot_j^{(4)} z) + x \odot_j^{(3)} N(y \odot_j^{(4)} z) - N(x \odot_j^{(3)} (y \odot_j^{(4)} z))) \\
&-N(x \odot_j^{(3)} (N(y) \odot_j^{(4)} z + y \odot_j^{(4)} N(z) - N(y \odot_j^{(4)} z)))\Big) \\
= &\sum_{j=1}^{r} -N(N(x) \odot_j^{(3)} (y \odot_j^{(4)} z) - N(x \odot_j^{(3)} (y \odot_j^{(4)} z)) \\
&+x \odot_j^{(3)} (N(y) \odot_j^{(4)} z) + x \odot_j^{(3)} (y \odot_j^{(4)} N(z))).
\end{aligned}
$$

This verifies the last relation. □

Applying Theorem 6.3.2 repeatedly, we get

Corollary 6.3.3. *Let $(D, \{P_i\}_i)$ be an algebra for a binary quadratic NS operad $\mathcal{P}(V, R)$ with k commuting linear operators P_i, such as Rota-Baxter or Nijenhuis operators. We obtain on D an algebra structure for the operad $\mathcal{P}(V, R)\blacksquare\ (\blacksquare_i \mathcal{P}(V_i, R_i))$ where $\mathcal{P}(V_i, R_i)$ is the operad corresponding to the operator P_i, $1 \leqslant i \leqslant k$.*

6.3.2 Examples

The following are some of the constructions that can be obtained from Theorem 6.3.2 and Corollary 6.3.3.

(a) A Rota-Baxter operator of weight zero on an algebra gives a dendriform algebra [3];
(b) A Rota-Baxter operator of non-zero weight on an algebra gives a tridendriform algebra [47, 126];
(c) A Rota-Baxter operator of weight zero on a dendriform algebra gives a quadri-algebra [6];
(d) A pair of commuting Rota-Baxter operator of weight zero on an algebra gives a quadri-algebra [6];
(e) A Rota-Baxter operator of non-zero weight on a tridendriform algebra gives an ennea-algebra [126];
(f) A pair of commuting Rota-Baxter operators of non-zero weight on an algebra gives an ennea-algebra [126];
(g) A Nijenhuis operator on an algebra gives an NS-algebra [127];
(h) A Nijenhuis operator on a tridendriform algebra gives a dendriform-Nijenhuis algebra [127];
(i) A set of three pairwise commuting Rota-Baxter operators of weight zero on an algebra gives an octo-algebra [128].

6.4 Some related literature and problems

This chapter is based on [52]. For further discussion of binary quadratic ns operads and splitting of associativity, see [54, 135]. General references on operads are [140, 147].

For the splitting of more general operads, Manin products and Rota-Baxter operators on operads, see [15, 168, 170].

The concept of unit actions of operads has been introduced by Loday [135] and has been used to construct Hopf algebras on the free algebras. A classification of binary quadratic ns operads with unit actions and the relationship between products of operads and their unit actions are investigated in [54]. More general unit actions on operads can be found in [115].

See [15, 168] for the relationship between Rota-Baxter algebras and dendriform algebras in the context of operads.

Bibliography

[1] E. Abe, *Hopf Algebras*, Cambridge University Press, Cambridge, 1980.

[2] M. Aguiar, Prepoisson algebras, *Lett. Math. Phys.*, **54** (2000), 263 – 277.

[3] M. Aguiar, Infinitesimal Hopf algebras, *Contemporary Mathematics*, **267** (2000), 1 – 29.

[4] M. Aguiar, N. Bergeron and F. Sottile, Combinatorial Hopf algebras and generalized Dehn-Sommerville relations, *Compos. Math.*, **142** (2006), 1 – 30.

[5] M. Aguiar and S. Hsiao, Canonical characters on quasi-symmetric functions and bivariate Catalan numbers, *Electron. J. Combin.*, **11** (2004 – 2005), R15: 34 pp.

[6] M. Aguiar and J. -L. Loday, Quadri-algebras, *J. Pure Applied Algebra*, **191** (2004), 205 – 221.

[7] M. Aguiar and W. Moreira, Combinatorics of the free Baxter algebra, *Electron. J. Combin.*, **13** (2006), R17: 38 pp.

[8] M. Aguiar and F. Sottile, Structure of the Loday-Ronco Hopf algebra of trees, *J. Algebra*, **295** (2006), 473 – 511.

[9] M. Aguiar and F. Sottile, Cocommutative Hopf algebras of permutations and trees, *J. Algebraic Combin.*, **22** (2005), 451 – 470.

[10] H. An and C. Bai, From Rota-Baxter algebras to pre-Lie algebras, *J. Phys. A:Math. Theor.*, **41** (2008), 015201 (19pp).

[11] G.E. Andrew, L. Guo, W. Keigher and K. Ono, Baxter algebras and Hopf algebras, *Trans. Amer. Math. Soc.*, **355** (2003), 4639 – 4656.

[12] F. V. Atkinson, Some aspects of Baxter's functional equation, *J. Math. Anal. and Applications*, **7** (1963), 1 – 30.

[13] F. Baader, and T. Nipkow, *Term Rewriting and All That*, Cambridge University Press, Cambridge, 1998.

[14] C. Bai, A unified algebraic approach to the classical Yang-Baxter equation, *J. Phys. A: Math. Theor.*, **40** (2007), 11073 – 11082.

[15] C. Bai, O. Bellier, L. Guo and X. Ni, Splitting of operations, Manin products and Rota-Baxter operators, to appear in *Int. Math. Res. Not.*, arXiv: 1106.6080.

[16] C. Bai, L. Guo and X. Ni, Nonabelian generalized Lax pairs, the classical Yang-Baxter equation and PostLie algebras, *Comm. Math. Phys.*, **297** (2010), 553 – 596.

[17] C. Bai, L. Guo and X. Ni, Nonabelian generalized Lax pairs, the classical Yang-Baxter equation and PostLie algebras, *Int. Math. Res. Not. IMRM.* (2012); dov: 10.1093/imrn/rnr266.

[18] C. Bai, L. Guo and X. Ni, $\mathcal{O}$-operators on associative algebras, associative Yang-Baxter equations and dendriform algebras (a survey), In "Proceeding of International Workshop on Quantized Algerba and Physics", (2012), 10 – 51, World Scientific Publication Company.

[19] C. Bai, L. Guo and X. Ni, Generalizations of the classical Yang-Baxter equation and $\mathcal{O}$-operators, *J. Math. Phys*, **52** (2011), 063515.

[20] C. Bai, L. Guo and X. Ni, $\mathcal{O}$-operators on associative algebras and associative Yang-Baxter equations, to appear in *Pacific J. Math.*

[21] C. Bai, L. Guo and X., Ni, $\mathcal{O}$-operators on associative algebras and dendriform algebras, arXiv: 1003.2432.

[22] C. Bai, L. Liu and X. Ni, Some results on L-dendriform algebras, *J. Geom. Phys.*, **60** (2010), 940 – 950.

[23] K. Barron, Y. Huang and J. Lepowsky, Factorization of formal exponentials and uniformalization, *J. Algebra*, **228** (2000), 551 – 579.

[24] G. Baxter, An analytic problem whose solution follows from a simple algebraic identity, *Pacific J. Math.*, **10** (1960), 731 – 742.

[25] L. A. Bokut, Y. Chen and J. Qiu, Gröbner-Shirshov bases for associative algebras with multiple operators and free Rota-Baxter algebras, *J. Pure Appl. Alg.*, **214** (2010), 89 – 110.

[26] M. Bordemann, Generalized Lax pairs, the modified classical Yang-Baxter equation, and affine geometry of Lie groups, *Comm. Math. Phys.*, **135** (1990), 201 – 216.

[27] J. M. Borwein and D. M. Bradley, Resolution of some open problems concerning multiple zeta evaluations of arbitrary depth, *Compos. Math.*, **139** (2003), 85 – 100.

[28] J. M. Borwein, D. J. Broadhurst, D. M. Bradley and P. Lisoněk, Special values of multiple polylogarithms, *Trans. Amer. Math. Soc.*, **353** (2001), 907 – 941.

[29] D. M. Bradley, Multiple q-zeta values, *J. Algebra.* **283** (2005), 752 – 798.

[30] W. Cao, An algebraic study of averaging operators, Ph.D. Thesis, Rutgers University at Newark, 2000.

[31] L. Carlitz, Extended Stirling and exponential numbers, *Duke Math. J.*, **32** (1965), 205 – 224.

[32] L. Carlitz, Some partition problems related to the Stirling numbers of the second kind, *Acta Arith.*, **10** (1964/1965), 409 – 422.

[33] P. Cartier, On the structure of free Baxter algebras, *Adv. Math.*, **9** (1972), 253 – 265.

[34] P. Cartier, Fonctions polylogarithmes, nombres polyzêtas et groupes prounipotents, *Astrisque*, **282** (2002), 137 – 173, (Sm. Bourbaki no. 885).

[35] F. Chapoton, Un théorème de Cartier-Milnor-Moore-Quillen pour les bigèbres dendriformes et les algèbres braces, *J. Pure Appl. Alg.*, **168** (2002), 1 – 18.

[36] Ch. A. Charalambides and J. Singh: A review of the Stirling numbers, their generalizations and statistical applications, *Commun. Statist. -Theory Meth.*, **17** (1988), 2533 – 2595.

[37] Y. Chen and Q. Mo, Embedding dendriform dialgebra into its universal enveloping Rota-Baxter algebra, *Proc. AMS SOC.* **139** (2011), 4207 – 4216.

[38] C. Chu and L. Guo, Localization of Rota-Baxter algebras, preprint.

[39] J. C. Collins, *Renormalization*, Cambridge University Press, Cambridge, 1984.

[40] L. Comtet, *Advanced Combinatorics,* D. Reidel Publishing Co., Dordrecht, Holland, 1974.

[41] A. Connes, D. Kreimer, Hopf algebras, Renormalization and Noncommutative Geometry, *Comm. Math. Phys.*, **199** (1998), 203 – 242.

[42] A. Connes and D. Kreimer, Renormalization in quantum field theory and the Riemann-Hilbert problem. I. The Hopf algebra structure of graphs and the main theorem, *Comm. Math. Phys.*, **210** (2000), no. 1, 249 – 273.

[43] A. Connes and D. Kreimer, Renormalization in quantum field theory and the Riemann-Hilbert problem. II. The β-function, diffeomorphisms and the renormalization group., *Comm. Math. Phys.*, **216** (2001), no. 1, 215 – 241.

[44] A. Connes and M. Marcolli, *From Physics to Number Theory via Noncommutative Geometry, Part II: Renormalization, the Riemann-Hilbert correspondence, and motivic Galois theory,* in "*Frontiers in Number Theory, Physics, and Geometry, II*", Springer-Verlag, 2006, 617 – 713.

[45] A. Connes and M. Marcolli, *Noncommutative Geometry, Quantum Fields and Motives*, Ame. Math. Soc., Providence, 2008.

[46] S. L. de Bragança, Finite dimensional Baxter algebras, *Studies in Applied Math.* **LIV** (1975), 75 – 89.

[47] K. Ebrahimi-Fard, Loday-type algebras and the Rota-Baxter relation, *Lett. Math. Phys.*, **61** 139, (2002).

[48] K. Ebrahimi-Fard, On the associative Nijenhuis relation, *Electron. J. Combin.* **11** (1), (2004), R38.

[49] K. Ebrahimi-Fard, J. M. Gracia-Bondia and F. Patras, Lie theoretic approach to renormalization, *Comm. Math. Phys.*, **276** (2007), 519 – 549.

[50] K. Ebrahimi-Fard, J. M. Gracia-Bondia and F. Patras, Rota-Baxter algebras and new combinatorial identities, *Lett. Math. Phy.*, **81** (2007), 61 – 75.

[51] K. Ebrahimi-Fard, J. M. Gracia-Bondia, L. Guo and J. C. Várilly, Combinatorics of renormalization as matrix calculus, *Physics Letters B*, **19** (2006), 552 – 558.

[52] K. Ebrahimi-Fard and L. Guo, On the products and dual of binary, quadratic, regular operads, *J. Pure Appl. Alg.*, **200** (2005), 293 – 317.

[53] K. Ebrahimi-Fard and L. Guo, Mixable shuffles, quasi-shuffles and Hopf algebras, *J. Algebraic Combin.*, **24** (2006), 83 – 101.

[54] K. Ebrahimi-Fard and L. Guo, Unit actions on operads and Hopf algebras, *Theory Appl. Categ.* **18** (2007) 348 – 371.

[55] K. Ebrahimi-Fard and L. Guo, Rota-Baxter algebras in renormalization of perturbative quantum field theory, in: "Universality and Renormalization", I. Binder and D. Kreimer, editors, *Fields Institute Communicatins*, **50**, AMS 2007, 47 – 105.

[56] K. Ebrahimi-Fard and L. Guo, Rota-Baxter algebras and dendriform algebras, *J. Pure Appl. Alg.*, **212** (2008), 320 – 339.

[57] K. Ebrahimi-Fard and L. Guo, Multiple zeta values and Rota-Baxter algebras, *Integers: Electronic Journal of Combinatorial Number Theory*, **8** (2008), A4.

[58] K. Ebrahimi-Fard and L. Guo, Free Rota-Baxter algebras and rooted trees, *J. Algebra and Its Applications*, **7** (2008), 167 – 194

[59] K. Ebrahimi-Fard and L. Guo, On matrix representation of renormalization in prerturbative quantum field theory, arXiv: hep-th/0508155.

[60] K. Ebrahimi-Fard, L. Guo and D. Kreimer, Integrable renormalization I: the ladder case, *J. Math. Phys*, **45** (2004), 3758 – 3769.

[61] K. Ebrahimi-Fard, L. Guo and D. Kreimer, Integrable renormalization II: the general case, *Annales Henri Poincare*, **6** (2005), 369 – 395.

[62] K. Ebrahimi-Fard, L. Guo and D. Kreimer, Spitzer's identity and the algebraic Birkhoff decomposition in pQFT, *J. Phys. A: Math. Gen.*, **37** (2004), 11037 – 11052.

[63] K. Ebrahimi-Fard, L. Guo and D. Manchon, Birkhoff type decompositions and the Baker-Campbell-Hausdorff recursion, *Comm. Math. Phys.*, **267** (2006), 821 – 845.

[64] K. Ebrahimi-Fard and P. Lerous, Generalized shuffles related to Nijenhuis and TD-algebras, *Comm. Algebra*, **37** (2009) 3065 – 3094.

[65] K. Ebrahimi-Fard and D. Manchon, Dendriform Equations, *J. Algebra*, **322** (2009), 4053 – 4079.

[66] K. Ebrahimi-Fard and D. Manchon, A Magnus- and Fer-type formula in dendriform algebras, *Foundations of Computational Mathematics*, **9** (2009), 295.

[67] K. Ebrahimi-Fard and D. Manchon, On matrix differential equations in the Hopf algebra of renormalization, *Adv. Theor. Math. Phys.*, **10** (2006), 879 – 913.

[68] K. Ebrahimi-Fard and D. Manchon, The combinatorics of Bogoliubov's recursion in renormalization, CIRM 2006 workshop "Renormalization and Galois Theory", Org. F. Fauvet, J.-P. Ramis.

[69] K. Ebrahimi-Fard, D. Manchon and F. Patras, New identities in dendriform algebras, *J. Algebra*, **320** (2008), 708 – 727.

[70] K. Ebrahimi-Fard, D. Manchon and F. Patras, A noncommutative Bohnenblust-Spitzer identity for Rota-Baxter algebras solves Bogoliubov's recursion, *J. Noncomm. Geometry*, **3** (2009), 181 – 222.

[71] K. Ebrahimi-Fard and F. Patras, A Zassenhaus-type algorithm solves the Bogoliubov recursion, In "Lie Theory and Its Applications in Physics VII" ed. V. K. Dobrev et al., Heron Press, Sofia, 2008.

[72] K. Ebrahimi-Fard and F. Patras, Exponential renormalization, *Annales Henri Poincare*, **11** (2010), 943 – 971.

[73] K. Ebrahimi-Fard and F. Patras, Exponential Renormalization II: Bogoliubov's R-operation and momentum subtraction schemes, arXiv:1104.3415.

[74] R. Ehrenborg, On posets and Hopf algebras, *Adv. Math.*, **119** (1996), 1 – 25.

[75] H. Figueroa, J. M. Gracia-Bondía, Combinatorial Hopf algebras in quantum field theory I, *Reviews of Mathematical Physics*, **17** (2005), 881 – 976.

[76] L. Foissy, Les algèbres de Hopf des arbres enracinés décorés II, *Bull. Sci. Math.*, **126**, (2002), 249 – 288.

[77] A. Frabetti, Dialgebra homology of associative algebras, *C. R. Acad. Sci. Paris,* **325**, (1997), 135 – 140.

[78] A. Frabetti, Leibniz homology of dialgebras of matrices, *J. Pure Appl. Alg.*, **129**, (1998), 123 – 141.

[79] J. M. Freeman, On the classification of operator identities, *Studies in Appl. Math.*, **51** (1972), 73 – 84.

[80] V. Ginzburg and M. Kapranov, Koszul duality for operads, *Duke Math. J.*, **76** (1994), 203 – 272.

[81] A. G. Goncharov, Galois symmetries of fundamental groupoids and noncommutative geometry, *Duke Math. J.*, **128** (2005), 209 – 284.

[82] R. Grossman and R. G. Larson, Hopf-algebraic structure of families of trees, *J. Algebra*, **126** (1989), 184 – 210.

[83] L. Guo, Properties of free Baxter algebras, *Adv. Math.*, **151** (2000), 346 – 374.

[84] L. Guo, Baxter algebras and the umbral calculus, *Adv. Appl. Math.*, **27** (2001), 405 – 426.

[85] L. Guo, Ascending chain conditions in free Baxter algebras, *Int. J. Algebra Comput.*, **12** (2002), 601 – 622.

[86] L. Guo, Baxter algebra and differential algebra, in "Differential Algebra and Related Topics", World Scientific Publishing Company, 2002, 281 – 305.

[87] L. Guo, Baxter algebras, Stirling numbers and partitions, *J. Algebra Appl.*, **4** (2005), 153 – 164.

[88] L. Guo, Operated semigroups, Motzkin paths and rooted trees, *J. Algebraic Combinatorics*, **29** (2009), 35 – 62, arXiv:0710.0429.

[89] L. Guo, Algebraic Birkhoff Decomposition and its applications, In: *Automorphic Forms and the Langlands Program*, Higher Education Press and International Press, 2009, 35 – 62.

[90] L. Guo, WHAT IS a Rota-Baxter algebra, *Notices of Amer. Math. Soc.*, **56** (2009), 1436 – 1437.

[91] L. Guo and W. Keigher, *On integration algebras,* preprint.

[92] L. Guo and W. Keigher, Baxter algebras and shuffle products, *Adv. Math.*, **150** (2000), 117 – 149.

[93] L. Guo and W. Keigher, On free Baxter algebras: completions and the internal construction, *Adv. Math.*, **151** (2000), 101 – 127.

[94] L. Guo and W. Keigher, On differential Rota-Baxter algebras, *J. Pure Appl. Algebra*, **212** (2008), 522 – 540.

[95] L. Guo and Z. Liu, Rota-Baxter operators on generalized power series, *J. Algebra and Its Applications*, **8** (2009) 557 – 564.

[96] L. Guo, S. Paycha, B. Xie and B. Zhang, Double shuffle relations and renormalization of multiple zeta values, *The Geometry of Algebraic Cycles (Columbus, 2008)*, R. Akhtar et. al (eds.), Clay Math. Proc. 9, Amer. Math. Soc., Providence, 2010.

[97] L. Guo, S. Paycha and B. Zhang, Renormalization by Birkhoff-Hopf factorization and by generalized evaluators: a case study, in *Noncommutative Geometry Arithmetic and Related Topics*, Johns Hopkins University Press, (2011), 183 – 211.

[98] L. Guo and W. Y. Sit, Enumeration and generating functions of Rota-Baxter words, *Math. Comput. Sci.*, **4** (2010), 313 – 337.

[99] L. Guo and W. Y. Sit, Enumeration and generating functions of differential Rota-Baxter words, *Math. Comput. Sci.*, **4** (2010), 339 – 358.

[100] L. Guo, W. Sit and R. Zhang, On Rota's problem for linear operators in associative algebras, *Proc. ISSAC 2011*, 147 – 154.

[101] L. Guo, W. Sit and R. Zhang, Classification of differential type operators and Gröbner-Shirshov bases, To appear in *J. Symbolic Comput.*

[102] L. Guo and B. Xie, Structure theorems of mixable shuffle algebras and free commutative Rota-Baxter algebras, to appear in *Comm. Alg.* arXiv: 0807: 2267[math.RA].

[103] L. Guo and B. Xie, Explicit double shuffle relations and a generalization of Euler's decomposition formula, arXiv: 0808.2618[math.NT].

[104] L. Guo and B. Xie, Weighted sum formula for multiple zeta values, *J. Number Theory*, **129** (2009), 2747 – 2765.

[105] L. Guo and B. Xie, Shuffle relation of fractions from multiple zeta values and their product formula, *Ramanujan J. Math.*, **25** (2011), 307 – 317.

[106] L. Guo and B. Zhang, Renormalization of multiple zeta values, *J. Algebra*, **319** (2008), 3770 – 3809.

[107] L. Guo and B. Zhang, Differential Algebraic Birkhoff Decomposition and renormalization of multiple zeta values, *J. Number Theory*, **128** (2008), 2318 – 2339.

[108] L. Guo and B. Zhang, Polylogarithms and multiple zeta values from free Rota-Baxter algebras, *Sci. China Math.*, **53** (2010), 2239 – 2258.

[109] M. Hazewinkel, Generalized overlapping shuffle algebras *J. Math. Sci. (New York)*, **106** (2001), 3168 – 3186.

[110] M. E. Hoffman, Multiple harmonic series, *Pacific J. Math.*, **152** (1992), no. 2, 275 – 290.

[111] M. E. Hoffman, The algebra of multiple harmonic series, *J. Algebra*, **194** (1997), no. 2, 477 – 495.

[112] M. E. Hoffman, Quasi-shuffle products, *J. Algebraic Combin.*, **11**, (2000), no. 1, 49 – 68.

[113] M. E. Hoffman, Algebraic aspects of multiple zeta values, in "Zeta Functions, Topology and Quantum Physics", T. Aoki et. al. (eds.), *Developments in Math.* **14**, Springer, New York (2005), 51 – 74.

[114] R. Holtkamp, Comparison of Hopf algebras on trees, *Arch. Math. (Basel)*, **80** (2003), 368 – 383.

[115] R. Holtkamp, On Hopf algebra structures over operads, *Adv. Math.*, **207** (2006), 544 – 565.

[116] L. C. Hsu, Garry L. Mullen and P. J.-S. Shiue, Dickson-Stirling numbers, *Proc. Edinburgh Math. Soc.*, **40** (1997), 409 – 423.

[117] T. W. Hungerford, *Algebra*, Springer-Verlag, New York, 1974.

[118] C. Kassel, *Quantum Groups,* Springer-Verlag, New York, 1995.

[119] J. F. C. Kingman, Spitzer's identity and its use in probability theory, *J. London Math. Soc.*, **37** (1962), 309 – 316.

[120] E. R. Kolchin, *Differential Algebras and Algebraic Groups,* Academic Press, New York, 1973.

[121] Y. Kosmann-Schwarzbach and F. Magri, Poisson-Nijenhuis structures, *Ann. Inst. H. Poincaré Phys. Théor.*, **53** (1990), no. 1, 35 – 81.

[122] D. Kreimer, On the Hopf algebra structure of perturbative quantum field theories, *Adv. Theor. Math. Phys.*, **2** (1998), 303.

[123] D. Kreimer, Shuffling quantum field theory, *Lett. Math. Phys.*, **51** (2000), 179 – 191.

[124] J. P. S. Kung, Baxter algebras, entry in Springer Online Encyclopedia of Mathematics, M. Hazewinkel, ed. http://reference.springerlink.com/

[125] P. Lei and L. Guo, Nijenhuis algebra, NS algebras and N-dendriform algebra, *Front. Math. China,* (2012), to appear.

[126] P. Leroux, Ennea-algebras, *J. Algebra,* **281** (2004), 287 – 302.

[127] P. Leroux, Construction of Nijenhuis operators and dendriform trialgebras, *Int. J. Math. Math. Sci.*, **52** (2004), no. 40, 2595 – 2615.

[128] P. Leroux, On some remarkable operads constructed from Baxter operators, preprint, Nov. 2003, arXiv: math.QA/0311214.

[129] X. Li, D. Hou and C. Bai, Rota-Baxter operators on pre-Lie algebras, *J. Nonlinear Math. Phys.*, **14** (2007), 269 – 289.

[130] L. Liu, X. Ni and C. Bai, L-quadri-algebras (in Chinese), *Sci. Sin. Math.*, **42** (2011), 105 – 124.

[131] J.-L. Loday, Une version non commutative des algèbre de Lie: les algèbres de Leibniz, *Ens. Math.*, **39** (1993), 269 – 293.

[132] J. -L. Loday, Algèbres ayant deux opérations associatives (digèbres), *C. R. Acad. Sci. Paris,* **321** (1995), 141 – 146.

[133] J. -L. Loday, La renaissance des opérades, Séminaire Bourbaki, Vol. 1994/95. Astérisque No. 237 (1996), Exp. No. 792, 3, 47 – 74.

[134] J. -L. Loday, Dialgebras, in Dialgebras and related operads, *Lecture Notes in Math.*, **1763** (2002), 7 – 66.

[135] J. -L. Loday, Scindement d'associativité et algèbres de Hopf, the Proceedings of the Conference in honor of Jean Leray, Nantes (2002), *Séminaire et Congrès (SMF)*, **9** (2004), 155 – 172.

[136] J. -L. Loday, On the algebra of quasi-shuffles, *Manuscripta Mathematica,* **123** (2007), 79 – 83.

[137] J. -L. Loday and M. O. Ronco, Hopf algebra of the planar binary trees, *Adv. Math.*, **139** (1998), 293 – 309.

[138] J. -L. Loday and M. O. Ronco, Trialgebras and families of polytopes, in "Homotopy Theory: Relations with Algebraic Geometry, Group Cohomology, and Algebraic K-theory", *Contemporary Mathematics* **346** (2004), 369 – 398.

[139] J. -L. Loday and M. O. Ronco, Algèbre de Hopf colibres, *C. R. Acad. Sci. Paris*, **337** (2003), 153 – 158.

[140] J. -L. Loday and B. Vallette, *Algebraic Opeards*, *Grundlehren Math. Wiss.*, **346**, Springer, Heideberg, 2012.

[141] I. G. Macdonald, *Symmetric Functions and Hall Polynomials*, 2nd ed., Oxford University Press, Oxford, 1999.

[142] Y. I. Manin, Some remarks on Koszul algebras and quantum groups, *Ann. Inst. Fourier*, **37** (1987), 191 – 205.

[143] Y. I. Manin, Quantum groups and non-commutative geometry, *Pub. Centre Rech. Math. Montréal*, 1988.

[144] D. Manchon, Hopf algebras, from basics to applications to renormalization, Comptes-rendus des Rencontres mathmatiques de Glanon 2001. arXiv: math. QA/0408405

[145] D. Manchon and S. Paycha, Shuffle relations for regularised integrals of symbols, *Comm. Math. Phys.*, **270** (2007), 13 – 31.

[146] D. Manchon and S. Paycha, Nested sums of symbols and renormalized multiple zeta values, *Int. Math. Res. Not. IMRN* (2010), 4628 – 4687.

[147] M. Markl, S. Shnider and J. Stasheff, *Operads in Algebra, Topology and Physics,* Amer. Math. Soc., Providence, 2002.

[148] H. Z. Munthe-Kaas, G.R.W. Quispel and A. Zanna, Generalized polar decompositions on Lie groups with involutive automorphisms, *Found. Comput. Math.*, **1** (2001), no. 3, 297 – 324.

[149] T. Pirashvili, Sets with two associative operations, *C.E.J.M.*, **2** (2003), 169 – 183.

[150] C. Reutenauer, *Free Lie Algebras*, Oxford University Press, Oxford, 1993.

[151] J. Riordan, *Combinatorial Identities*, Wiley, New York, 1968.

[152] J. F. Ritt, Differential Equations from the Algebraic Standpoint, *Amer. Math. Sco. Colloq. Pub.*, **14** (1934), Amer. Math. Soc., New York.

[153] J. F. Ritt, Differential Algebra, *Amer. Math. Sco. Colloq. Pub.*, **33** (1950), Amer. Math. Soc., New York.

[154] M. R. Ronco, Eulerian idempotents and Milnor-Moore theorem for certain non-commutative Hopf algebras, *J. Algebra*, **254** (2002), 152 – 172.

[155] M. Rosenkranz and G. Regensburger, Solving and factoring boundary problems for linear ordinary differential equations in differential algebra, *J. Symbolic Comput.*, **43** (2008), 515 – 544.

[156] G. -C. Rota, Baxter algebras and combinatorial identities I, II, *Bull. Amer. Math. Soc.*, **75** (1969), 325 – 329, 330 – 334.

[157] G. -C. Rota, Baxter operators, an introduction, In: "Gian-Carlo Rota on Combinatorics, Introductory papers and commentaries", Joseph P.S. Kung, editor, Birkhäuser, Boston, 1995.

[158] G. -C. Rota, Ten mathematics problems I will never solve, Invited address at the joint meeting of the American Mathematical Society and the Mexican Mathematical Society, Oaxaca, Mexico, December 6, 1997, *DMV Mittellungen Heft* **2**, 1998, 45 – 52.

[159] G. -C. Rota, Letter dated July 26, 1998.

[160] G. -C. Rota, D. A. Smith, Fluctuation theory and Baxter algebras, *Istituto Nazionale di Alta Matematica,* **IX** (1972), 179 – 201.

[161] M. A. Semenov-Tian-Shansky, What is a classical r-matrix? *Funct. Ana. Appl.*, **17** (1983), no. 4, 259 – 272.

[162] M. Singer and M. van der Put, *Galois Theory of Linear Differential Equations*, Springer, New York, 2003.

[163] F. Spitzer, A combinatorial lemma and its application to probability theory, *Trans. Amer. Math. Soc.*, **82** (1956), 323 – 339.

[164] R. P. Stanley, *Enumerative Combinatorics: Vol. 2*, Cambridge University Press, Cambridge, New York, 1999.

[165] J. Stirling, Methodus Differentialis sine Tractatus de Summatione et Interpolatione Serierum Infinitarium, Londini (English translation by F. Lolliday with title: The Differential Method, London, 1749).

[166] M. Sweedler, *Hopf Algebras*, W.A. Benjamin, Inc., New York, 1969.

[167] K. Uchino, Quantum analogy of Poisson geometry, related dendriform algebras and Rota-Baxter operators, *Lett. Math. Phys.*, **85** (2008), 91 – 109.

[168] K. Uchino, Derived bracket construction and Manin products, *Lett. Math. Phys.*, **93** (2010), 37 – 53.

[169] B. Vallette, Homology of generalized partition posets, *J. Pure Appl. Algebra*, **208** (2007), 699 – 725.

[170] B. Vallette, Manin products, Koszul duality, Loday algebras and Deligne conjecture, *J. Reine Angew. Math.*, **620** (2008), 105 – 164.

[171] W. Vogel, Die kombinatorische Lösung einer Operator-Gleichung, *Z. f. Wahrscheinlichkeitstheorie und verw. Gebiete*, **2** (1963), 122 – 134.

[172] J. G. Wendel, A brief proof of a theorem of Baxter, *Math. Scand.*, **11** (1962), 107 – 108.

[173] R. Winkel, Sequences of symmetric polynomials and combinatorial properties of tableaux, *Adv. Math.*, **134** (1998), 46 – 89.

[174] D. Zagier, Values of zeta functions and their applications, First European Congress of Mathematics, Vol. II (Paris, 1992), 497 – 512, *Progr. Math.*, **120**, Birkhuser, Basel, 1994.

[175] A. Zanna, Recurrence relations and convergence theory of the generalized polar decomposition on Lie groups, *Math. Comp.*, **73** (2004), no. 246, 761 – 776

[176] Y. Zhang, C. Bai and L. Guo, Totally compatible associative and Lie dialgebras, tridendriform algebras and Post Lie algebras, preprint.

[177] J. Zhao, Renormalization of multiple q-zeta values, *Acta Math. Sinica*, **24** (2008), 1593 – 1616.

[178] C. Zhou, Free TD-algebras, in "*Operads and Universal Algebra*", World Scientific Publishing Company, 2012, 199 – 215.

Index

E

F

G

H

I

L

M

N

O

P

Q

R

Made in the USA
Monee, IL
08 November 2026

68136722R20131